Lecture Notes in Control and Information Sciences 311

Editors: M. Thoma · M. Morari

F. Lamnabhi-Lagarrigue · A. Loría
E. Panteley (Eds.)

Advanced Topics in Control Systems Theory

Lecture Notes from FAP 2004

With 12 Figures

Springer

Editors

Dr. Françoise Lamnabhi-Lagarrigue
Dr. Antonio Loría
Dr. Elena Panteley

Laboratoire des Signaux et Systèmes
Centre National de la Recherche Scientifique (CNRS)
SUPELEC
3 rue Joliot Curie
91192 Gif-sur-Yvette
France

British Library Cataloguing in Publication Data
Advanced topics in control systems theory : lecture notes from FAP 2004. - (Lecture notes in control and information sciences ; 311)
1. Automatic control 2. Automatic control - Mathematical models
3. Control theory 4. Systems engineering
I. Lamnabhi-Lagarrigue, F. (Francoise), 1953- II. Loria, Antonio
III. Panteley, Elena
629.8'312
ISBN 1852339233

Library of Congress Control Number: 2004117782

Lecture Notes in Control and Information Sciences ISSN 0170-8643
ISBN 1-85233-923-3
Springer Science+Business Media
springeronline.com

Typesetting: Data conversion by the authors.
Final processing by PTP-Berlin Protago-TEX-Production GmbH, Germany
Cover-Design: design & production GmbH, Heidelberg
Printed in Germany
69/3141 Yu-543210 Printed on acid-free paper SPIN 11334774

To our lovely daughters,
AL & EP.

Preface

Advanced topics in control systems theory is a byproduct of the European school "Formation d'Automatique de Paris" (Paris Graduate School on Automatic Control) which took place in Paris through February and March 2004. The school benefited of the valuable participation of 17 European renowned control researchers and about 70 European PhD students. While the program consisted of the modules listed below, the contents of the present monograph collects selected notes provided by the lecturers and is by no means exhaustive.

Program of FAP 2004:

P1 Nonlinear control of electrical and electromechanical systems
A. Astolfi, R. Ortega

P2 Algebraic analysis of control systems defined by partial differential equations
J-F. Pommaret

P3 Nonlinear flatness-based control of complex electromechanical systems E. Delaleau

P4 Modeling and control of chemical and biotechnological processes
Jan van Impe, D. Dochain,

P5 Modeling and boundary control of infinite dimensional systems
B. Maschke, A.J. van der Schaft, H. Zwart

P6 Linear systems, algebraic theory of modules, structural properties
H. Bourles, M. Fliess

P7 Lyapunov-based control: state and output feedback
L. Praly, A. Astolfi, A. Loría

P8 Nonlinear control and mechanical systems
B. Bonnard

P9 Tools for analysis and control of time-varying systems
J. M. Coron, A. Loría

P10 Control of oscillating mechanical systems, synchronization and chaos
J. Levine, H. Nijmeijer

In particular, the lecture notes included in the subsequent chapters stem from modules P1, P2, P5, P6, P7 and P8. The material, which covers a wide range of topics from control theory, is organized in six chapters: two chapters on Lyapunov-like methods for control design and stability analysis, one chapter on nonlinear optimal control, one chapter on modeling of Hamiltonian infinite-ddimensional systems and two chapters on algebraic methods.

Each module listed above was taught over 21hrs within one week. Therefore, the contents of the present monograph may be used in support to either a one-term general advanced course on non linear control theory, thereby devoting a few lectures to each topic, or it may be used in support to more focused intensive courses at graduate level. The academic requirement for the class student or the reader in general is a basic knowledge on control theory (linear and non linear).

Advanced topics in control systems theory also constitutes an ideal start for researchers in control theory who wish to broaden their general culture or to get involved in fields different to their expertise, while avoiding a thorough book-keeping. Indeed, the monograph presents in a concise but pedagogical manner diverse aspects of modern control theory.

This book is the first of a series of yearly volumes, which shall prevail beyond the lectures taught in class during each FAP season. Further information on FAP, in particular, on the scientific program for the subsequent years is updated in due time on our URL `http://www.supelec.lss/cts/fap`.

FAP is organized within the context of the European teaching network "Control Training Site" sponsored by the European Community through the Marie Curie program. The editors of the present text greatefully acknowledge such sponsorship. We also take this oportunity to acknowledge the French national center for scientific research (C.N.R.S.) which provides us with a working environment and ressources probably unparalleled in the world.

Gif-sur-Yvette, France.
September 2004

Françoise Lamnabhi-Lagarrigue,
Antonio Loría,
Elena Panteley.

Contents

List of Figures

List of Contributors

A. Astolfi
Department of Electrical and Electronic Engineering, Imperial College,
Exhibition Road, London SW7 2BT, UK.
astolfi@imperial.ac.uk

B. Bonnard
Département de Mathématiques, Laboratoire Analyse Appliquée et Optimisation
B.P 47870, 21078 Dijon Cedex, France.
bbonnard@u-bourgogne.fr

H. Bourlès
SATIE, ENS de Cachan et CNAM,
61 Ave du Président Wilson,
94230 Cachan, France.
henri.bourles@satie.ens_cachan.fr

M. Jabeur
Département de Mathématiques, Laboratoire Analyse Appliquée et Optimisation
B.P 47870, 21078 Dijon Cedex, France.
mjabeur@u-bourgogne.fr

G. Janin
Département de Mathématiques, Laboratoire Analyse Appliquée et Optimisation
B.P 47870, 21078 Dijon Cedex, France.
gabriel.janin@ensta.org

D. Karagiannis
Department of Electrical and Electronic Engineering, Imperial College,
Exhibition Road, London SW7 2BT, UK.
d.karagiannis@imperial.ac.uk

A. Loría
Laboratoire des Signaux et Systèmes, Supélec,
3, Rue Joliot Curie, 91192 Gif-sur-Yvette, France.
loria@lss.supelec.fr

B. Maschke
Laboratoire d'Automatique et de Génie des Procédés,
Université Claude Bernard Lyon-1,
CPE Lyon - Bâtiment 308 G, 43,
Bd du 11 Novembre 1918, F-69622 Villeurbanne cedex, France.
maschke@lagep.univ-lyon1.fr

R. Ortega
Laboratoire des Signaux et Systèmes, Supélec,
3, Rue Joliot Curie, 91192
Gif-sur-Yvette, France.
Romeo.Ortega@lss.supelec.fr

E. Panteley
Laboratoire des Signaux et Systèmes, Supélec,
3, Rue Joliot Curie, 91192
Gif-sur-Yvette, France.
panteley@lss.supelec.fr

J. Pommaret
CERMICS/Ecole Nationale des Ponts et Chaussées, 6/8 ave Blaise Pascal, Cité Descartes,
77455 MARNE-la-Vallée CEDEX 2, France.
pommaret@cermics.enpc.fr

A. van der Schaft
Systems, Signals and Control Department, Dept. of Applied Mathematics, University of Twente,
P.O. Box 217, 7500 AE Enschede, The Netherlands.
twarjan@math.utwente.nl

1

Nonlinear Adaptive Stabilization via System Immersion: Control Design and Applications

D. Karagiannis[1], R. Ortega[2], and A. Astolfi[1]

[1] Department of Electrical and Electronic Engineering, Imperial College, Exhibition Road, London SW7 2AT, UK. Email: d.karagiannis@imperial.ac.uk, a.astolfi@imperial.ac.uk

[2] Laboratoire des Signaux et Systèmes, Supélec, Plateau de Moulon, 91192 Gif-sur-Yvette, France. Romeo.Ortega@lss.supelec.fr

This chapter presents a review of a recently developed methodology for the (adaptive) stabilization of nonlinear systems via state and output feedback. The method relies upon the notion of system immersion and is well-suited in applications where a controller for a reduced-order model is known and we would like to robustify it with respect to higher-order dynamics. This is achieved by *immersing* the dynamics of the controlled plant into the desired dynamics of the reduced-order model. The method is illustrated with several practical and academic examples.

1.1 Introduction

The problem of (adaptive) stabilization of nonlinear systems has been an active area of research in the past years with several constructive methodologies being introduced, such as feedback linearization [3, 7], sliding-mode control [10], backstepping [6] and passivity-based control [8], see also the monograph [4]. Most of the existing methods rely on the use of (control) Lyapunov functions, *i.e.* the control law and/or the adaptive law are designed so that a candidate Lyapunov function—typically a quadratic function of the states and the (parameter) estimation error—is rendered negative definite. For systems with certain "triangular" structures, this approach has proven particularly successful [6, 7].

More recently in [1] the concepts used in the theory of output regulation [2] have been exploited to develop a novel framework for solving nonlinear stabilization and adaptive control problems. This new approach makes use of two classical tools from nonlinear regulator theory and geometric nonlinear

F. Lamnabhi-Lagarrigue et al. (Eds.): Adv. Top. in Cntrl. Sys. Theory, LNCIS 311, pp. 1–21, 2005

control: (system) immersion and (manifold) invariance. For this reason the method is referred to as *immersion and invariance* (I&I).

The basic idea in this methodology is to *immerse* the plant dynamics into a stable (lower-order) target system. To illustrate this, consider the system

$$\dot{x} = f(x, u) \tag{1.1}$$

with $x \in \mathbb{R}^n$ and $u \in \mathbb{R}^m$ and the problem of finding, whenever possible, a state feedback control law $u = v(x)$ such that the closed-loop system is (globally) asymptotically stable. This is equivalent to finding a target dynamical system

$$\dot{\xi} = \alpha(\xi) \tag{1.2}$$

with $\xi \in \mathbb{R}^p$ and $p < n$, which is (globally) asymptotically stable, a mapping[3] $x = \pi(\xi)$ and a control law $v(x)$ such that

$$\pi(\xi(0)) = x(0) \tag{1.3}$$

$$\pi(0) = 0 \tag{1.4}$$

and

$$f(\pi(\xi), v(\pi(\xi))) = \frac{\partial \pi}{\partial \xi}\alpha(\xi). \tag{1.5}$$

If the above conditions hold, then any trajectory $x(t)$ of the closed-loop system

$$\dot{x} = f(x, v(x)) \tag{1.6}$$

is the *image* through the mapping $\pi(\cdot)$ of a trajectory $\xi(t)$ of the target system (1.2). From (1.4), this implies that $x(t)$ converges to the origin. Thus the stabilization problem for the system (1.1) can be recasted as a problem of solving the partial differential equation (1.5) with the boundary conditions (1.3)-(1.4).

A geometric interpretation of (1.3)–(1.5) is the following. Consider the closed-loop system (1.6) and a manifold in the n-dimensional state-space defined by

$$\mathcal{M} = \{x \in \mathbb{R}^n \mid x = \pi(\xi),\ \xi \in \mathbb{R}^p\}.$$

From (1.5), the manifold $\mathcal{M}$ is *invariant* with internal dynamics (1.2), hence all trajectories $x(t)$ that start on the manifold remain there and asymptotically converge to the point $x = \pi(0)$, which is the origin by (1.4). Moreover, the condition (1.3) guarantees that the initial state of (1.6) lies on the manifold $\mathcal{M}$.

The above formulation is impractical for two reasons. First, from (1.3), the mapping $\pi(\cdot)$ will, in general, depend on the initial conditions. Second, even if

[3] Note that, since the dimension of ξ is strictly lower than the dimension of x, the mapping $x = \pi(\xi)$ is an immersion.

the target system (1.2) is *globally* asymptotically stable, we cannot conclude that the trajectories of (1.6) are bounded, without additional conditions on the mapping $\pi(\cdot)$.

These obstacles can be removed by modifying the control law $u = v(x)$ so that, for all initial conditions, the trajectories of the system (1.1) remain bounded and asymptotically *converge* to the manifold $\mathcal{M}$, *i.e.* $\mathcal{M}$ is rendered attractive. The attractivity of the manifold can be expressed in terms of the distance

$$z = \mathrm{dist}(x, \mathcal{M}),$$

which should be driven to zero. Notice that the variable z, which is referred to as the *off-the-manifold coordinate*, is not uniquely defined. This provides an additional degree of freedom in the control design.

1.2 Nonlinear Stabilization via System Immersion

The present section reviews the basic theoretical results of [1], namely a set of sufficient conditions for the construction of globally asymptotically stabilizing state feedback control laws for general nonlinear systems.

Theorem 1.1. *[1] Consider the system (1.1) with an equilibrium point $x^* \in \mathbb{R}^n$ to be stabilized and assume we can find mappings $\alpha(\cdot)$, $\pi(\cdot)$, $c(\cdot)$, $\phi(\cdot)$ and $\psi(\cdot)$ such that the following hold.*

(A1) (Target system) The system (1.2) has a globally asymptotically stable equilibrium at $\xi^* \in \mathbb{R}^p$ and $x^* = \pi(\xi^*)$.

(A2) (Immersion) For all $\xi \in \mathbb{R}^p$

$$f(\pi(\xi), c(\pi(\xi))) = \frac{\partial \pi}{\partial \xi}\alpha(\xi). \tag{1.7}$$

(A3) (Implicit manifold) The following set identity holds

$$\mathcal{M} = \{x \in \mathbb{R}^n \mid \phi(x) = 0\} = \{x \in \mathbb{R}^n \mid x = \pi(\xi),\ \xi \in \mathbb{R}^p\}. \tag{1.8}$$

(A4) (Manifold attractivity and trajectory boundedness) All trajectories of the system

$$\dot{z} = \frac{\partial \phi}{\partial x} f(x, \psi(x, z)) \tag{1.9}$$

$$\dot{x} = f(x, \psi(x, z)) \tag{1.10}$$

are bounded and satisfy

$$\lim_{t \to \infty} z(t) = 0. \tag{1.11}$$

Then x^* is a globally asymptotically stable equilibrium of the closed-loop system

$$\dot{x} = f(x, \psi(x, \phi(x))).$$

Remark 1.1. The result in Theorem 1.1 implies that the stabilization problem for the system (1.1) can be divided into two subproblems. First, given the target system (1.2) find, if possible, a manifold $\mathcal{M}$ described implicitly by $\{x \in \mathbb{R}^n \mid \phi(x) = 0\}$ and in parameterized form by $\{x \in \mathbb{R}^n \mid x = \pi(\xi),$ for some $\xi \in \mathbb{R}^p\}$, which can be rendered invariant with internal dynamics a copy of the target dynamics. Second, design a control law $u = \psi(x, z)$ that drives to zero the off-the-manifold coordinate $z = \phi(x)$ and keeps the system trajectories bounded.

Remark 1.2. The convergence condition (1.11) can be relaxed, *i.e.* to prove asymptotic stability of the equilibrium x^* it suffices to require

$$\lim_{t \to \infty} [f(x(t), \psi(x(t), z(t))) - f(x(t), \psi(x(t), 0))] = 0. \qquad (1.12)$$

In other words, it is not necessary to reach the manifold $\mathcal{M}$, in order to stabilize the equilibrium x^*.

Remark 1.3. If we can find a partition of $x = \mathrm{col}(x_1, x_2)$ with $x_1 \in \mathbb{R}^p$ and $x_2 \in \mathbb{R}^{n-p}$ and a corresponding partition of $\pi(\cdot) = \mathrm{col}(\pi_1(\cdot), \pi_2(\cdot))$ such that $x_1 = \pi_1(\xi)$ is a global change of coordinates, then (A3) is satisfied with $z = \phi(x) = x_2 - \pi_2(\pi_1^{-1}(x_1))$. As a result, instead of considering the trajectories of the extended system (1.9)-(1.10) in (A4), it suffices to study the trajectories of the system with state (x_1, z).

We conclude this section by recalling a definition introduced in [1], which will be used in the rest of the chapter.

Definition 1.1. *The system (1.1) is said to be* I&I-stabilizable *with target dynamics (1.2), if the assumptions (A1)-(A4) of Theorem 1.1 are satisfied.*

1.3 Adaptive Control via System Immersion

In this section we show how the general theory of Section 1.2 can be used to design adaptive stabilizing controllers for nonlinear systems with parametric uncertainties. The first step is to assume that a "full-information" control law (*i.e.* a control law that depends on the unknown parameters) is available.[4]

[4] Obviously, if it is not available, we could design one using Theorem 1.1.

Then the problem is reduced to finding an adaptive law such that the closed-loop system is immersed into the system that would result if we applied the full-information controller.

To illustrate this approach, consider again the system (1.1), where the vector field $f(\cdot)$ may depend on an unknown parameter vector $\theta \in \mathbb{R}^q$. Assume that there exists a parameterized control law $u = \upsilon(x, \theta)$ such that the closed-loop system

$$\dot{x} = f(x, \upsilon(x, \theta)) \tag{1.13}$$

is globally asymptotically stable. Consider now the system

$$\begin{aligned} \dot{x} &= f(x, \upsilon(x, \hat{\theta} + \beta_1(x))) \\ \dot{\hat{\theta}} &= \beta_2(x, \hat{\theta}) \end{aligned} \tag{1.14}$$

with extended state $(x, \hat{\theta}) \in \mathbb{R}^n \times \mathbb{R}^q$ and "controls" β_1 and β_2. The adaptive stabilization problem boils down to selecting the functions $\beta_1(\cdot)$ and $\beta_2(\cdot)$ so that the system (1.14) is asymptotically immersed into the system (1.13). This leads us to the following definition.

Definition 1.2. *The system (1.1) is said to be* adaptively I&I stabilizable, *if the system (1.14) is I&I stabilizable with target dynamics (1.13).*

The above formulation implies that we can use Theorem 1.1 to solve the adaptive control problem, provided that a full-information control law is known. Note that the control law applied in (1.14) does not follow the certainty-equivalence philosophy. In fact, the unknown parameter vector θ is replaced by the term $\hat{\theta} + \beta_1(x)$. The term $\beta_1(x)$ introduces a "proportional" action in addition to the classical "integral" action given by the second equation in (1.14). For this reason, this adaptive scheme is sometimes referred to as *nonlinear PI* adaptation.

1.3.1 Systems Linear in the Unknown Parameters

In this section, to demonstrate the application of the foregoing theory and for simplicity, we will consider linearly parameterized, control affine systems of the form

$$\dot{x} = f_0(x) + f_1(x)\theta + g(x)u \tag{1.15}$$

with state $x \in \mathbb{R}^n$ and input $u \in \mathbb{R}^m$, where $\theta \in \mathbb{R}^q$ is an unknown constant vector.

Proposition 1.1. *Consider the system (1.15) with an equilibrium point x^* to be stabilized and assume the following hold.*

(B1) There exists a full-information control law $u = v(x, \theta)$ such that the closed-loop system

$$\dot{x} = f^*(x) := f_0(x) + f_1(x)\theta + g(x)v(x, \theta) \tag{1.16}$$

has a globally asymptotically stable equilibrium at x^*.

(B2) There exists a mapping $\beta_1(\cdot)$ such that all trajectories of the system

$$\begin{aligned} \dot{z} &= -\frac{\partial \beta_1}{\partial x} f_1(x) z \\ \dot{x} &= f^*(x) + g(x)\left(v(x, \theta + z) - v(x, \theta)\right) \end{aligned} \tag{1.17}$$

are bounded and satisfy

$$\lim_{t \to \infty} \left[v(x(t), \theta + z(t)) - v(x(t), \theta)\right] = 0.$$

Then the system (1.15) with assumptions (B1)-(B2) is adaptively I&I stabilizable.

Proof. We will verify that for the extended system

$$\begin{aligned} \dot{x} &= f_0(x) + f_1(x)\theta + g(x)v(x, \hat{\theta} + \beta_1(x)) \\ \dot{\hat{\theta}} &= \beta_2(x, \hat{\theta}) \end{aligned}$$

the conditions of Theorem 1.1 hold with (1.12) replacing (1.11) as pointed out in Remark 1.2. First, (A1) is automatically satisfied from (B1) for $\dot{\xi} = f^*(\xi)$. Second, for the immersion condition (A2), we are looking for mappings $\pi_1(\cdot)$, $\pi_2(\cdot)$, $c_1(\cdot)$ and $c_2(\cdot)$ with

$$x = \pi_1(\xi), \qquad \hat{\theta} = \pi_2(\xi)$$

that solve the equations

$$\begin{aligned} \frac{\partial \pi_1}{\partial \xi} f^*(\xi) &= f_0(\pi_1(\xi)) + f_1(\pi_1(\xi))\theta + g(\pi_1(\xi))v(\pi_1(\xi), \pi_2(\xi) + c_1(\pi_1(\xi))) \\ \frac{\partial \pi_2}{\partial \xi} f^*(\xi) &= c_2(\pi_1(\xi), \pi_2(\xi)). \end{aligned}$$

A solution to these equations is given by $\pi_1(\xi) = \xi$, $\pi_2(\xi) = \theta - c_1(\xi)$, for any function $c_1(\cdot)$, with $c_2(\cdot)$ defined by the last identity. Setting $\beta_1(\xi) = c_1(\xi)$, the implicit manifold in (A3) is described by the equation

$$\phi(x, \hat{\theta}) = \hat{\theta} - \theta + \beta_1(x) = 0.$$

It remains to prove that (A4) holds. To this end, note that the dynamics of the off-the-manifold coordinate $z = \phi(x, \hat{\theta})$ are given by the equation

$$\dot{z} = \beta_2(x, \hat{\theta}) + \frac{\partial \beta_1}{\partial x}\left[f_0(x) + f_1(x)\left(\hat{\theta} + \beta_1(x) - z\right) + g(x)v(x, \hat{\theta} + \beta_1(x))\right].$$

Selecting the adaptation law

$$\beta_2(x, \hat{\theta}) = -\frac{\partial \beta_1}{\partial x}\left[f_0(x) + f_1(x)\left(\hat{\theta} + \beta_1(x)\right) + g(x)v(x, \hat{\theta} + \beta_1(x))\right]$$

yields the first equation in (1.17), while the second equation is obtained directly from (1.15) using (B1) and the definition of z. Hence, by assumption (B2), the condition (A4) with (1.11) replaced by (1.12) holds. ■

Example 1.1. Consider the stabilization to zero of the unstable first-order linear system

$$\dot{x} = \theta x + u, \tag{1.18}$$

where $\theta > 0$ is an unknown constant and note that (B1) is satisfied with $u = -kx - \theta x$, for any $k > 0$, hence the I&I adaptive control law is given by

$$u = -kx - \left(\hat{\theta} + \beta_1(x)\right)x, \qquad \dot{\hat{\theta}} = \beta_2(x, \hat{\theta}). \tag{1.19}$$

Defining

$$\beta_2(x, \hat{\theta}) = -\frac{\partial \beta_1}{\partial x}\left[\left(\hat{\theta} + \beta_1(x)\right)x + u\right] = \frac{\partial \beta_1}{\partial x}kx$$

and $z = \hat{\theta} - \theta + \beta_1(x)$, as in the proof of Proposition 1.1, yields the error system

$$\begin{aligned} \dot{z} &= -\frac{\partial \beta_1}{\partial x}xz \\ \dot{x} &= -kx - xz, \end{aligned} \tag{1.20}$$

which is of the form (1.17). Selecting

$$\beta_1(x) = \frac{\gamma}{2}x^2$$

ensures that all trajectories of the system (1.20) are bounded and

$$\lim_{t \to \infty}[x(t)z(t)] = 0,$$

i.e. the condition (B2) holds. As a result, the closed-loop system (1.18)-(1.19) has a globally asymptotically stable equilibrium at $x = 0$.

1.3.2 Systems in Feedback Form

The assumption (B2), on which the result in Proposition 1.1 relies, is quite restrictive in the sense that there is no systematic way of treating the *cascade* system (1.17), unless the function $v(x, \theta)$ is linear in θ or satisfies a Lipschitz condition.[5] In this section we depart from this approach and do not assume that a control law $u = v(x, \theta)$ is known. Instead, we design a dynamic controller directly using Theorem 1.1.

Consider a class of systems described by equations of the form

$$\begin{aligned} \dot{x}_1 &= x_2 \\ &\vdots \\ \dot{x}_{p-1} &= x_p \\ \dot{x}_p &= x_{p+1} + \phi^T(x_1, \dots, x_p)\theta \\ \dot{x}_{p+1} &= x_{p+2} \\ &\vdots \\ \dot{x}_n &= x_{n+1} = u \end{aligned} \tag{1.21}$$

with states $x_i \in \mathbb{R}$, $i = 1, \dots, n$ and input $u \in \mathbb{R}$, where $\theta \in \mathbb{R}^q$ is a vector of unknown parameters. We will show that an adaptive, globally asymptotically stabilizing control law for the system (1.21) can be obtained by applying Theorem 1.1. In particular, we have the following result.

Proposition 1.2. *Consider the system (1.21) and the adaptation law*

$$\dot{\hat{\theta}} = \beta_2(x, \hat{\theta}) \tag{1.22}$$

with $\hat{\theta} \in \mathbb{R}^q$. The system (1.21)-(1.22) with inputs u and β_2 is I&I stabilizable with target dynamics

$$\begin{aligned} \dot{\xi}_1 &= \xi_2 \\ &\vdots \\ \dot{\xi}_{p-1} &= \xi_p \\ \dot{\xi}_p &= -K^T\xi, \end{aligned} \tag{1.23}$$

where $\xi = [\xi_1, \dots, \xi_p]^T$ and K is a constant vector.

Proof. To begin with, note that the vector K can be selected so that the target system (1.23) is globally asymptotically stable, hence (A1) holds. Consider now the immersion condition (A2) and the mappings

$$\begin{aligned} x_i &= \pi_i(\xi), \qquad i = 1, \dots, n \\ \hat{\theta} &= \pi_{n+1}(\xi). \end{aligned}$$

Setting $\pi_i(\xi) = \xi_i$, for $i = 1, \dots, p$, the equations (1.7) reduce to

[5] See [1, Section IV-C] for more detail.

$$\begin{aligned}
\pi_{p+1}(\xi) &= -K^T\xi - \phi^T(\xi)\theta \\
\pi_{p+2}(\xi) &= \dot{\pi}_{p+1}(\xi) \\
&\vdots \\
\pi_n(\xi) &= \dot{\pi}_{n-1}(\xi) \\
c_1(\pi(\xi)) &= \dot{\pi}_n(\xi) \\
c_2(\pi(\xi)) &= \dot{\pi}_{n+1}(\xi).
\end{aligned}$$

Selecting $\pi_{n+1}(\xi) = \theta - \beta_1(\xi)$, from the first identity we obtain the solution

$$\pi_{p+1}(\xi) = -K^T\xi - \phi^T(\xi)\left(\pi_{n+1}(\xi) + \beta_1(\xi)\right),$$

while from the remaining identities we can *recursively* define $\pi_{p+2}(\cdot), \ldots, \pi_n(\cdot)$, $c_1(\cdot)$ and $c_2(\cdot)$. Note now that, from (A3), the off-the-manifold coordinates

$$z = x - \pi(\xi)$$

are given by

$$\begin{aligned}
z_1 &= x_{p+1} + K^T\left[x_1, \ldots, x_p\right]^T + \phi^T(x_1, \ldots, x_p)\left(\hat{\theta} + \beta_1(x_1, \ldots, x_p)\right) \\
z_2 &= x_{p+2} - \pi_{p+2}(x_1, \ldots, x_p) \\
&\vdots \\
z_{n-p} &= x_n - \pi_n(x_1, \ldots, x_p) \\
z_{n-p+1} &= \hat{\theta} - \theta + \beta_1(x_1, \ldots, x_p).
\end{aligned}$$

As a result, the system (1.21)-(1.22) can be described in the $(z, x, \hat{\theta})$ coordinates by the equations[6]

$$\begin{aligned}
\dot{z}_1 &= z_2 \\
&\vdots \\
\dot{z}_{n-p-1} &= z_{n-p} \\
\dot{z}_{n-p} &= u(x,\hat{\theta}) - \frac{\partial \pi_n}{\partial \hat{\theta}}\beta_2(x,\hat{\theta}) - \sum_{j=1}^{p}\frac{\partial \pi_n}{\partial x_j}x_{j+1} \\
&\quad - \frac{\partial \pi_n}{\partial x_p}\phi^T(x_1, \ldots, x_p)\left(\hat{\theta} + \beta_1(x_1, \ldots, x_p) - z_{n-p+1}\right) \\
\dot{z}_{n-p+1} &= \beta_2(x,\hat{\theta}) + \sum_{j=1}^{p}\frac{\partial \beta_1}{\partial x_j}x_{j+1} \\
&\quad + \frac{\partial \beta_1}{\partial x_p}\phi^T(x_1, \ldots, x_p)\left(\hat{\theta} + \beta_1(x_1, \ldots, x_p) - z_{n-p+1}\right)
\end{aligned}$$

[6] Notice that the last $n-p+1$ equations are redundant, since they are globally diffeomorphic to the first $n-p+1$ equations (see Remark 1.3).

$$
\begin{aligned}
\dot{x}_1 &= x_2 \\
&\vdots \\
\dot{x}_{p-1} &= x_p \\
\dot{x}_p &= -K^T \left[x_1, \dots, x_p\right]^T - \phi^T(x_1, \dots, x_p) z_{n-p+1} + z_1 \\
\dot{x}_{p+1} &= x_{p+2} \\
&\vdots \\
\dot{x}_n &= u(x, \hat{\theta}) \\
\dot{\hat{\theta}} &= \beta_2(x, \hat{\theta}).
\end{aligned}
$$

Consider now the control laws

$$
\begin{aligned}
\beta_2(x, \hat{\theta}) &= -\sum_{j=1}^{p} \frac{\partial \beta_1}{\partial x_j} x_{j+1} - \frac{\partial \beta_1}{\partial x_p} \phi^T(x_1, \dots, x_p) \left(\hat{\theta} + \beta_1(x_1, \dots, x_p)\right) \\
u(x, \hat{\theta}) &= -\Lambda^T \left[z_1, \dots, z_{n-p}\right]^T + \frac{\partial \pi_n}{\partial \hat{\theta}} \beta_2(x, \hat{\theta}) + \sum_{j=1}^{p} \frac{\partial \pi_n}{\partial x_j} x_{j+1} \\
&\quad + \frac{\partial \pi_n}{\partial x_p} \phi^T(x_1, \dots, x_p) \left(\hat{\theta} + \beta_1(x_1, \dots, x_p)\right) \\
&\quad - \frac{1}{2\epsilon} \left(\frac{\partial \pi_n}{\partial x_p}\right)^2 \left[z_1, \dots, z_{n-p}\right]^T P \left[0, \dots, 1\right]^T,
\end{aligned}
$$

where $\epsilon > 0$ is an arbitrary constant, the vector Λ is chosen so that the matrix

$$
A = \left[\begin{array}{cc} 0 & I_{n-p-1} \\ \hline \multicolumn{2}{c}{-\Lambda^T} \end{array}\right]
$$

is Hurwitz and P satisfies the Lyapunov equation $A^T P + PA = -I_{n-p}$. Defining the function

$$
\beta_1(x_1, \dots, x_p) = \gamma \int_0^{x_p} \phi(x_1, \dots, x_{p-1}, \chi) d\chi \tag{1.24}
$$

with $\gamma > 0$ yields the closed-loop system

$$
\begin{aligned}
\dot{z}_1 &= z_2 \\
&\vdots \\
\dot{z}_{n-p-1} &= z_{n-p} \\
\dot{z}_{n-p} &= -\Lambda^T [z_1, \dots, z_{n-p}]^T + \frac{\partial \pi_n}{\partial x_p} \phi^T(x_1, \dots, x_p) z_{n-p+1} \\
&\quad - \frac{1}{2\epsilon} \left(\frac{\partial \pi_n}{\partial x_p} \right)^2 [z_1, \dots, z_{n-p}]^T P [0, \dots, 1]^T \\
\dot{z}_{n-p+1} &= -\gamma \phi(x_1, \dots, x_p) \phi^T(x_1, \dots, x_p) z_{n-p+1} \\
\dot{x}_1 &= x_2 \\
&\vdots \\
\dot{x}_{p-1} &= x_p \\
\dot{x}_p &= -K^T [x_1, \dots, x_p]^T - \phi^T(x_1, \dots, x_p) z_{n-p+1} + z_1.
\end{aligned}
$$

The proof is completed by considering the Lyapunov function

$$
V = [z_1, \dots, z_{n-p}] P [z_1, \dots, z_{n-p}]^T + \frac{1+\epsilon}{2\gamma} |z_{n-p+1}|^2,
$$

whose time-derivative along the trajectories of the closed-loop system given above satisfies

$$
\dot{V} \leq -[z_1, \dots, z_{n-p}]^T [z_1, \dots, z_{n-p}] - \left(\phi^T(x_1, \dots, x_p) z_{n-p+1} \right)^2.
$$

Hence z is bounded and the signals $z_1, \dots, z_{n-p}$ and $\phi^T(x_1, \dots, x_p) z_{n-p+1}$ converge to zero. As a result, the system (1.21) is globally asymptotically stable. ∎

1.4 Output Feedback Stabilization

In this section we extend the results of Section 1.3 to the *output feedback* stabilization problem, where we seek a dynamic stabilizing control law that does not depend on the unmeasured states.

Consider the system

$$
\begin{aligned}
\dot{\eta} &= f(\eta, y, u) \\
\dot{y} &= h(\eta, y, u)
\end{aligned}
\tag{1.25}
$$

with state $(\eta, y) \in \mathbb{R}^n \times \mathbb{R}^r$ and input $u \in \mathbb{R}^m$ and suppose that only the state y is available for measurement. Note that the system (1.25) may also include unknown parameters, *i.e.* equations of the form $\dot{\eta}_i = 0$. Following the ideas of Section 1.3 we assume that there exists a full-information control law $u = v(\eta, y)$ such that all trajectories of the closed-loop system

$$\begin{aligned}\dot{\eta} &= f(\eta, y, v(\eta, y)) \\ \dot{y} &= h(\eta, y, v(\eta, y))\end{aligned} \tag{1.26}$$

are bounded and

$$\lim_{t \to \infty} y(t) = y^*. \tag{1.27}$$

Note that we only require that the output y converges to a set-point and that η remains bounded. This is because it may not be possible to drive the whole state (η, y) to a desired equilibrium. This is the case, for instance, when η contains unknown parameters. Consider now the system

$$\begin{aligned}\dot{\eta} &= f(\eta, y, v(\hat{\eta} + \beta_1(y), y)) \\ \dot{y} &= h(\eta, y, v(\hat{\eta} + \beta_1(y), y)) \\ \dot{\hat{\eta}} &= \beta_2(y, \hat{\eta})\end{aligned} \tag{1.28}$$

with extended state $(\eta, y, \hat{\eta})$ and "controls" β_1 and β_2. The output feedback stabilization problem boils down to finding functions $\beta_1(\cdot)$ and $\beta_2(\cdot)$ such that the system (1.28) is immersed into the system (1.26).

1.4.1 Linearly Parameterized Systems

As in Section 1.3.1, we will first consider linearly parameterized, control affine systems of the form

$$\begin{aligned}\dot{\eta} &= f_0(y) + f_1(y)\eta + g_1(y)u \\ \dot{y} &= h_0(y) + h_1(y)\eta + g_2(y)u\end{aligned} \tag{1.29}$$

with state $(\eta, y) \in \mathbb{R}^n \times \mathbb{R}^r$, output y and input $u \in \mathbb{R}^m$.

Proposition 1.3. *Consider the system (1.29) and assume the following hold.*

(C1) There exists a full-information control law $u = v(\eta, y)$ such that all trajectories of the closed-loop system

$$\begin{aligned}\dot{\eta} &= f^*(\eta, y) := f_0(y) + f_1(y)\eta + g_1(y)v(\eta, y) \\ \dot{y} &= h^*(\eta, y) := h_0(y) + h_1(y)\eta + g_2(y)v(\eta, y)\end{aligned} \tag{1.30}$$

are bounded and satisfy (1.27).

(C2) There exists a mapping $\beta_1(\cdot)$ such that all trajectories of the system

$$\begin{aligned}\dot{z} &= \left(f_1(y) - \frac{\partial \beta_1}{\partial y} h_1(y) \right) z \\ \dot{\eta} &= f^*(\eta, y) + g_1(y)\left(v(\eta + z, y) - v(\eta, y)\right) \\ \dot{y} &= h^*(\eta, y) + g_2(y)\left(v(\eta + z, y) - v(\eta, y)\right)\end{aligned} \tag{1.31}$$

are bounded and satisfy

$$\lim_{t \to \infty} \left[v(\eta(t) + z(t), y(t)) - v(\eta(t), y(t)) \right] = 0.$$

Then there exists a dynamic output feedback control law described by equations of the form

$$u = v(\hat{\eta} + \beta_1(y), y), \qquad \dot{\hat{\eta}} = \beta_2(y, \hat{\eta}) \tag{1.32}$$

such that all trajectories of the closed-loop system (1.25)-(1.32) are bounded and satisfy (1.27).

Proof. As in the proof of Proposition 1.1, we will construct a function $\beta_2(y, \hat{\eta})$ so that the closed-loop system (1.29)-(1.32) is transformed into (1.31). To this end, let

$$z = \hat{\eta} - \eta + \beta_1(y),$$

which represents the distance of the system trajectory from the manifold $\phi(\eta, y, \hat{\eta}) = \hat{\eta} - \eta + \beta_1(y) = 0$, and note that the dynamics of the variable z are given by

$$\begin{aligned} \dot{z} = {} & \beta_2(y, \hat{\eta}) - f_0(y) - f_1(y)\left(\hat{\eta} + \beta_1(y) - z\right) - g_1(y) v(\hat{\eta} + \beta_1(y), y) \\ & + \frac{\partial \beta_1}{\partial y}\left[h_0(y) + h_1(y)\left(\hat{\eta} + \beta_1(y) - z\right) + g_2(y) v(\hat{\eta} + \beta_1(y), y)\right]. \end{aligned}$$

Selecting the adaptation law

$$\begin{aligned} \beta_2(y, \hat{\eta}) = {} & f_0(y) + f_1(y)\left(\hat{\eta} + \beta_1(y)\right) + g_1(y) v(\hat{\eta} + \beta_1(y), y) \\ & - \frac{\partial \beta_1}{\partial y}\left[h_0(y) + h_1(y)\left(\hat{\eta} + \beta_1(y)\right) + g_2(y) v(\hat{\eta} + \beta_1(y), y)\right] \end{aligned}$$

yields the first equation of (1.31), while the other two equations are obtained from (1.29) with $u = v(\hat{\eta} + \beta_1(y), y) = v(\eta + z, y)$ by adding and subtracting the term $v(\eta, y)$. Hence, by assumptions (C1)-(C2), all trajectories of the closed-loop system (1.25)-(1.32) are bounded and satisfy (1.27). ■

Example 1.2. Consider the stabilization to zero of the second-order nonlinear system

$$\begin{aligned} \dot{\eta} &= \eta + y \\ \dot{y} &= \eta\left(y^2 + 1\right) + u \end{aligned} \tag{1.33}$$

with input u and output y, where η is an unknown state. It is interesting to note that the zero dynamics are described by the equation $\dot{\eta} = \eta$, hence the system (1.33) is not minimum-phase. To begin with, note that (C1) is satisfied with the function

$$v(\eta, y) = -\eta(y^2 + 1) - (1 + k_1 + k_2)(\eta + y) - k_1 k_2 \eta$$

with $k_1 > 0$, $k_2 > 0$. Consider now the dynamic control law (1.32) and the change of coordinates $\tilde{y} = y + (1 + k_1)\eta$. Selecting

$$\beta_2(y) = \frac{\partial \beta_1}{\partial y}\left[(1 + k_1 + k_2)\left(\hat{\eta} + \beta_1 + y\right) + k_1 k_2\left(\hat{\eta} + \beta_1\right)\right] + \hat{\eta} + \beta_1 + y$$

yields the error system

$$\begin{aligned}\dot{z} &= -\left[\frac{\partial \beta_1}{\partial y}\left(y^2+1\right)-1\right] z \\ \dot{\eta} &= -k_1\eta + \tilde{y} \\ \dot{\tilde{y}} &= -k_2\tilde{y} - \left(y^2+1\right) z - \left(1+k_1\right)\left(1+k_2\right) z.\end{aligned} \tag{1.34}$$

A suitable selection for the function β_1 is given by

$$\beta_1(y) = \gamma\left(\frac{y^3}{3}+y\right) + (1+\gamma)\arctan(y),$$

which is such that

$$\frac{\partial \beta_1}{\partial y}\left(y^2+1\right) - 1 = \gamma\left(y^2+1\right)^2 + \gamma.$$

This ensures that the signals $\left(y^2+1\right) z$ and $\left(1+k_1\right)\left(1+k_2\right) z$ are in $\mathcal{L}_2$, hence all trajectories of the system (1.34) are bounded and η and y converge to zero.

1.4.2 Control Design Using a Separation Principle

The result in Proposition (1.3) relies on our ability to stabilize the cascade system (1.31) by assigning the function $\beta_1(\cdot)$. Note that, by construction, when $z = 0$, the (η, y)-subsystem in (1.31) is globally stable with y converging to the desired equilibrium. Therefore, it is natural to ask whether we can neglect the (η, y)-subsystem and concentrate on the stabilization of the z-subsystem alone. In the case of Proposition (1.3) this is not possible because, even if z converges to zero exponentially, the term $v(\eta + z, y) - v(\eta, y)$ may destabilize the (η, y)-subsystem. This clearly poses a restriction on the function $v(\eta, y)$.

The present section provides an answer to this problem by establishing a condition under which the design of the full-information control law $u = v(\eta, y)$ can be *decoupled* from the design of the function $\beta_1(\cdot)$ and so the stabilization of the z-subsystem can be considered independently. In this sense it can be considered as a nonlinear counterpart of the well-known separation principle used in linear systems. Interestingly, this result can be applied to systems that are not affine in the input. Moreover, the result can be extended to systems that are nonlinear in the unmeasured states.[7]

We consider systems described by equations of the form

$$\begin{aligned}\dot{\eta} &= f_0(y, u) + f_1(y, u)\eta \\ \dot{y} &= h_0(y, u) + h_1(y, u)\eta\end{aligned} \tag{1.35}$$

with state $(\eta, y) \in \mathbb{R}^n \times \mathbb{R}^r$, output y and input $u \in \mathbb{R}^m$. As before, we are seeking a dynamic output feedback control law

[7] See [5] for more detail.

$$u = v(\hat{\eta} + \beta_1(y), y), \qquad \dot{\hat{\eta}} = \beta_2(y, \hat{\eta}) \tag{1.36}$$

such that all trajectories of the closed-loop system (1.35)-(1.36) are bounded and

$$\lim_{t \to \infty} y(t) = y^*. \tag{1.37}$$

Proposition 1.4. *Consider the system (1.35) and assume the following hold.*

(D1) There exists a full-information control law

$$u = v(\eta, y) \tag{1.38}$$

such that all trajectories of the closed-loop system (1.35)-(1.38) are bounded and satisfy (1.37). Moreover, the system (1.35) with

$$u = v(\eta + z, y)$$

is globally bounded-input bounded-state stable with respect to the input z.

(D2) There exists a mapping $\beta_1(\cdot)$ such that the system

$$\dot{z} = \left(f_1(y, u) - \frac{\partial \beta_1}{\partial y} h_1(y, u) \right) z \tag{1.39}$$

is uniformly globally stable for any y and u and $z(t)$ is such that, for any fixed η and y,

$$\lim_{t \to \infty} [v(\eta + z(t), y)] = v(\eta, y).$$

Then there exists a dynamic output feedback control law described by equations of the form (1.36) such that all trajectories of the closed-loop system (1.35)-(1.36) are bounded and satisfy (1.37).

The proof is similar to the one of Proposition 1.3, hence it is omitted.

Remark 1.4. Assumption (D2) can be replaced by the following (stronger) condition.

(D3) There exists a mapping $\beta_1(\cdot)$ such that the system (1.39) is uniformly globally stable for any y and u and $z(t)$ is such that, for any y and u,

$$\lim_{t \to \infty} z(t) = 0.$$

If (D3) holds, then $\hat{\eta}$ can be used to construct an *asymptotic* estimate of the unmeasured states η, which is given by $\hat{\eta} + \beta_1(y)$. However, to achieve the desired control goal, only the reconstruction of the full-information control law $v(\eta, y)$ is necessary.

1.5 Applications

1.5.1 Aircraft Wing Rock Suppression

In this section we apply the result in Section 1.3.2 to the problem of wing rock elimination in high-performance aircrafts. This example has been adopted from [6, Section 4.6], where a classical controller, based on the adaptive backstepping method, has been proposed.

Consider the system

$$\begin{aligned}\dot{x}_1 &= x_2\\ \dot{x}_2 &= x_3 + \phi(x_1, x_2)^T\theta\\ \dot{x}_3 &= \frac{1}{\tau}u - \frac{1}{\tau}x_3,\end{aligned} \tag{1.40}$$

where the states x_1, x_2 and x_3 represent the roll angle, roll rate and aileron deflection angle respectively, τ is the aileron time constant, u is the control input, $\theta \in \mathbb{R}^5$ is an unknown constant vector and

$$\phi(x_1, x_2)^T = \left[1,\, x_1,\, x_2,\, |x_1|x_2,\, |x_2|x_2\right].$$

The control objective is to regulate x_1 to zero. Note that, despite the presence of extra terms in the dynamics of x_3, the result in Section 1.3.2 still applies. The target dynamics are defined as

$$\dot{\xi} = \begin{bmatrix} 0 & 1 \\ -k_1 & -k_2 \end{bmatrix}\xi \tag{1.41}$$

with $k_1 > 0$ and $k_2 > 0$. From equation (1.24), the function β_1 is given by

$$\beta_1(x_1, x_2) = \gamma\left[x_2,\, x_1x_2,\, \tfrac{1}{2}x_2^2,\, \tfrac{1}{2}|x_1|x_2^2,\, \tfrac{1}{3}|x_2|x_2^2\right]$$

with $\gamma > 0$, while the control laws are defined as

$$\begin{aligned}\beta_2(x,\hat{\theta}) &= -\frac{\partial\beta_1}{\partial x_1}x_2 - \frac{\partial\beta_1}{\partial x_2}\left[x_3 + \phi^T(x_1,x_2)\left(\hat{\theta} + \beta_1(x_1,x_2)\right)\right]\\ \frac{1}{\tau}u &= \frac{1}{\tau}x_3 - \left(k_1 + \frac{\partial\phi^T}{\partial x_1}\hat{\theta} + \frac{\partial\phi^T\beta_1}{\partial x_1}\right)x_2 - \phi^T(x_1,x_2)\beta_2(x,\hat{\theta})\\ &\quad - \left(k_2 + \frac{\partial\phi^T}{\partial x_2}\hat{\theta} + \frac{\partial\phi^T\beta_1}{\partial x_2}\right)\left[x_3 + \phi^T(x_1,x_2)\left(\hat{\theta} + \beta_1(x_1,x_2)\right)\right]\\ &\quad -\lambda z_1 - \frac{1}{2\epsilon}\left(k_2 + \frac{\partial\phi^T}{\partial x_2}\hat{\theta} + \frac{\partial\phi^T\beta_1}{\partial x_2}\right)^2 z_1\end{aligned}$$

with $\lambda > 0$.

Figure 1.1 shows the trajectory of the closed-loop system from the initial state $x(0) = [0.4, 0, 0]$ for the data provided in [6, Section 4.6], namely

$$\theta = [0, -26.67, 0.76485, -2.9225, 0]$$

and $\tau = 1/15$. The design parameters are $k_1 = 25$, $k_2 = 10$, $\lambda = 5$, $\gamma = 100$ and $\epsilon = 5000$. We see that the proposed adaptive scheme recovers the performance of the full-information controller. The speed of response can be further increased (or reduced) by tuning the parameter γ.

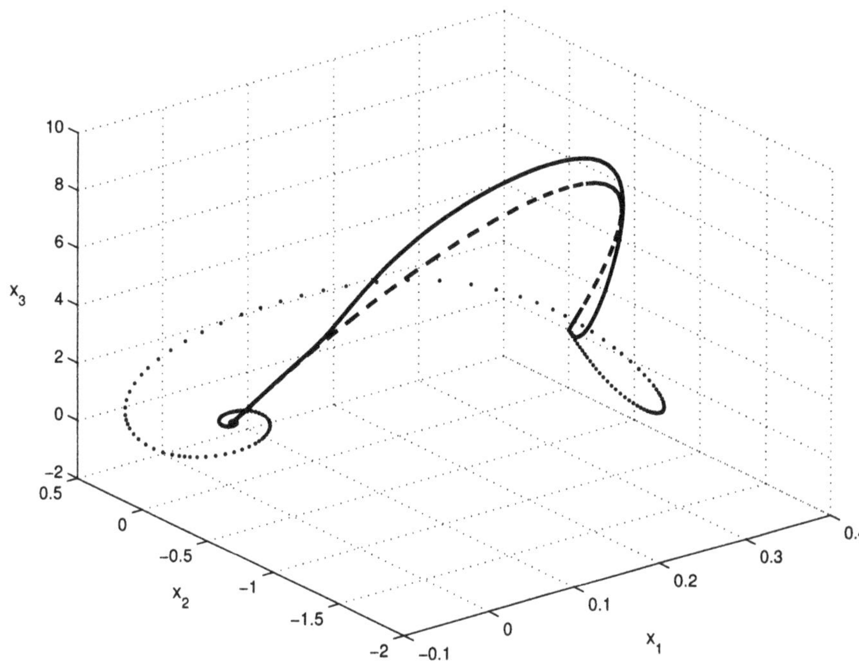

Fig. 1.1. State-space trajectory of the wing rock system. Dashed line: Full-information controller. Dotted line: Adaptive backstepping controller. Solid line: Proposed controller.

1.5.2 Output Voltage Regulation of Boost Converters

In this section we consider the problem of regulating the output voltage of a DC–DC boost converter with the circuit topology shown in Figure 1.2. The averaged model of the system can be described by equations of the form (1.35), namely

$$\begin{aligned} \dot{\eta}_1 &= 0 \\ \dot{\eta}_2 &= -\frac{1}{L}uy + \frac{1}{L}\eta_1 \\ \dot{y} &= -\frac{1}{RC}y + \frac{1}{C}u\eta_2, \end{aligned} \tag{1.42}$$

where the states η_1, η_2 and y represent the input voltage, inductor current and output voltage respectively, L, C and R are positive constants and $u \in [\epsilon, 1]$, with $0 < \epsilon < 1$, is the modulating signal of the PWM circuit controlling the switch and acts as the control input. The control objective is to regulate the output voltage y to a set-point y^* using output feedback only. Note that, due to the constraint $\epsilon \leq u \leq 1$, the set-point y^* must be such that

$$\epsilon \leq \frac{\eta_1}{y^*} \leq 1.$$

To begin with, note that the full-information control law

$$u = v(\eta, y) = \mathrm{sat}_{[\epsilon,1]}(\frac{\eta_1}{y^*})$$

is such that the assumption (D1) of Proposition 1.4 holds. Consider now a dynamic output feedback control law of the form (1.36), namely

$$\begin{aligned} u &= \mathrm{sat}_{[\epsilon,1]}(\frac{\begin{bmatrix} 1 & 0 \end{bmatrix} (\hat{\eta} + \beta_1(y))}{y^*}) \\ \dot{\hat{\eta}} &= \beta_2(y, \hat{\eta}) \end{aligned} \tag{1.43}$$

and the error variable $z = \hat{\eta} - \eta + \beta_1(y)$, whose time-derivative is given by

$$\begin{aligned} \dot{z} = \beta_2(y, \hat{\eta}) - \begin{bmatrix} 0 & 0 \\ \frac{1}{L} & 0 \end{bmatrix} (\hat{\eta} + \beta_1(y) - z) + \begin{bmatrix} 0 \\ \frac{1}{L} uy \end{bmatrix} \\ + \frac{\partial \beta_1}{\partial y} \left(-\frac{1}{RC} y + \begin{bmatrix} 0 & \frac{1}{C} u \end{bmatrix} (\hat{\eta} + \beta_1(y) - z) \right). \end{aligned}$$

Selecting the adaptation law

$$\begin{aligned} \beta_2(y, \hat{\eta}) = \begin{bmatrix} 0 & 0 \\ \frac{1}{L} & 0 \end{bmatrix} (\hat{\eta} + \beta_1(y)) - \begin{bmatrix} 0 \\ \frac{1}{L} uy \end{bmatrix} \\ - \frac{\partial \beta_1}{\partial y} \left(-\frac{1}{RC} y + \begin{bmatrix} 0 & \frac{1}{C} u \end{bmatrix} (\hat{\eta} + \beta_1(y)) \right) \end{aligned}$$

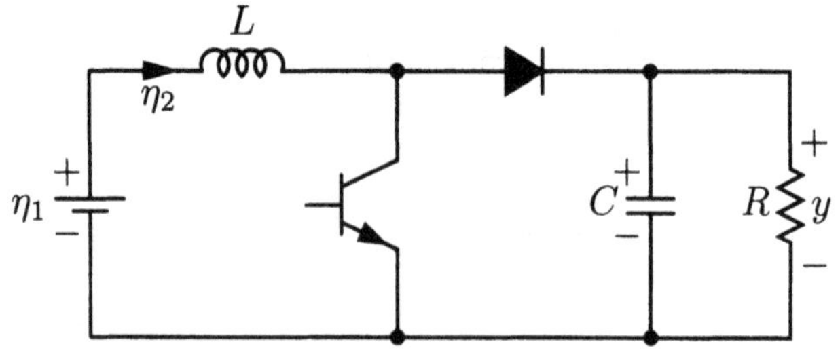

Fig. 1.2. Diagram of the DC–DC boost converter.

and the mapping $\beta_1(y) = \left[\lambda_1\ \lambda_2\right]^T y$, with $\lambda_1 > 0$, $\lambda_2 > 0$, yields the error dynamics

$$\dot{z} = \begin{bmatrix} 0 & -\frac{\lambda_1}{C}u \\ \frac{1}{L} & -\frac{\lambda_2}{C}u \end{bmatrix} z. \tag{1.44}$$

It remains to select the constants λ_1 and λ_2 so that the system (1.44) satisfies the assumption (D2) of Proposition 1.4. To this end, let $\bar{u} = (1+\epsilon)/2$ and rewrite the system (1.44) as

$$\dot{z} = \begin{bmatrix} 0 & -\frac{\lambda_1}{C}\bar{u} \\ \frac{1}{L} & -\frac{\lambda_2}{C}\bar{u} \end{bmatrix} z + \begin{bmatrix} \frac{\lambda_1}{C}\bar{u} \\ \frac{\lambda_2}{C}\bar{u} \end{bmatrix} w \tag{1.45}$$

with

$$w = \frac{\bar{u} - u}{\bar{u}} r, \qquad r = [0\ 1]z,$$

i.e. as a linear system, which is asymptotically stable (since $\lambda_1 > 0$ and $\lambda_2 > 0$), in closed loop with a time-varying gain $(\bar{u} - u)/\bar{u}$, which is such that

$$|\frac{\bar{u} - u}{\bar{u}}| \leq \frac{1-\epsilon}{1+\epsilon}.$$

Note now that the H_∞ norm of the system (1.45) with output r is given by

$$\gamma(\mu) = \frac{1}{\sqrt{1 + 2\mu^2\sqrt{1 + 2/\mu} - 2\mu^2 - 2\mu}}, \qquad \mu = \frac{\lambda_1}{\lambda_2^2}\frac{C}{L\bar{u}}.$$

Hence, a simple application of the small-gain theorem[8] shows that the system (1.44) is asymptotically stable provided

$$\gamma(\mu)\frac{1-\epsilon}{1+\epsilon} < 1.$$

Since $\gamma(\mu) \geq 1$ for any $\mu \geq 0$, this condition is equivalent to

$$\epsilon > \frac{\gamma(\mu) - 1}{\gamma(\mu) + 1}. \tag{1.46}$$

Note that the function on the right-hand side of (1.46) is zero for $\mu = 0$ and it is monotonically increasing and tends to one for $\mu \to \infty$, hence, for any ϵ, it is always possible to select $\mu > 0$, thus $\lambda_1 > 0$ and $\lambda_2 > 0$, such that (1.46) holds. As a result, the system (1.44) is uniformly globally asymptotically stable. Moreover, since the condition (D3) in Remark 1.4 holds, an asymptotic estimate of the unmeasured state η is given by

[8] See [9, Proposition 3.4.7].

$$\hat{\eta} + \beta_1(y) = \begin{bmatrix} \hat{\eta}_1 + \lambda_1 y \\ \hat{\eta}_2 + \lambda_2 y \end{bmatrix} = \begin{bmatrix} \eta_1 + z_1 \\ \eta_2 + z_2 \end{bmatrix}.$$

The closed-loop system (1.42)-(1.43) has been simulated using the parameters $L = 20m$H, $C = 20\mu$F and $R = 30\Omega$. The desired output voltage is $y^* = 30$V and the design parameters are $\epsilon = 0.1$ and $\lambda_1 = \lambda_2 = 0.01$, which satisfy the condition (1.46). It is assumed that the input voltage is $\eta_1 = 15$V at $t = 0$s and it changes to $\eta_1 = 10$V at $t = 0.02$s. In Figure 1.3 the input voltage η_1 and the inductor current η_2 together with their estimates, the output voltage y and the control signal u are displayed. We see that the output voltage y tracks the desired value y^*, despite partial state measurement and the change in the input voltage. In addition, the input voltage and the inductor current estimates converge to the true values.

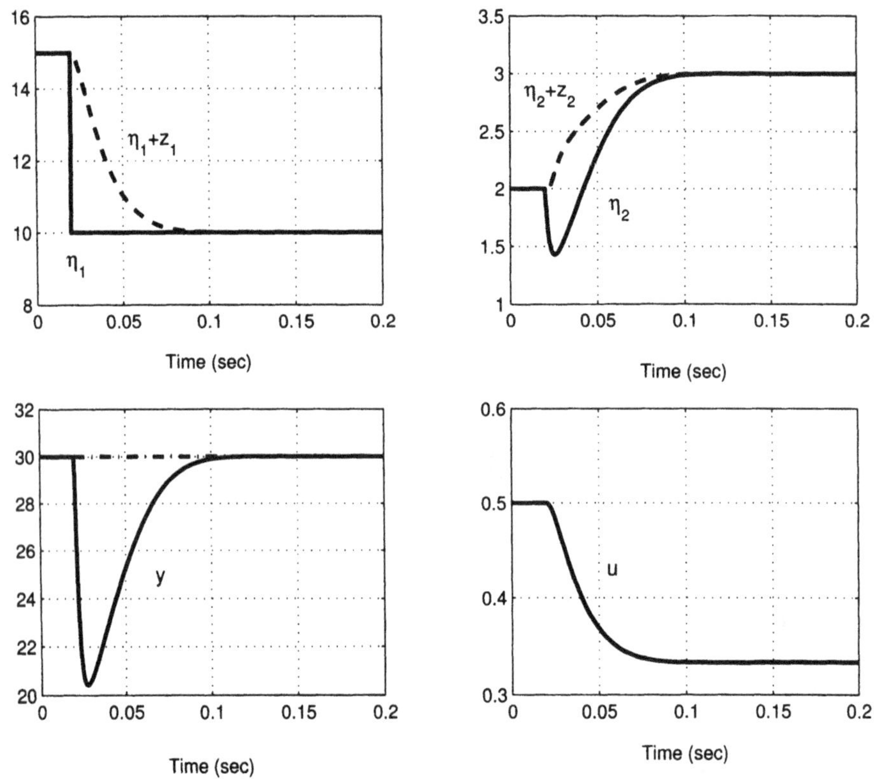

Fig. 1.3. State and control histories of the boost converter.

1.6 Conclusions

The problem of designing globally stabilizing adaptive control laws for general nonlinear systems has been addressed from a new perspective using the notion of system immersion. The proposed methodology is applicable to both state and output feedback control problems and treats unmeasured parameters and unknown states in a unified way. The method has been illustrated with several academic and practical examples, highlighting the potential of the proposed approach in solving nonlinear control problems with partial state and/or parameter information.

References

1. A. Astolfi and R. Ortega (2003) Immersion and invariance: a new tool for stabilization and adaptive control of nonlinear systems. *IEEE Trans. Automatic Control*, 48(4):590–606.
2. C.I. Byrnes, F. Delli Priscoli, and A. Isidori (1997) *Output regulation of uncertain nonlinear systems*. Birkhäuser.
3. A. Isidori (1995) *Nonlinear Control Systems*. Springer-Verlag, 3rd edition.
4. A. Isidori (1999) *Nonlinear Control Systems II*. Springer-Verlag.
5. D. Karagiannis, A. Astolfi, and R. Ortega (2003) Two results for adaptive output feedback stabilization of nonlinear systems. *Automatica*, 39(5):857–866.
6. M. Krstić, I. Kanellakopoulos, and P. Kokotović (1995) *Nonlinear and Adaptive Control Design*. John Wiley and Sons.
7. R. Marino and P. Tomei (1995) *Nonlinear Control Design: Geometric, Adaptive and Robust*. Prentice Hall.
8. R. Ortega, A. Loría, P.J. Nicklasson, and H. Sira-Ramírez (1998) *Passivity-based Control of Euler-Lagrange Systems*. Springer-Verlag.
9. A. van der Schaft (2000) *L_2-Gain and Passivity Techniques in Nonlinear Control*. Springer-Verlag, 2nd edition.
10. V.I. Utkin (1992) *Sliding Modes in Control and Optimization*. Springer-Verlag.

2

Cascaded Nonlinear Time-Varying Systems: Analysis and Design

Antonio Loría and Elena Panteley

CNRS–LSS, Supélec, 3 rue Joliot Curie, 91192 Gif s/Yvette, France.
loria@lss.supelec.fr, panteley@lss.supelec.fr.

The general topic of study is Lyapunov stability of nonlinear time-varying cascaded systems. Roughly speaking these are systems in "open loop" as illustrated in the figure below.

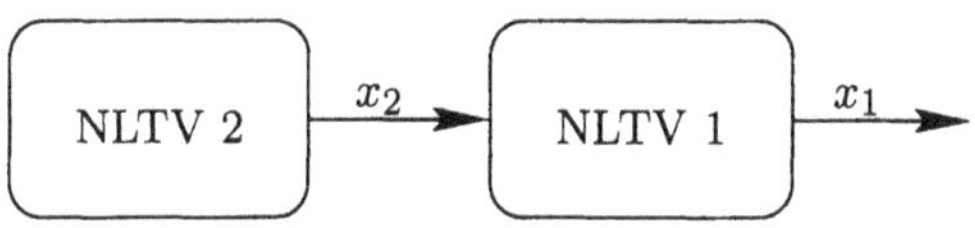

Fig. 2.1. Block-diagram of a cascaded system.

The results that we present here are not original, they have been published in different scientific papers so the adequate references are provided in the Bibliography. The chapter gathers the material of lectures given by the first author within the "Formation d'Automatique de Paris" 2004.

The material is organised in three main sections. In the first, we introduce the reader to several definitions and state our motivations to study time-varying systems. We also state the problems of design and analysis of cascaded control systems. The second part contains the main stability results. All the theorems and propositions in this section are on conditions to guarantee Lyapunov stability of cascades. No attention is paid to the control design problem. The third section contains some selected practical applications where control design aiming at obtaining a cascaded system in closed loop reveals to be better than classical Lyapunov-based designs such as Backstepping.

The technical proofs of the main stability results are omitted here but the interested readers are invited to see the references cited in the Bibliography at the end of the chapter.

Throughout this chapter we use the following nomenclature.

F. Lamnabhi-Lagarrigue et al. (Eds.): Adv. Top. in Cntrl. Sys. Theory, LNCIS 311, pp. 23–64, 2005

Notation. The solution of a differential equation $\dot{x} = f(t,x)$, where $f : \mathbb{R}_{\geq 0} \times \mathbb{R}^n \to \mathbb{R}^n$, with initial conditions $(t_\circ, x_\circ) \in \mathbb{R}_{\geq 0} \times \mathbb{R}^n$ and $x_\circ = x(t_\circ)$, is denoted by $x(\cdot\,; t_\circ, x_\circ)$ or simply by $x(\cdot)$. We say that the system $\dot{x} = f(t,x)$ is uniformly globally stable (UGS) if the trivial solution $x(\cdot\,; t_\circ, x_\circ) \equiv 0$ is UGS (cf. Definition 2.4). Respectively for uniform global asymptotic stability (UGAS, cf. Definition 2.5). These properties will be precisely defined later. $|\cdot|$ stands for the Euclidean norm of vectors and induced norm of matrices, and $\|\cdot\|_p$, where $p \in [1,\infty]$, denotes the $\mathcal{L}_p$ norm of time signals. In particular, for a measurable function $\phi : \mathbb{R}_{\geq t_\circ} \to \mathbb{R}^n$, $\|\phi\|_p$ denotes $(\int_{t_\circ}^{\infty} |\phi(t)|^p \, dt)^{1/p}$ for $p \in [1,\infty)$ and $\|\phi\|_\infty$ denotes the quantity $\operatorname{ess\,sup}_{t \geq t_\circ} |\phi(t)|$. We denote by B_r the open ball $B_r := \{x \in \mathbb{R}^n \, : \, |x| < r\}$. We denote by $\dot{V}_{(\#)}(t,x) := \frac{\partial V}{\partial t} + \frac{\partial V}{\partial x} f(t,x)$ the time derivative of the (differentiable) Lyapunov function $V(t,x)$ along the solutions of the differential equation $\dot{x} = f(t,x)$ labelled by $(\#)$. When clear from the context we use the compact notation $V(t,x(t)) = V(t)$. We also use $L_\psi V = \frac{\partial V}{\partial x} \psi$ for a vector field $\psi : \mathbb{R}_{\geq 0} \times \mathbb{R}^q \to \mathbb{R}^n$.

2.1 Preliminaries on Time-Varying Systems

The first subject of study in this report is to establish sufficient (and for some cases, necessary) conditions to guarantee *uniform* global asymptotic stability (UGAS) (cf. Definition 2.5) of the origin, for nonlinear ordinary differential equations (ODE)

$$\dot{x} = f(t,x) \qquad x(t_\circ) =: x_\circ \, . \tag{2.1}$$

Most of the literature for nonlinear systems in the last decades has been devoted to time-*invariant* systems nonetheless, the importance of nonautonomous systems should not be underestimated; these arise for instance as closed-loop systems in nonlinear trajectory tracking control problems, that is, where the goal is to design a control input $u(t,x)$ for the system

$$\dot{x} = f(x,u) \qquad x(t_\circ) =: x_\circ \tag{2.2a}$$

$$y = h(x) \tag{2.2b}$$

such that the output y follows asymptotically a desired time-varying reference $y_d(t)$. For a "feasible" trajectory $y_d(t) = h(x_d(t))$ some "desired" state trajectory $x_d(t)$, satisfying $\dot{x}_d = f(x_d, u)$, system (2.2) in closed loop with the control input $u = u(x, x_d(t), y_d(t))$ may be written as

$$\dot{\tilde{x}} = \tilde{f}(t,\tilde{x})\,, \qquad \tilde{x}(t_\circ) = \tilde{x}_\circ\,, \tag{2.3a}$$

$$\tilde{y} = \tilde{h}(t,\tilde{x}) \tag{2.3b}$$

where $\tilde{x} = x - x_d$, $\tilde{y} = y - y_d$ and $\tilde{f}(t,x) := f(\tilde{x} + x_d(t),\, u(\tilde{x} + x_d(t), x_d(t), y_d(t))$. The so stated tracking control problem, applies to many physical systems, e.g.

mechanical and electromechanical, for which there is a large body of literature (see [39] and references therein).

Another typical situation in which closed-loop systems of the form (2.3) arise, is in regulation problems (that is, when the desired set-point (x_d, y_d) is *constant*) such that the open-loop plant is not stabilisable by continuous time-invariant feedbacks $u = u(x)$. This is the case, for instance, of some driftless (e.g. nonholonomic) systems, $\dot{x} = g(x)u$. See e.g. [4, 22].

A classical approach to analyse the *stability* of the *nonautonomous* system (2.1) is to search a so-called Lyapunov function with certain properties (see e.g. [66, 20]). Consequently, for the tracking control design problem previously described one looks for so-called Control Lyapunov Function (CLF) for system (2.2) so that the control law u is derived from the CLF (see e.g. [23, 47, 57]). In general, finding an adequate LF or CLF is very hard and one has to focus on systems with specific structural properties. This gave rise to approaches such as the so-called integrator backstepping [23] and feedforwarding [30, 57].

The second subject of study in this chapter, is a specific structure of systems, which are wide enough to cover a large number of applications while simple enough to allow criteria for stability which are easier to verify than finding an LF for the closed-loop system. These are *cascaded* systems. We distinguish between two problems: stability *analysis* and control *design*. For the design problem, we do not offer a general methodology as in [23, 57] however, we show through different applications, that simple (from a mathematical viewpoint) controllers can be obtained by aiming at giving the closed-loop system a cascaded structure.

2.1.1 Stability Definitions

There are various types of asymptotic stability that can be pursued for time-varying nonlinear systems. As we shall see in this section, from a *robustness* viewpoint, the most useful are *uniform* (global) asymptotic stability and *uniform* (local) exponential stability (ULES). For clarity of exposition and self-containedness let us recall a few definitions (cf. e.g. [19, 66, 10]) as we use them throughout the chapter.

Definition 2.1. *A continuous function* $\alpha : \mathbb{R}_{\geq 0} \to \mathbb{R}_{\geq 0}$ *is said to belong to* class $\mathcal{K}$ *if it is strictly increasing and* $\alpha(0) = 0$. *It is said to be of* class $\mathcal{K}_\infty$ *if moreover* $\alpha(s) \to \infty$ *as* $s \to \infty$.

Definition 2.2. *A continuous function* $\beta : \mathbb{R}_{\geq 0} \times \mathbb{R}_{\geq 0} \to \mathbb{R}_{\geq 0}$ *is said to belong to* class $\mathcal{KL}$ *if, for each fixed* s $\beta(\cdot, s)$ *is of class* $\mathcal{K}$ *and for each fixed* r, $\beta(r, \cdot)$ *is strictly decreasing and* $\beta(r, s) \to 0$ *as* $s \to \infty$.

For system (2.1), we define the following.

Definition 2.3 (Uniform boundedness). *We say that the solutions of (2.1) are uniformly (resp. globally) bounded (UB, resp., UGB) if there exist a class $\mathcal{K}$ (resp. $\mathcal{K}_\infty$) function α and a number $c > 0$ such that*

$$|x(t;t_\circ,x_\circ)| \le \alpha(|x_\circ|) + c \qquad \forall\, t \ge t_\circ\,. \tag{2.4}$$

Definition 2.4 (Uniform stability). *The origin of system (2.1) is said to be uniformly stable (US) if there exist a constant $r > 0$ and $\gamma \in \mathcal{K}$ such that, for each $(t_\circ, x_\circ) \in \mathbb{R}_{\ge 0} \times B_r$,*

$$|x(t;t_\circ,x_\circ)| \le \gamma(|x_\circ|) \qquad \forall\, t \ge t_\circ\,. \tag{2.5}$$

If the bound (2.5) holds for all $(t_\circ, x_\circ) \in \mathbb{R}_{\ge 0} \times \mathbb{R}^n$ and with $\gamma \in \mathcal{K}_\infty$, then the origin is uniformly globally stable (UGS).

Remark 2.1.

1. The formulation of uniform stability given above is equivalent to the classical one, i.e., "the system (2.1) is uniformly stable in the sense defined above if and only if for each ε there exists $\delta(\varepsilon)$ such that $|x_\circ| \le \delta$ implies that $|x(t)| \le \varepsilon$ for all $t \ge t_\circ$. Necessity becomes evident if we take $\delta(s) = \gamma^{-1}(s)$ for sufficiently small[1].
2. It is clear from the above that uniform global boundedness is a necessary condition for uniform global stability, that is, in the case that $\gamma \in \mathcal{K}_\infty$ we have that (2.5) implies (2.4). However, a system may be (*locally*) uniformly stable and yet, have unbounded solutions.
3. Another common characterization of UGS and which we use in some proofs is the following (see e.g. [23]): "the system is UGS if it is US and uniformly globally bounded (UGB)". Indeed, observe that US implies that there exists $\gamma \in \mathcal{K}$ such that (2.5) holds. Then, using (2.4) and letting $b \le r$, one can construct $\bar\gamma \in \mathcal{K}_\infty$ such that $\bar\gamma(s) \ge \alpha(s) + c$ for all $s \ge b > 0$ and $\bar\gamma(s) \ge \alpha(s)$ for all $s \le b \le r$. Hence (2.5) holds with $\bar\gamma$.

Definition 2.5 (Uniform asymptotic stability). *The origin of system (2.1) is said to be uniformly asymptotically stable (UAS) if it is uniformly stable and uniformly attractive, i.e., there exists $r > 0$ and for each $\sigma > 0$ there exists $T > 0$, such that*

$$|x_\circ| \le r \implies |x(t;t_\circ,x_\circ)| \le \sigma \qquad \forall t \ge t_\circ + T\,. \tag{2.6}$$

If moreover the origin of system is UGS and the bound (2.6) holds for each $r > 0$ then the origin is uniformly globally asymptotically stable (UGAS).

[1] "Sufficiently small" here means that s is taken to belong to the domain of γ^{-1}. Recall that in general, *class* $\mathcal{K}$ functions are not invertible on $\mathbb{R}_{\ge 0}$. For instance, $\tanh(|\cdot|) \in \mathcal{K}$ but $\tanh^{-1} : (-1, 1) \to \mathbb{R}_{\ge 0}$.

Remark 2.2 (class-$\mathcal{KL}$ characterization of UGAS). It is known (see, e.g., [10, Section 35] and [26, Proposition 2.5]) that the two properties characterizing uniform global asymptotic stability hold if and only if there exists a function $\beta \in \mathcal{KL}$ such that, for all $(t_\circ, x_\circ) \in \mathbb{R}_{\geq 0} \times \mathbb{R}^n$, all solutions satisfy

$$|x(t; t_\circ, x_\circ)| \leq \beta(|x_\circ|, t - t_\circ) \qquad \forall\, t \geq t_\circ\,. \tag{2.7}$$

The local counterpart is that the origin of system (2.1) is UAS if there exist a constant $r > 0$ and $\beta \in \mathcal{KL}$ such that for all $(t_\circ, x_\circ) \in \mathbb{R}_{\geq 0} \times B_r$.

Definition 2.6 (Exponential convergence). *The trajectories of system (2.1) are said to be exponentially convergent if there exists $r > 0$ such that* for each *pair of initial conditions $(t_\circ, x_\circ) \in \mathbb{R}_{\geq 0} \times B_r$, there exist γ_1 and $\gamma_2 > 0$ such that each solution $x(t; t_\circ, x_\circ)$ of (2.1), satisfies*

$$|x(t; t_\circ, x_\circ)| \leq \gamma_1 |x_\circ| e^{-\gamma_2 (t - t_\circ)}\ . \tag{2.8}$$

The trajectories of the system are said to be globally exponentially convergent if the constants γ_i exist for each pair of initial conditions $(t_\circ, x_\circ) \in \mathbb{R}_{\geq 0} \times \mathbb{R}^n$.

Remark 2.3. Sometimes, in the literature, the dependence of γ_1 and γ_2 on the initial conditions is overlooked and, inappropriately, (uniform) global exponential stability is concluded. Motivated by such abuse of terminology we introduce below the following definition of exponential stability in which we emphasize the uniformity with respect to initial conditions.

Definition 2.7 (Uniform exponential stability). *The origin of the system (2.1) is said to be uniformly (locally) exponentially stable (ULES) if there exist constants γ_1, γ_2 and $r > 0$ such that for all $(t_\circ, x_\circ) \in \mathbb{R}_{\geq 0} \times B_r$*

$$|x(t; t_\circ, x_\circ)| \leq \gamma_1 |x_\circ| e^{-\gamma_2 (t - t_\circ)} \qquad \forall t \geq t_\circ. \tag{2.9}$$

If for each $r > 0$ there exist γ_1, γ_2 such that condition (2.9) holds for all $(t_\circ, x_\circ) \in \mathbb{R}_{\geq 0} \times B_r$ then, the system is said to be uniformly semiglobally exponentially stable (USGES)[2].

Finally, the origin of system (2.1) is said to be uniformly globally exponentially stable (UGES) if there exist γ_1, $\gamma_2 > 0$ such that (2.9) holds for all $(t_\circ, x_\circ) \in \mathbb{R}_{\geq 0} \times \mathbb{R}^n$.

2.1.2 Why *Uniform* Stability?

One of the main interests of the *uniform* forms of stability, is *robustness* with respect to bounded disturbances. If the time-varying system (2.1) with $f(t, \cdot)$

[2] This property has been called also, uniform exponential stability in any ball –cf. [25].

locally Lipschitz uniformly in t, is ULAS or ULES then the system is locally Input-to-State Stable (ISS); that is, for this system, there exist $\beta \in \mathcal{KL}$, $\gamma \in \mathcal{K}$ and a number δ such that, for all $t \geq t_\circ \geq 0$ (see e.g. [19, Definition 5.2])

$$\max\{|x_\circ|, \|u\|_\infty\} \leq \delta \implies |x(t; t_\circ, x_\circ, u)| \leq \beta(|x_\circ|, t - t_\circ) + \gamma(\|u\|_\infty). \tag{2.10}$$

This fact can be verified invoking [19, Lemma 5.4], and the converse Lyapunov theorems in [24, 26]. The importance of this implication is that, in particular, local ISS implies *total stability*, which can be defined as follows.

Definition 2.8 (Total stability[3]). *The origin of of $\dot{x} = f(t, x, 0)$, is said to be totally stable if for the system $\dot{x} = f(t, x, u)$ small bounded inputs $u(t, x)$ and small initial conditions $x_\circ = x(t_\circ)$ yield small state trajectories for all $t \geq t_\circ$, i.e., if for each $\varepsilon > 0$ there exists $\delta > 0$ such that*

$$\max\{|x_\circ|, \|u\|_\infty\} \leq \delta \qquad \implies \qquad |x(t; t_\circ, x_\circ, u)| \leq \varepsilon \quad \forall t \geq t_\circ \geq 0 . \tag{2.11}$$

In contrast to this, weaker forms of asymptotic stability for time-varying systems *do not* imply total stability. More precisely:

Proposition 2.1. *Consider the system (2.1) and assume that $f(t, \cdot)$ is locally Lipschitz uniformly in t, and the origin is UGS. The following conditions are* not *sufficient for total stability:*

1. *The origin is globally attractive*
2. *The trajectories of the system are exponentially convergent and $f(t, x)$ is globally Lipschitz in x, uniformly in t.*

Proof. We present an interesting example of an UGS nonlinear time-varying system which satisfies items 1 and 2 of the proposition above; yet, is not totally stable.

Example 2.1. [44] Consider system (2.1) with

$$f(t, x) = \begin{cases} -a(t)\mathrm{sgn}(x) & \text{if} \quad |x| \geq a(t) \\ -x & \text{if} \quad |x| \leq a(t) \end{cases} \tag{2.12}$$

and $a(t) = \dfrac{1}{t+1}$. This system has the following properties:

[3] The definition provided here is a modern version of the notion of total stability, originally introduced in [29] and which is more suitable for the purposes of these notes.

1. The function $f(t,x)$ is globally Lipschitz in x, uniformly in t and the system is UGS with linear gain equal to one i.e., with $\gamma(s) = s$.
2. For each $r > 0$ and $t_\circ \geq 0$ there exist strictly positive constants κ and λ such that for all $t \geq t_\circ$ and all $|x(t_\circ)| \leq r$
$$|x(t)| \leq \kappa |x(t_\circ)| e^{-\lambda(t-t_\circ)} \tag{2.13}$$
3. The origin is *not* totally stable. Furthermore, there always exist a bounded (arbitrarily small) additive perturbation, and $(t_\circ, x_\circ) \in \mathbb{R}_{\geq 0} \times \mathbb{R}^n$ such that the trajectories of the system grow *unboundedly* as $t \to \infty$.

The proof of these properties is provided in [44]. See also [35] for examples of linear time-varying systems proving the claim in Proposition 2.1. ∎

This lack of total stability for GAS (but not UGAS) and LES (but not ULES) nonautonomous systems, added to the unquestionable interest on total stability in time-varying systems arising in practical applications, is our main motivation to study sufficient conditions that guarantee UGAS and ULES for nonlinear nonautonomous systems.

As it has been mentioned before the stability analysis problem and hence, control design for time-varying systems is in general very hard to solve. By restricting the class of NLTV systems to cascades, we can establish simple-to-verify conditions for UGAS and UGES. The importance of these results is evident from their application to specific control design problems that we address.

2.2 Cascaded Systems

2.2.1 Introduction

To put the topic of cascades in perspective, let us consider the two linear time-invariant systems

$$\dot{x}_1 = A_1 x_1 \tag{2.14a}$$
$$\dot{x}_2 = A_2 x_2 \tag{2.14b}$$

where A_1 and A_2 are Hurwitz matrices of not necessarily equal dimension. Reconsider now the system (2.14a) as a system with an input and let these systems be interconnected in *cascade*, that is, redefine (2.14a) to be

$$\dot{x}_1 = A_1 x_1 + B x_2 \tag{2.15}$$

where B is a matrix of appropriate dimensions. It follows immediately from the property of *superposition*, which is inherent to linear systems, that if each of the systems in (2.14) is exponentially stable the *cascade* (2.15) - (2.14b) is also exponentially stable. (Note that no other assumptions are imposed).

This property comes even more "handy" if we are confronted to designing an observer-based controller for the system $\dot{x} = Ax + Bu$ where (A, B) is controllable. Assume we can measure the output $y = Cx$ and that the pair (A, C) is observable. Then, as it is studied in many textbooks on control systems theory, this problem is easily solved by means of the control law $u := -K\hat{x}$, where $\hat{x}$ is the estimate of x and K is such that $(A - BK)$ is Hurwitz, and the observer

$$\dot{\hat{x}} = A\hat{x} - LC\bar{x}\,, \quad \bar{x} := x - \hat{x} \tag{2.16}$$

with L such that $A - LC$ is Hurwitz. Indeed, global exponential stability of the overall closed-loop system

$$\dot{x} = (A - BK)x + BK\bar{x} \tag{2.17a}$$
$$\dot{\bar{x}} = (A - LC)\bar{x} \tag{2.17b}$$

follows if also $(A - LC)$ is Hurwitz, from the so called *separation principle* which is a direct consequence of the property of superposition. An alternative reasoning to infer GES for (2.17) starts with observing that this system has a so called *cascaded* structure as it does (2.15) - (2.14b). Roughly speaking, GES is concluded since both subsystems (2.17a), (2.17b) *separately* are GES and the interconnection term along the trajectories of the system, $BK\bar{x}(t)$, is obviously exponentially decaying.

Holding this simple viewpoint, it is natural to wonder whether a similar reasoning would hold to infer stability and convergence properties of cascaded *nonlinear (time-varying) systems*:

$$\dot{x}_1 = f_1(t, x_1) + g(t, x)x_2 \tag{2.18a}$$
$$\dot{x}_2 = f_2(t, x_2) \tag{2.18b}$$

For now, let us simply assume that the functions $f_1(\cdot,\cdot)$, $f_2(\cdot,\cdot)$ and $g(\cdot,\cdot)$ are such that the solutions exist and are unique on bounded intervals. The answer to this general question is far from obvious even for autonomous partially linear and linear time-varying systems, and has been object of study during the last 15 years, at least. In particular, obtaining sufficient conditions for a nonlinear separation principle is of special interest. While a short literature review is provided in Section 2.3.1, let us briefly develop on our motivations and goals in the study of cascaded systems.

We identify two problems:

(Design) For the cascaded nonlinear time-varying system:

$$\dot{x}_1 = f_1(t, x_1, x_2) \tag{2.19a}$$
$$\dot{x}_2 = f_2(t, x_2, u) \tag{2.19b}$$

where $x_1 \in \mathbb{R}^n$, $x_2 \in \mathbb{R}^m$ and $u \in \mathbb{R}^l$, the function $f_1(t, x_1, x_2)$ is continuously differentiable in (x_1, x_2) uniformly in t, and measurable in t, find a control input $u = u(t, x_1, x_2)$ or $u = u(t, x_2)$ such that the cascade interconnection be uniformly globally asymptotically stable (UGAS) or uniformly globally stable (UGS).

(Analysis) Consider the cascade (2.19a), (2.18b). Assume that the perturbing system (2.19b) is uniformly globally exponentially stable (UGES), or UGAS. Assume further that the zero-input dynamics of the perturbed system (2.18b),

$$\dot{x}_1 = f_1(t, x_1, 0),$$

is UGAS (respectively UGS).
Find sufficient conditions under which the cascade (2.19) is UGAS (respectively UGS).

While we study both problems nevertheless, we take a deeper insight into the second one. Concerning control design, we do not establish a general methodology as done for instance in [38, 23, 57]. However we show, through diverse applications in Section 2.4, that when the structural properties of the system in question allow it one can design relatively simple controllers and observers by simply aiming at obtaining a closed-loop system with a cascaded structure.

2.2.2 Peaking: A Technical Obstacle to Analysis

Let us come back to the linear cascade (2.15) - (2.14b). As discussed above, this system is GES if both subsystems in (2.14) are GES. As it shall become clearer later this follows essentially because, for each fixed x_2 and large values of x_1 the drift term $f(x_1, x_2) := A_1 x_1$ (assuming A_1 Hurwitz) dominates over the decaying "perturbation" $g(x_1, x_2) := B x_2$ (this is because $x_2 \to 0$). In general we do not have such property for nonlinear systems and, as a matter of fact, the growth rate of the functions $f(t, \cdot)$ and $g(t, x)$ for each fixed x_2 and t plays a major role in the stability properties of the cascade. For illustration let us consider the following example which we cite from the celebrated paper [63] which first analysed in detail the so-called *peaking phenomenon*:

Example 2.2 (Peaking).

$$\begin{aligned}
\dot{x}_1 &= -\left(1 - x_{2_2}\right) x_1^3 \\
\dot{x}_{2_1} &= x_{2_2} \\
\dot{x}_{2_2} &= -2a x_{2_2} - a^2 x_{2_1}, \quad a > 0.
\end{aligned}$$

Clearly, the *linear* x_2-subsystem is GES. Indeed, the explicit solution for x_{2_2} is $x_{2_2}(t) = a^2 t e^{-at}$ and it is illustrated in Fig. 2.2 below. Moreover, the subsystem $\dot{x}_1 = -0.5 x_1^3$ is obviously globally asymptotically stable (GAS).

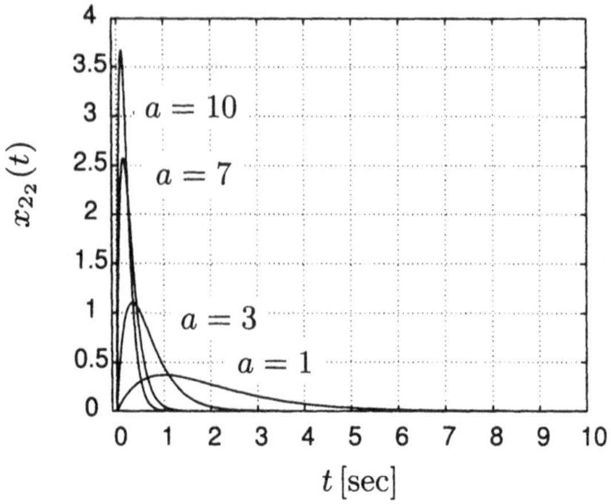

Fig. 2.2. The peaking phenomenon.

Due to the term $-x_1^3$ one would conjecture that the faster the perturbing input $x_{2_2}(t)$ vanishes the better. However, as it is clear from the expression and the graph of $x_{2_2}(t)$ (cf. Fig. 2.2) the price paid for fast convergence is to have (relatively) large *peaks* during the transient. These peaks may destabilize the system. To see this, consider the explicit solution with initial conditions $(t_\circ, x_1(t_\circ), x_2(t_\circ)) = (0, x_{1_\circ}, 0)$ i.e., $x_1(t; t_\circ, x_\circ)$, which satisfies[4]

$$x_1(t)^2 = x_\circ^2 \left[1 + 2x_\circ^2 \left(t + (1 + at)e^{-at} - 1\right)\right]^{-1} . \tag{2.20}$$

It must be clear from the expression above that for *any* values of a and $x_{1\circ}$ there exists $t_e < \infty$ such that the sum in the square brackets becomes zero at $t = t_e$. An approximation of the escape time t_e may be obtained as follows: approximate $e^{-at} \sim (1 - at)$ and substitute in (2.20), then

$$t_e \sim \frac{1}{x_{1\circ}^2} + \frac{1}{a^2}$$

that is, it becomes smaller as a and $x_{1\circ}$ become larger.

While the example above makes clear that the analysis and design problems are far from trivial, in the forecoming section we present an example which illustrates the advantages of a cascades point of view in control design.

[4] This follows observing that the solution of $\dot{x} = -b(t)x^3$ satisfies $x_1(t)^2 = x_\circ^2 \left[1 + 2x_\circ^2 \int_0^t b(s)ds\right]^{-1}$

2.2.3 Control Design from a Cascades Point of View

A Synchronization Example

Let us consider the problem of synchronizing the two pendula showed in Fig. 2.3 . That is, we consider the problem of making the "slave" pendulum oscillate at the frequency of the "master", while assuming that *no control action* is available *at the joints.* Instead, we desire to achieve our task by modifying, on line, the length of the slave pendulum and thereby its oscillating frequency, which is given by

$$\omega_1 = \sqrt{\frac{l_1}{9.81}}\,.$$

The dynamic equations are:

$$\begin{array}{ll} \text{(slave)} & \ddot{y} + 2\zeta_1\omega_1\dot{y} + \omega_1^2 y = a_1\cos\omega_1 t\,, \\ \text{(master)} & \ddot{y}_d + 2\zeta_2\omega_2\dot{y}_d + \omega_2^2 y_d = a_2\cos\omega_2 t \end{array}$$

where the control input corresponds to the change in ω_1, i.e.,

$$\dot{\omega}_1 = u \tag{2.21}$$

As it has been shown in [28] if $\omega_2 > 0$, $\zeta_1 = \zeta_2$, $a_1 = a_2$, using a cascades approach it is easy to prove that the linear control law $u = -k\tilde{\omega}$, with $k > 0$ makes that

$$\lim_{t\to\infty}\tilde{\omega}(t) = 0 \qquad \lim_{t\to\infty}\tilde{y}(t) = 0$$

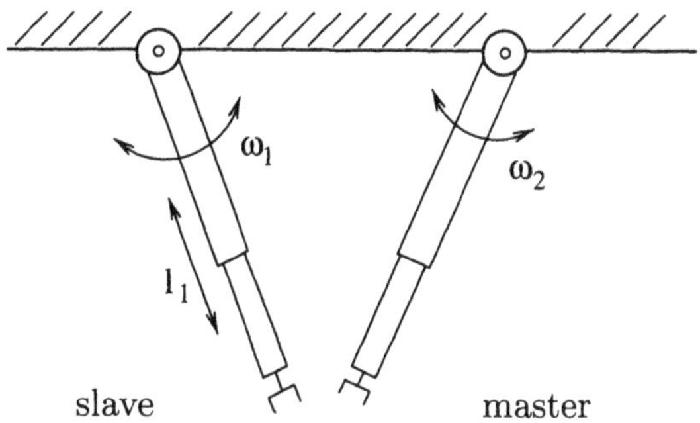

Fig. 2.3. Synchronisation of two pendula.

where $\tilde{\omega}(t) := \omega_1(t) - \omega_2$ and $\tilde{y}(t) := y(t) - y_d(t)$. To see this, we first remark that, defining

$$v := \ddot{y}_d + 2\zeta\omega_2\dot{y}_d + \omega_2^2 y_d - a\cos\omega_2 t = 0$$

and $z :=$ col$[\tilde{y},\, \dot{\tilde{y}}]$, the two pendula dynamic equations

$$\ddot{y} + 2\zeta\omega_1\dot{y} + \omega_1^2 y = a\cos\omega_1 t + v$$
$$\ddot{y}_d + 2\zeta\omega_2\dot{y}_d + \omega_2^2 y_d = a\cos\omega_2 t$$

together with the control law $u = -k\tilde{\omega}$ and eq. (2.21), are equivalent to

$$\begin{cases} \dot{z}_1 = z_2 \\ \dot{z}_2 = -2\zeta\omega_2 z_2 - \omega_2^2 z_1 + g_2(t, z, \tilde{\omega}) \end{cases}$$
$$\dot{\tilde{\omega}} = -k\tilde{\omega}$$

where $g_2(t, z, \tilde{\omega}) = 2\zeta\tilde{\omega}z_2 + 2\zeta\tilde{\omega}\dot{y}_d(t) - \tilde{\omega}^2(z_1 + y_d(t)) - 2\tilde{\omega}\omega_2 y_d + a(cos\omega_1 t - \cos\omega_2 t)$. Then, notice that this system is of the cascaded form

$$\dot{z} = f_1(z) + g(t,\, z,\, \tilde{\omega}) \tag{2.22a}$$
$$\dot{\tilde{\omega}} = -k\tilde{\omega}\,, \qquad k > 0, \tag{2.22b}$$

where $g(t, z, \tilde{\omega}) :=$ col$[0,\, g_2(t, z, \tilde{\omega})]$ and clearly, $\dot{z} = f_1(z)$ is exponentially stable. It occurs that, since the "growth" of the function $g(t, \cdot\,, \tilde{\omega})$ is linear for each fixed $\tilde{\omega}$ uniformly in t that is, for each fixed $\tilde{\omega}$ there exists $c > 0$ such that $|g(t, z, \tilde{\omega})| \leq c\,|z|$ for all $t \geq 0$, the cascade (2.22) is globally asymptotically stable (for details, see [28]).

Roughly speaking, the difference between this system and the one in Example 2.2 is that the exponentially stable dynamics $\dot{z} = f_1(z)$ dominates over the interconnection term, no matter how large the input $\tilde{\omega}(t)$ gets. In Section 2.3 we establish in a formal way sufficient conditions in terms of the growth rates for cascades including (2.22).

From Feedback Interconnections to Cascades

The objective of the application that we discuss here is to illustrate that, under certain conditions, a *feedback interconnected* system can be viewed as a cascade (thereby, "neglecting the feedback interconnection!) if you twist your eyes. This property will be used in the design applications presented in Section 2.4. To introduce the reader to this technique we briefly discuss below the problem of tracking control of rigid-joint robot manipulators driven by AC motors. This problem was originally solved in [45].

To illustrate the main idea exploited in that reference, let

$$\dot{x}_1 = \phi_1(x_1) + \tau, \tag{2.23}$$

represent the rigid-joint robot dynamics with state x_1 and let τ be the (control) input torque. Assume that this torque is provided by an AC motor with dynamics

$$\dot{x}_2 = \phi_2(x_2, u), \tag{2.24}$$

and u is a control input to the AC motor. The control goal is to find u such that the robot generalized coordinates follow a specified time-varying reference $x_{1d}(t)$. That is, we wish to design $u(t, x_1, x_2)$ such that the closed-loop system be UGAS. The rational behind the approach undertaken in [45] can be summarized in two steps:

1. Design a control law $\tau_d(t, x_1)$ to render UGAS the robot closed-loop dynamics,
$$\dot{x}_1 = \phi_1(x_1) + \tau_d(t, x_1) \tag{2.25}$$
at the "equilibrium" $x_1 \equiv x_{1d}(t)$.
2. Design $u = u(t, x_1, x_2)$ so that the AC drive dynamics
$$\dot{x}_2 = \phi_2(x_2, u(t, x_1, x_2)) \tag{2.26}$$
be uniformly globally exponentially stable[5] (UGES), uniformly in x_1 at the equilibrium $x_2 \equiv x_{2d}(t)$. Equivalently, so that the output error $\tau - \tau_d$ decay exponentially fast with the same rate for any value of $x_1(t)$ and any initial conditions.

Provided that we are able to show that system (2.25)-(2.26) is forward complete, the dynamics (2.26) may be considered along the trajectories $x_1(t)$ and hence, we can write (2.26) as $\dot{x}_2 = f(t, x_2)$ with $f_2(t, x_2) := \phi_2(x_2, u(t, x_1(t), x_2))$ which is well defined for all $t \geq t_\circ \geq 0$ and all $x_2 \in \mathbb{R}^{n_2}$. The closed-loop equation (2.25), represents the ideal case when the drives, provide the desired torque τ_d. Using this, we can rewrite the closed-loop equation (2.23) as $\dot{x}_1 = f_1(t, x_1, x_2)$ with $f_1(t, x_1, x_2) := \phi_1(x_1) + \tau_d(t, x_1) + \tau(t, x_1, x_2) - \tau_d(t, x_1)$. By design, $x_2 \equiv x_{2d}(t) \Rightarrow \tau \equiv \tau_d$. This reveals the cascaded structure of the overall closed-loop system[6].

This example suggests that the global stabilization of nonlinear systems which allow a cascades decomposition, may be achieved by ensuring UGAS for both subsystems separately. The question remaining is to know whether the stability properties of both subsystems separately, remains valid under the cascaded interconnection (2.18b), (2.19a). The latter motivates us to study the *stability analysis problem* exposed above.

[5] As we will show in this paper, in some cases, UGAS suffices.

[6] To explain the rationale of this cascades design approach, we have abbreviated $x_1(t) = x_1(t; t_0, x_0)$, however, due to the uniformity of the GES property, in x_1, it is valid to consider the closed-loop as a cascade. See [45] for details.

2.3 Stability of Cascades

2.3.1 Brief Literature Review

The stability analysis problem for nonlinear *autonomous* systems

$$\Sigma_1' : \dot{x}_1 = f_1(x_1, x_2) \tag{2.27}$$
$$\Sigma_2' : \dot{x}_2 = f_2(x_2) \tag{2.28}$$

where $x_1 \in \mathbb{R}^n$, $x_2 \in \mathbb{R}^m$ and the functions $f_1(\cdot,\cdot)$, $f_2(\cdot)$ are sufficiently smooth in their arguments, was addressed for instance in [59]. The author showed GAS of the origin of the cascaded system Σ_1', Σ_2' under the assumption that the respective origins of subsystems $\dot{x}_1 = f_1(x_1, 0)$ and (2.28) are GAS and the "Converging Input - Bounded State" (CIBS) property holds:

CIBS: For each input $x_2(\cdot)$ on $[0,\infty)$ such that $\lim_{t\to\infty} x_2(t) = 0$, and for each initial state $x_{1_\circ}$, the solution of (2.27) with $x_1(0) = x_{1_\circ}$ exists for all $t \geq 0$ and it is bounded.

Independently, based on Krasovskii-LaSalle's invariance principle, the authors of [55] showed that the origin of the composite system is GAS assuming that all solutions are bounded (in short, BS) and that both subsystems, (2.28) and $\dot{x}_1 = f_1(x_1, 0)$, are GAS at the origin.

$$\text{Fact 1: GAS} + \text{GAS} + \text{BS} \Rightarrow \text{GAS}.$$

For *autonomous* systems this fact is a fundamental result which has been used by many authors to prove GAS for the cascade (2.27), (2.28). The natural question which arises next, is "*how do we guarantee boundedness of the solutions?*". One way is to use the now well known property of Input-to-State stability (ISS), introduced in [58]. For convenience we cite below the following characterization of ISS (see e.g. [62]).

ISS: The system $\Sigma_1' : \dot{x}_1 = f_1(x_1, x_2)$ with input x_2, is Input-to-State Stable if and only if there exists a positive definite proper function $V(x_1)$, and two class $\mathcal{K}$ functions α_1 and α_2 such that, the implication

$$\{|x_1| \geq \alpha_1(|x_2|)\} \Longrightarrow \left\{ \frac{\partial V}{\partial x_1} f_1(x_1, x_2) \leq -\alpha_2(|x_1|) \right\} \tag{2.29}$$

holds for each $x_1 \in \mathbb{R}^n$ and $x_2 \in \mathbb{R}^m$.

Example 2.3. Consider the system (cf. [60])

$$\dot{x}_1 = -x_1^3 + x_1^2 x_2 \tag{2.30}$$

with input $x_2 \in \mathbb{R}$ and the Lyapunov function candidate $V(x_1) = \frac{1}{2}x_1^2$. The time derivative of V is $\dot{V} = -x_1^4 + x_1^3 x_2$ clearly, if $|x_1| \geq 2|x_2|$ then $\dot{V} \leq -\frac{1}{2}|x_1|^4$.

Unfortunately, proving the ISS property as a condition to imply CIBS may appear, in some cases, very restrictive. For instance consider the one-dimensional system

$$\dot{x}_1 = -x_1 + x_1 x_2 \tag{2.31}$$

which is not ISS with respect to the input $x_2 \in \mathbb{R}$. While it may be already intuitive from the last two examples, we will see formally in this chapter that what makes the difference is that the terms in (2.31) have the *same* linear growth rate in the variable x_1, while in (2.30) the term x_1^3 dominates over $x_1^2 x_2$ for each fixed x_2 and "large" values of x_1.

Concerned by the control design problem, i.e., to stabilize the cascaded system Σ_1', Σ_2' by using feedback of the state x_2 only, the authors of [51] studied the case when Σ_2' is a linear controllable system. Assuming $f_1(x_1, x_2)$ in (2.27) to be continuously differentiable, rewrite (2.27) as

$$\dot{x}_1 = f_1^*(x_1) + g(x_1, x_2)x_2 \,. \tag{2.32}$$

In [51] the authors introduced the *linear* growth condition

$$|g(x_1, x_2)x_2| \leq \theta(|x_2|)\,|x_1| \tag{2.33}$$

where θ is $\mathcal{C}^1$, non-decreasing and $\theta(0) = 0$, together with the assumption that $x_2 = 0$ is GES, to prove boundedness of the solutions. Using such a condition one can deal with systems which are not ISS with respect to the input x_2.

From these examples one may conjecture that, in order to prove CIBS for the system (2.32) with decaying input $x_2(\cdot)$, some growth restrictions should be imposed on the functions $f_1^*(\cdot)$ and $g(\cdot,\cdot)$. For instance, for the NL system (2.32) one may impose a linear growth condition such as (2.33) or the ISS property with respect to the input x_2. As we will show later, for the latter it is "needed" that the function $f_1^*(x_1)$ grows faster than $g(x_1, x_2)$ as $|x_1| \to \infty$.

In the papers [30] and [12], the authors addressed the problem of global stabilisability of feedforward systems, by a systematic recursive design procedure, which leads to the construction of a Lyapunov function for the complete system. In order to prove that all solutions remain bounded under the cascaded interconnection, the authors of [12] used the linear growth restriction

$$|g(x_1, x_2)x_2| \leq \theta_1(|x_2|)\,|x_1| + \theta_2(|x_2|) \tag{2.34}$$

where $\theta_1(\cdot)$, $\theta_2(\cdot)$ are $\mathcal{C}^1$ and $\theta_i(0) = 0$, together with the growth rate condition on the Lyapunov function $V(x_1)$ for the zero-dynamics $\dot{x}_1 = f_1(x_1, 0)$: $\left|\frac{\partial V}{\partial x_1}\right| |x_1| \leq cV$ for $|x_1| \geq c_2$ (which holds e.g. for polynomial functions

$V(x_1)$) and a condition of exponential stability for Σ_2. In [30] the authors used the assumption on the existence of continuous nonnegative functions ρ, $\kappa : \mathbb{R}_{>0} \to \mathbb{R}_{>0}$, such that $\left|\frac{\partial V}{\partial x_1} g(x) x_2\right| \leq \kappa(x_2)[1+\rho(V)]$ and $\frac{1}{1+\rho(V)} \notin \mathcal{L}_1$ and $\kappa(x_2) \in \mathcal{L}_1$. The choice of κ is restricted depending on the type of stability of Σ_2'. In other words, there is a tradeoff between the decay rate of $x_2(t)$ and the growth of $g(x)$.

Nonetheless, all the results mentioned above apply only to *autonomous* nonlinear systems whereas in these notes we are interested in trajectory tracking control problems and time-varying stabilization therefore, nonautonomous systems deserve particular attention. Some of the first efforts made to extend the ideas exposed above for *time-varying* nonlinear cascaded systems are contained in [15, 42, 41, 31].

In [15] the stabilization problem of a robust (vis-a-vis dynamic uncertainties) controller was considered, while in [42, 41] we established sufficient conditions for UGAS of cascaded nonlinear non autonomous systems based on a similar *linear* growth condition as in (2.34), and an *integrability* assumption on the input $x_2(\cdot)$ thereby, relaxing the exponential-decay condition used in other references. In [31] the results of [30] are extended to the nonautonomous case.

2.3.2 Nonautonomous Cascades: Problem Statement

We study cascaded NLTV systems of the form

$$\Sigma_1 : \ \dot{x}_1 = f_1(t, x_1) + g(t, x) x_2 \tag{2.35a}$$
$$\Sigma_2 : \ \dot{x}_2 = f_2(t, x_2) \tag{2.35b}$$

where $x_1 \in \mathbb{R}^n$, $x_2 \in \mathbb{R}^m$, $x := \mathrm{col}[x_1,\ x_2]$. We assume that the functions $f_1(\cdot,\cdot)$, $f_2(\cdot,\cdot)$ and $g(\cdot,\cdot)$ are continuous in their arguments, locally Lipschitz in x, uniformly in t, and $f_1(\cdot,\cdot)$ is continuously differentiable in both arguments. We also assume that there exists a nondecreasing function $G(\cdot)$ such that

$$|g(t, x)| \leq G(|x|)\,. \tag{2.36}$$

One of the main observations in this section is that Fact 1 above holds for nonlinear time-varying systems and as a matter of fact, the implication holds in both senses. That is, uniform global boundedness (UGB) is a *necessary* condition for UGAS of cascades. See Lemma 2.1.

We also present several statements of sufficient conditions for UGB. These statements are organised in the following two sections. In Section 2.3.4 we present a theorem and some lemmas which rely on a *linear* growth condition (in x_1) of the interconnection term $g(t, x)$ and the fundamental assumption that the perturbing input $x_2(\cdot)$ is *integrable*. In Section 2.3.5 we enunciate sufficient conditions to establish UGAS for three classes of cascades: roughly speaking, we consider systems such that, for each fixed x_2, the following hold uniformly in t:

(i) the function $f_1(t, x_1)$ grows faster than $g(t, x)$ as $|x_1| \to \infty$,
(ii) both functions $f_1(t, x_1)$ and $g(t, x)$ grow at similar rate as $|x_1| \to \infty$,
(iii) the function $g(t, x)$ grows faster than $f_1(t, x_1)$ as functions of x_1.

In each case, we present sufficient conditions to guarantee that a UGAS nonlinear time-varying system

$$\dot{x}_1 = f_1(t, x_1) \tag{2.37}$$

remains UGAS when it is perturbed by the output of another UGAS system of the form Σ_2, that is, we establish sufficient conditions to ensure UGAS for the system (2.35).

2.3.3 Basic Assumptions and Results

Since we consider cascades for which the origin of (2.37) is UGAS it follows from converse Lyapunov theorems (see e.g. [24, 20, 26]) that there exists a positive definite proper Lyapunov function $V(t, x_1)$ with a negative definite bound on the total time derivative. Thus consider the assumption below which we divide in two parts for ease of reference.

Assumption 1
a) The system (2.37) is UGAS.
b) There exists a known $\mathcal{C}^1$ Lyapunov function $V(t, x_1)$, α_1, $\alpha_2 \in \mathcal{K}_\infty$, a positive semidefinite *function $W(x_1)$ a continuous non-decreasing function $\alpha_4(\cdot)$, such that*

$$\alpha_1(|x_1|) \le V(t, x_1) \le \alpha_2(|x_1|) \tag{2.38}$$

$$\dot{V}_{(2.37)}(t, x_1) \le -W(x_1) \tag{2.39}$$

$$\left|\frac{\partial V}{\partial x_1}\right| \le \alpha_4(|x_1|). \tag{2.40}$$

Remark 2.4. We point out that to verify Assumption 1a) it is enough to have a Lyapunov function with only negative semidefinite time derivative. Yet, we have the following.

Proposition 2.2. *Assumption 1a implies the existence of a Lyapunov function $\mathcal{V}(t, x_1)$, functions $\bar{\alpha}_1$, $\bar{\alpha}_2 \in \mathcal{K}_\infty$ and $\bar{\alpha}_4 \in \mathcal{K}$ such that,*

$$\bar{\alpha}_1(|x_1|) \le \mathcal{V}(t, x_1) \le \bar{\alpha}_2(|x_1|) \tag{2.41}$$

$$\dot{\mathcal{V}}_{(2.37)}(t,x_1) \le -\mathcal{V}(t,x_1) \tag{2.42}$$

$$\left|\frac{\partial \mathcal{V}}{\partial x_1}\right| \le \bar{\alpha}_4(|x_1|) \,. \tag{2.43}$$

Sketch of proof. The inequalities in (2.41), as well as the existence of $\bar{\alpha}_3 \in \mathcal{K}$ such that,

$$\dot{\mathcal{V}}_{(2.37)}(t,x_1) \le -\bar{\alpha}_3(|x_1|) \,, \tag{2.44}$$

follow from [26, Theorem 2.9]. The property (2.43) follows along the lines of proofs of [19, Theorems 3.12, 3.14] and [24], using the assumption that $f_1(t,x_1)$ is continuously differentiable and locally Lipschitz. Finally, (2.42) follows using (2.44) and [48, Proposition 13]. See also [64].

We stress the importance of formulating Assumption 1b) with the less restrictive conditions (2.38), (2.39) since, for some applications, UGAS for (2.37) may be established with a Lyapunov function $V(t,x_1)$ with a negative semidefinite derivative. For autonomous systems, e.g., using invariance principles (such as Krasovskii-LaSalle's) or, for non-autonomous systems, via Matrosov's theorem [49]. See Section 2.4 and [25] for some examples.

We also remark that the same observations apply to [42, Theorem 2] where we overlooked this important issuc, imposing the unnecessarily restrictive assumption of negative definiteness on $\dot{V}_{(2.37)}(t,x_1)$. This is implicitly assumed in the proof of that Theorem. In Section 2.3.4 we present a theorem which includes the same result.

Further, we assume that

(Assumption 2) *the subsystem Σ_2 is UGAS.*

Let us stress some direct consequences of Assumption 2 in order to introduce some notation. Firstly, it means that there exists $\beta \in \mathcal{KL}$ such that, for all $(t_\circ, x_\circ) \in \mathbb{R}_{\ge 0} \times \mathbb{R}^n$,

$$|x_2(t;t_\circ,x_{2\circ})| \le \beta(|x_{2\circ}|\,,t-t_\circ), \quad \forall\, t \ge t_\circ \tag{2.45}$$

and hence, for each $r > 0$

$$|x_2(t;t_\circ,x_{2\circ})| \le c := \beta(r,0) \qquad \forall\, |x_{2\circ}| < r \,. \tag{2.46}$$

Secondly, note that due to [30, Lemma B.1] (2.36) implies that there exist continuous functions $\theta_1 : \mathbb{R}_{\ge 0} \mapsto \mathbb{R}_{\ge 0}$ and $\alpha_5 : \mathbb{R}_{\ge 0} \mapsto \mathbb{R}_{\ge 0}$ such that $|g(t,x)| \le \theta_1(|x_2|)\alpha_5(|x_1|)$. Hence, under Assumption 2, we have that for each $r > 0$ and for all $t_\circ \ge 0$

$$|g(t,x(t;t_\circ,x_\circ))| \le c_g(r)\alpha_5(|x_1(t;t_\circ,x_\circ)|), \quad \forall\ |x_{2\circ}| < r\,,\ \forall t \ge t_\circ \tag{2.47}$$

where $c_g(\cdot)$ is the class $\mathcal{K}$ function defined by $c_g(\cdot) := \theta_1(\beta(\cdot,0))$.

Having laid the basic assumptions, we are ready to present some stability theorems. The following lemma extends the fact that GAS + GAS + BS $\Rightarrow$ GAS, to the nonautonomous case. This is probably the most fundamental result of this chapter and therefore, we provide the proof of it.

Lemma 2.1 (UGAS + UGAS + UGB $\Leftrightarrow$ UGAS). *The cascade (2.35) is UGAS if and only if the systems (2.35b) and (2.37) are UGAS and the solutions of (2.35) are uniformly globally bounded (UGB).*

Proof. (*Sufficency*). By assumption (from UGB), for each $r > 0$ there exists $\bar{c}(r) > 0$ such that, if $|x_\circ| < r$ then $|x(t; t_\circ, x_\circ)| \leq \bar{c}(r)$. Consider the function $\mathcal{V}(t, x_1)$ as defined in Proposition 2.2. It's time derivative along the trajectories of (2.35a) yields, using (2.43), (2.42), (2.47), and defining $v(t) := \mathcal{V}(t, x_1(t))$,

$$\dot{v}_{(2.35a)}(t) \leq -v(t) + c(r)\,|x_2(t)| \ , \tag{2.48}$$

where $c(r) := c_g(r)\bar{\alpha}_4(\bar{c}(r))\alpha_5(\bar{c}(r))$. Therefore, using (2.45) and defining $v_\circ := v(t_\circ)$, we obtain that for all $t_\circ \geq 0$, $|x_\circ| < r$ and $v_\circ < \bar{\alpha}_2(r)$,

$$\dot{v}_{(2.35a)}(t; t_\circ, v_\circ) \leq -v(t; t_\circ, v_\circ) + \tilde{\beta}(r, t - t_\circ) \tag{2.49}$$

where $\tilde{\beta}(r, t - t_\circ) := c(r)\beta(r, t - t_\circ)$.

Let $\tau_\circ \geq t_\circ$ and multiply by $e^{(t-\tau_\circ)}$ on both sides of (2.49). Rearranging the terms we obtain

$$\frac{d}{dt}\left[v(t)e^{(t-\tau_\circ)}\right] \leq \tilde{\beta}(r, t - t_\circ)e^{(t-\tau_\circ)} \,, \quad \forall t \geq \tau_\circ \,. \tag{2.50}$$

Then, integrating on both sides from $\tau_\circ$ to t and multiplying by $e^{-(t-\tau_\circ)}$ we obtain that

$$v(t) \leq v(\tau_\circ)e^{-(t-\tau_\circ)} + \int_{\tau_\circ}^{t} \tilde{\beta}(r, s - t_\circ)e^{-(t-s)}ds \,, \quad \forall t \geq \tau_\circ \,. \tag{2.51}$$

Next, let $\tau_\circ = t_\circ$ (2.51) implies that

$$v(t) \leq v(t_\circ) + \tilde{\beta}(r, 0)\left(1 - e^{-(t-t_\circ)}\right) \leq \bar{\alpha}_2(r) + \tilde{\beta}(r, 0) \,, \quad \forall t \geq t_\circ \,. \tag{2.52}$$

Hence $|x_1(t)| \leq \bar{\alpha}_1^{-1}(\bar{\alpha}_2(r) + \tilde{\beta}(r, 0)) =: \gamma(r)$. Uniform global stability follows observing that $\gamma \in \mathcal{K}_\infty$ and that the subsystem Σ_2 is UGS by assumption. On the other hand, for each $0 < \varepsilon_1 < r$, let $T_1(\varepsilon_1, r) \geq 0$ be such that $\tilde{\beta}(r, T_1) = \varepsilon_1/2$ ($T_1 = 0$ if $\tilde{\beta}(r, 0) \leq \varepsilon_1/2$). Then, defining now $\tau_\circ = t_\circ + T_1$, (2.51) also implies that

$$v(t) \leq v(t_\circ + T_1)e^{-(t-t_\circ-T_1)} + \int_{t_\circ+T_1}^{t} \tilde{\beta}(r, T_1)e^{-(t-s)}ds \,, \quad \forall t \geq t_\circ + T_1 \,.$$

This, in vue of (2.52), implies that

$$v(t) \le \left[\bar{\alpha}_2(r) + \tilde{\beta}(r,0)\right] e^{-(t-t_\circ-T_1)} + \frac{\varepsilon_1}{2}, \quad \forall t \ge t_\circ + T_1 .$$

It follows that $v(t) \le \varepsilon_1$ for all $t \ge t_\circ + T$ with $T := T_1 + \ln\left[\frac{2[\bar{\alpha}_2(r)+\tilde{\beta}(r,0)]}{\varepsilon_1}\right]$. Finally, defining $\varepsilon := \bar{\alpha}_2(\varepsilon_1)$ we conclude that $|x_1(t)| \le \varepsilon$ for all $t \ge t_\circ + T$. The result follows observing that ε_1 is arbitrary, $\bar{\alpha}_2 \in \mathcal{K}_\infty$, and that Σ_2 is UGAS by assumption.

(*Necessity*). By assumption there exists $\beta \in \mathcal{KL}$ such that $|x(t)| \le \beta(|x_\circ|, t - t_\circ)$. UGB follows observing that $|x(t)| \le \beta(|x_\circ|, 0)$. Also, notice that the solutions $x(t)$ restricted to $x_2(t) \equiv 0$ satisfy $|x_1(t)| \le \beta(|x_{1\circ}|, t - t_\circ)$ which implies UGAS of (2.37). It is clear that (2.37) is UGAS only if (2.35b) is UGAS. ■

As discussed in the previous section, the next question is how to guarantee uniform global boundedness. This can be established by imposing extra *growth rate* assumptions. In particular, in Section 2.3.5, under Assumptions 1 and 2, we shall consider the three previously mentioned cases according to the growth rates of $f_1(t, x_1)$ and $g(t, x)$. For this, we make use of the following concepts.

Definition 2.9 (small order). *Let $\varrho : \mathbb{R}^n \to \mathbb{R}^n$, $\varphi : \mathbb{R}_{\ge 0} \times \mathbb{R}^n \to \mathbb{R}^n$ be continuous functions of their arguments. We denote $\varphi(t, \cdot) = o(\varrho(\cdot))$ (and say that "phi is of small order rho") if there exists a continuous function $\lambda : \mathbb{R}_{\ge 0} \to \mathbb{R}_{\ge 0}$ such that $|\varphi(t,x)| \le \lambda(|x|)\,|\varrho(x)|$ for all $(t,x) \in \mathbb{R}_{\ge 0} \times \mathbb{R}^n$ and $\lim_{|x|\to\infty} \lambda(|x|) = 0$.*

A direct consequence of this definition is that the following limit holds uniformly in t:

$$\lim_{|x|\to\infty} \frac{|\varphi(t,x)|}{|\varrho(x)|} = 0 .$$

Definition 2.10. *Let $\varrho(\cdot)$ and $\varphi(\cdot,\cdot)$ be continuous. We say that the function $\varrho(x)$ majorises the function $\varphi(t,x)$ if*

$$\limsup_{|x|\to\infty} \frac{|\varphi(t,x)|}{|\varrho(x)|} < +\infty \qquad \forall t \ge 0 .$$

Notice that, as a consequence of the definition above, it holds true that there exist finite positive constants η and λ such that, the following holds uniformly in t:

$$|x| \ge \eta \;\Rightarrow\; \frac{|\varphi(t,x)|}{|\varrho(x)|} < \lambda . \tag{2.53}$$

We may also refer to this property as "large order" or "order" and write $\varphi(t,\cdot) = \mathcal{O}(\varrho(\cdot))$ to say that "phi is of large order rho".

2.3.4 An Integrability Criterion

Theorem 2.1. *Let Assumption 1a hold and suppose that the trajectories of (2.35b) are uniformly globally bounded. If moreover, Assumptions 3 – 5 below are satisfied, then the solutions $x(t; t_\circ, x_\circ)$ of the system (2.35) are uniformly globally bounded. If furthermore, the origin of system (2.35b) is UGAS, then so is the origin of the cascade (2.35).*

Assumption 3 *There exist constants c_1, c_2, $\eta > 0$ and a Lyapunov function $V(t, x_1)$ for (2.37) such that $V : \mathbb{R}_{\geq 0} \times \mathbb{R}^n \to \mathbb{R}_{\geq 0}$ is positive definite, radially unbounded, $\dot{V}_{(2.37)}(t, x_1) \leq 0$ and*

$$\left|\frac{\partial V}{\partial x_1}\right| |x_1| \leq c_1 V(t, x_1) \qquad \forall\ |x_1| \geq \eta \tag{2.54}$$

$$\left|\frac{\partial V}{\partial x_1}\right| \leq c_2 \qquad \forall\ |x_1| \leq \eta \tag{2.55}$$

Assumption 4 *There exist two continuous functions θ_1, $\theta_2 : \mathbb{R}_{\geq 0} \to \mathbb{R}_{\geq 0}$, such that $g(t, x)$ satisfies*

$$|g(t, x)| \leq \theta_1(|x_2|) + \theta_2(|x_2|)\,|x_1| \tag{2.56}$$

Assumption 5 *There exists a class $\mathcal{K}$ function $\alpha(\cdot)$ such that, for all $t_\circ \geq 0$, the trajectories of the system (2.35b) satisfy*

$$\int_{t_\circ}^{\infty} |x_2(t; t_\circ, x_2(t_\circ))|\, dt \leq \alpha(|x_2(t_\circ)|)\,. \tag{2.57}$$

The following proposition is a local counterpart of the theorem above and establishes uniform semiglobal exponential stability of the cascade (see Def. 2.7), i.e., for each r there exists $\gamma_1(r) > 0$ and $\gamma_2(r) > 0$ such that for all $|x_\circ| \leq r$ the system satisfies (2.9). Notice that this concept differs from ULES in that the numbers γ_1 and γ_2 depend on the size of initial conditions however, the convergence is still uniform.

Even though this proposition may seem obvious to some readers, we write it separately from Theorem 2.1 for ease of reference in Section 2.4, where it will be an instrumental tool in tracking control design of nonholonomic carts.

Proposition 2.3. *If in addition to the assumptions in Theorem 2.1 the systems (2.35b) and (2.37) are USGES then the cascaded system (2.35) is USGES and UGAS. In particular if the subsystems are UGES the cascade is UGES.*

2.3.5 Growth Rate Theorems

In Theorem 2.1 we have imposed a linear growth condition on the interconnection term and used an integrability assumption on the solutions of (2.35b) to establish UGAS of the cascade. In this section we allow for different growth rates of the interconnection term, relative to the growth of the drift term.

Case 1: "The Function $F_1(T, X_1)$ Grows Faster Than $G(T, X)$"

The following theorem allows to deal with systems which are ISS[7] but which do not necessarily satisfy a linear growth condition such as (2.34). Roughly speaking, the stability induced by the drift $f_1(t, x_1)$ dominates over the "perturbations" induced by the trajectories $x_2(t)$ through the interconnection term $g(t, x)$.

Theorem 2.2. *If Assumptions 1 and 2 hold, and*

(Assumption 6) *for each fixed x_2 and t the function $g(t, x)$ satisfies*

$$\left|\frac{\partial V}{\partial x} g(t, x)\right| = o(W(x_1)), \quad as \ |x_1| \to \infty \tag{2.58}$$

where $W(x_1)$ is defined in Assumption 1;

then, the cascade (2.35) is UGAS.

Proposition 2.4. *If in addition to the assumptions of Theorem 2.2 there exists $\alpha_3 \in \mathcal{K}$ such that $W(x_1) \geq \alpha_3(|x_1|)$ then the system (2.35a) is ISS with respect to the input $x_2 \in \mathbb{R}^m$.*

Remark 2.5. If for a particular (autonomous) system we have that $W(x_1) = \left|\frac{\partial V}{\partial x_1}\right| |f_1(x_1)|$ then condition (2.58) reads simply "$g(t, x) = o(f_1(x_1))$". However, it must be understood that in general, such relation of order between the functions $f_1(t, x_1)$ and $g(t, x)$ is not implied by condition (2.58). This motivates the use of "quotes" in the phrase "the function $f_1(t, x_1)$ grows faster than $g(t, x)$".

Example 2.4. Consider the ISS system $\dot{x}_1 = -x_1^3 + x_1^2 x_2$ with input $x_2 \in \mathbb{R}$ with an ISS-Lyapunov function $V(x_1) = \frac{1}{2}x_1^2$ which satisfies Assumption 6 with $\alpha_4(s) = s$ and $\alpha_3(s) = s^4$.

[7] It is worth mentioning that the concept of ISS systems was originally proposed and is more often used in the context of autonomous systems, for definitions and properties of time-varying ISS systems see e.g. [23].

Case 2: "The Function $F_1(T, X_1)$ Majorizes $G(T, X)$"

We consider now systems like (2.35a), which are not necessarily ISS with respect to the input $x_2 \in \mathbb{R}^m$ but for which the following assumption on the growth rates of $V(t, x_1)$ and $g(t, x)$ holds.

Assumption 7 *There exist a continuous non-decreasing function $\alpha_6 : \mathbb{R}_{\geq 0} \to \mathbb{R}_{\geq 0}$, and a constant $a \geq 0$, such that $\alpha_6(a) > 0$ and*

$$\alpha_6(s) \geq \alpha_4(\alpha_1^{-1}(s))\alpha_5(\alpha_1^{-1}(s)) \tag{2.59}$$

where α_5 was defined in (2.47), and

$$\int_a^\infty \frac{ds}{\alpha_6(s)} = \infty\,. \tag{2.60}$$

Assumption 7 imposes a growth restriction, with respect to x_1, on the function $g(t, x)$. Indeed, notice that for (2.60) to hold it is seemingly needed that $\alpha_6(s) = \mathcal{O}(s)$ for large s, thereby imposing a restriction on $g(t, x)$ with respect to x_1 for each t and x_2. The condition in (2.60) guarantees (considering that the "inputs" $x_2(t)$ are absolutely continuous on $[0, \infty)$) that the solutions of the overall cascaded system $x(t\,; t_\circ, x_\circ)$ do not escape in finite time. The formal statement which support our arguments, can be found for instance in [53].

Remark 2.6. The assumptions on the growth rates of $g(t, x)$ and $V(t, x_1)$ considered in [12] are a particular case of Assumption 7. Also, this assumption is equivalent to the hypothesis made in Assumption A3.1 of [30], on the existence of a nonnegative function ρ such that $\frac{1}{1+\rho} \notin \mathcal{L}_2$.

Theorem 2.3. *Let Assumptions 1, 2 and 7 hold, and suppose that*

(Assumption 8) *The function $g(t, x)$ is majorized by the function $f_1(t, x_1)$ in the following sense: for each $r > 0$ there exist λ, $\eta > 0$ such that, for all $t \geq 0$ and all $|x_2| < r$*

$$\left|\frac{\partial V}{\partial x} g(t, x)\right| \leq \lambda W(x_1) \quad \forall\ |x_1| \geq \eta\,. \tag{2.61}$$

where $W(x_1)$ is defined in Assumption 1.

Then, the cascade (2.35) is UGAS.

Example 2.5. The system $\dot{x}_1 = -x_1 + x_1 x_2$ with input x_2, satisfies Assumptions 1 and 8 with a quadratic Lyapunov function $V(x_1) = \frac{1}{2}x_1^2$, $W(x_1) = x_1^2$ and $\alpha_4(s) = s$. Assumption 7 is also satisfied with $\alpha_5(s) = s$, and $\alpha_1 = \frac{1}{2}s^2$ hence from (2.59), with $\alpha_6(s) = 2s$.

It is worth remarking that the practical problems of tracking control of robot manipulators with induction motors [45], and controlled synchronization of two pendula [28] mentioned in Section 2.2.3 fit into the class of systems considered in Theorem 2.3.

Case 3: "The Function $F_1(T, X_1)$ Grows Slower Than $G(T, X)$"

Theorem 2.4. *Let Assumptions 1, 2 and 7 hold and suppose that*

(Assumption 9) *there exists $\alpha \in \mathcal{K}$ such that, the trajectory $x_2(t; t_\circ, x_2(t_\circ))$ of Σ_2 satisfies (2.57) for all $t_\circ \geq 0$.*

Then, the cascade (2.35) is UGAS.

Example 2.6. Let us define the saturation function sat : $\mathbb{R} \to \mathbb{R}$ as a $\mathcal{C}^2$ non-decreasing function that satisfies $\text{sat}(0) = 0$, $\text{sat}(\zeta)\zeta > 0$ for all $\zeta \neq 0$ and $|\text{sat}(\zeta)| < 1$. For instance, we can take $\text{sat}(\zeta) := \tanh(\omega\zeta)$, $\omega > 0$, or $\text{sat}(\zeta) = \frac{\zeta}{1+\zeta^p}$ with p being an even integer. Consider

$$\dot{x}_1 = -\text{sat}(x_1) + x_1 \ln(|x_1| + 1)x_2 \tag{2.62}$$

$$\dot{x}_2 = f_2(t, x_2) \tag{2.63}$$

where $x_1 \in \mathbb{R}$ and the system $\dot{x}_2 = f_2(t, x_2)$ is UGAS and satisfies Assumption 9. The zero input dynamics of (2.62), $\dot{x}_1 = -\text{sat}(x_1)$, is UGAS with Lyapunov function $V(x_1) = \frac{1}{2}x_1^2$ hence, let $\alpha_1(s) = \frac{1}{2}s^2$ and $\alpha_4(s) = s$, while the function $\alpha_5(s) = s\ln(s+1)$. Condition (2.60) holds with $\alpha_6(s) = [\ln(\sqrt{2s}+1)+1](2s+\sqrt{2s})$.

The last example of this section illustrates the importance of the integrability condition and shows that, in general, it does *not* hold that GAS + GAS + Forward Completeness[8] $\not\Rightarrow$ GAS.

Example 2.7. Consider the autonomous system

$$\dot{x}_1 = -\text{sat}(x_1) + x_1 x_2 \tag{2.64}$$

$$\dot{x}_2 = -x_2^3 \tag{2.65}$$

where the function $\text{sat}(x_1)$ is defined as follows: $\text{sat}(x_1) = \sin(x_1)$ if $|x_1| < \pi/2$, $\text{sat}(x_1) = 1$ if $x_1 \geq \pi/2$, and $\text{sat}(x_1) = -1$ if $x_1 \leq -\pi/2$. Even though Assumptions 1, 2 and 7 are satisfied and the system is forward complete, the trajectories may grow unbounded. The latter follows observing that the set $S := \{(x_1, x_2) : z \geq 0,\ x_1 \geq 2,\ 1/2 \geq x_2 \geq 0\}$ with $z = -\text{sat}(x_1) + x_1 x_2 - 1$, is positively invariant. On the other hand, the solution of (2.65), $x_2(t) = \left(2t + \frac{1}{x_{2\circ}^2}\right)^{-1/2}$, does not satisfy (2.57).

[8] That is, that the solutions $x(\cdot)$ be defined over the infinite interval.

Discussion

Each of the examples above, illustrates a class of systems that one can deal with using Theorems 2.2, 2.3 and 2.4. In this respect, it shall be noticed that even though the three theorems presented here cover a large group of dynamical non-autonomous systems, our conditions are *not* necessary. Hence, these results can be improved in several directions.

Firstly, for clarity of exposition, we have assumed that the interconnection term in (2.35a) can be factorised as $g(t,x)x_2$; in some cases, this may be unnecessarily restrictive. With an abuse of notation, let us redefine $g(t,x)x_2 =: g(t,x)$, i.e., consider a cascaded system of the form

$$\dot{x}_1 = f_1(t,x_1) + g(t,x) \tag{2.66}$$
$$\dot{x}_2 = f_2(t,x_2) \tag{2.67}$$

where $g(t,x)$ satisfies

$$|g(t,x)| \leq \alpha_5'(|x_1|)\gamma'(|x_2|) + \alpha_5''(|x_1|)\gamma''(|x_2|) \tag{2.68}$$

where α_5', α_5'', γ' and γ'' are nondecreasing functions such that $\gamma''(s) \to 0$ as $s \to 0$ and $\alpha_5''(|x_1|) \leq c_1\alpha_5'(|x_1|)$ for all $|x_1| \geq c$.

Secondly, notice that Assumption 2 does not impose any condition on the convergence rate of the trajectory (input) $x_2(t; t_\circ, x_{2\circ})$. Let $V_2(t,x_2)$ be a Lyapunov function for system (2.67), and consider the following Corollary which, under the assumptions of Theorems 2.3 and 2.4, establishes UGAS of the cascade.

Corollary 2.1. *Consider the cascaded system (2.66), (2.67), (2.68), and suppose that Assumptions 1 —with a Lyapunov function $V_1(t,x_1)$ –, 2 and 7 hold with $\alpha_6(V_1) = \alpha_4(\alpha_1^{-1}(V_1))\alpha_5'(\alpha_1^{-1}(V_1))$. Assume further that $\alpha_5''(|x_1|)$ is majorised by the function $\frac{W(x_1)}{\alpha_4(|x_1|)}$ and the function $\gamma'(|x_2|)$ satisfies either of the following:*

$$\gamma'(|x_2|) \leq \kappa(V_2(x_2))U(x_2) \tag{2.69}$$

where $\dot{V}_2(t,x_2) \leq -U(x_2)$ with $\kappa : \mathbb{R}_{\geq 0} \to \mathbb{R}_{\geq 0}$ continuous; or there exists $\phi \in \mathcal{K}$, s.t.

$$\int_{t_\circ}^{\infty} \gamma'(|x_2(t)|)dt \leq \phi(|x_{2\circ}|)\,. \tag{2.70}$$

Under these conditions, the cascaded system (2.66), (2.67) is UGAS.

Further relevant remarks on the relation of the bound (2.69) with the order of zeroes of the *input* $x_2(t)$ in (2.66) and (2.67) were given in [30].

A last important observation concerns the restrictions on the growth rate of $g(t,x)$ with respect to x_1. Coming back to Example 2.6, we have seen that

Theorem 2.4 and Corollary 2.1 apply to the cascaded system (2.66), (2.67) for which the interconnection term $g(t,x)$ grows slightly faster than linearly in x_1. Allowing for high order growth rates in x_1 is another interesting direction in which our contributions may be extended. This issue has already been studied for instance in [13] for *not*-ISS *autonomous* systems, where the authors established conditions (assuming that Σ_2' is a linear controllable system) under which global and semiglobal stabilization of the cascade are *impossible.*

In this respect, it is also worth mentioning that in [32] the authors study a *feedback* interconnected *autonomous* system

$$\dot{x}_1 = f_1(x_1) + g(x) \tag{2.71}$$

$$\dot{x}_2 = f_2(x_1, x_2) \tag{2.72}$$

under the assumption that

$$|g(x)| \le \theta_1(|x_2|)\,|x_1|^k + \theta_2(|x_2|) \quad k \ge 1\,. \tag{2.73}$$

Then, global asymptotic stability of (2.71), (2.72) can be proven by imposing the following condition on the derivative of the Lyapunov function $V_2(x_2)$, for (2.72):

$$\dot{V}_{2_{(2.72)}}(x_1, x_2) \le -\gamma_1(x_2) - \gamma_2(x_2)\,|x_1|^{k-1}$$

with γ_1 and γ_2 positive definite functions. See also [56] for other interesting results (on stabilization) of partially linear cascades, not satisfying the linear growth condition of the interconnection term.

Theorem 2.1 is a reformulation of the main results in [42]. other similar integrability conditions have been published at same time than the cited reference, in [61]. More recent results also based on similar integrability conditions and covering a similarly wide class of nonlinear systems have been reported in [1].

2.4 Applications in Control Design

Cascaded systems may appear in many control applications. Most remarkably, in some cases a system can be decomposed in two subsystems for which control inputs can be designed with the aim that the closed-loop system have a cascaded structure. In this direction the results in [45] suggest that the global stabilization of nonlinear systems which allow a cascades decomposition, may be achieved by ensuring UGAS for both subsystems separately. The question remaining is to know whether the stability properties of both subsystems separately, is still valid under the cascaded interconnection.

See also [25] and references therein, for an extensive study and a very complete work on cascaded-based control of applications including ships and nonholonomic systems. Indeed, in that reference some of the theorems presented here have been successfully used to design *simple* controllers. The term "simple" is used relative to the mathematical complexity of the expression of the control laws. Even though it is not possible to show it in general, there exists a good number of applications where controllers based on a cascaded approach are simpler than highly nonlinear Lyapunov-based control laws.

The mathematical simplicity is of obvious importance from a practical point of view since it is directly translated into "lighter" computational load in engineering implementations. See for instance [46] for an experimental comparative study of cascaded-based and backstepping-based controllers. See also [39].

In this section we present some practical applications of our theorems. These works were originally reported in [40, 43, 27]. It is not our intention to repeat these results here but to treat, in more detail than we did for our previous examples, two applications in control *synthesis*. In contrast to [30, 57, 23] we do not give a design methodology but we illustrate, with these examples, that the control design with a closed-loop cascaded structure in mind may considerably simplify both the control laws and the stability analysis.

2.4.1 Output Feedback Dynamic Positioning of a Ship

The problem that we discuss here was originally solved in [27] using the results previously proposed in [42].

Let us consider the following model of a surface marine vessel as in [8]:

$$M\dot{\nu} + D\nu = \tau + J^\top(y)b \tag{2.74}$$
$$\dot{\eta} = J^\top(y)\nu \tag{2.75}$$
$$\dot{\zeta} = \Omega\zeta \tag{2.76}$$
$$\dot{b} = Tb \tag{2.77}$$

where $M = M^\top > 0$ is the constant mass of the ship, $\eta \in \mathbb{R}^3$ is the position and orientation vector with respect to an Earth-fixed reference frame, similarly to the example above. The only available state is the noisy measurement $y = \eta + \zeta$ where ζ is a noise vector obeying the slowly convergent dynamics (2.76). It is supposed that the ship rotates only about the yaw axis (perpendicular to the water surface) hence the rotation matrix $J(y)$ is orthogonal, i.e. $J(y)^\top J(y) \equiv I$. The matrix $D \geq 0$ is a natural damping and the bias b represents the force of environmental perturbations, such as wind, waves, etc. The dynamics (2.77) is also assumed to be asymptotically stable, but slowly-converging.

The goal is to design an *output feedback* control law τ, to maintain the vessel stable at the origin, while filtering out noise and disturbances. The approach followed in [27] was to design a state observer based on the output measurement y and a state feedback controller of a classical PD-type as used in the robotics literature. As in the previous example, to avoid redundancy we give here only the main ideas where the cascades approach, via theorems like those presented here have been fundamental.

Let us firstly introduce the notation $x_1 = \text{col}[\nu,\ \eta,\ \zeta,\ b]$ for the position error state, that is, the desired set-point (hence $\nu \equiv 0$) is the origin $\text{col}[\eta,\ \zeta,\ b] = 0$, therefore the error and actual state are taken to be the same. With this in mind, notice that the system (2.74)–(2.77) is linear, except for the Jacobian matrix $J(y)$, thus the dynamic equations can be written in the compact form

$$\dot{x}_1 = A(y)x_1. \tag{2.78}$$

One can also verify that the closed-loop system (2.78) with the *state* feedback $\tau = -K(y)x_1$, or in expanded form,

$$\tau = -J^\top(y)b - K_d\nu - J^\top(y)K_p\eta \tag{2.79}$$

with K_p, $K_d > 0$, is GAS. This follows by writing the closed-loop equations in the compact form $\dot{x}_1 = A_c(x_1)x_1$ with $A_c(x_1) := (A(y) - BK(y))$ and using the Lyapunov function candidate $V_1(x_1) = x_1^\top P_c x_1$ (with P_c constant and positive definite) whose time derivative is negative *semi*definite.

On the other hand, in [9], an observer of the form

$$\dot{\hat{x}}_1 = A(y)\hat{x}_1 - L(y - \hat{y}) + B\tau\,, \tag{2.80}$$

where L is a design parameters matrix of suitable dimensions and $\hat{(\cdot)}$ stands for the "estimate of" $(\cdot)$, was proposed. This observer in closed loop with (2.78) yields

$$\underbrace{(\dot{x}_1 - \dot{\hat{x}}_1)}_{\dot{x}_2} = \underbrace{(A(y) - LC)}_{\bar{A}_o(y)}\underbrace{(x_1 - \hat{x}_1)}_{x_2} \tag{2.81}$$

where $C := [0,\ I,\ I,\ 0]$. In [9], using the Kalman-Yacubovich-Popov lemma, it is shown that (2.81) can be made globally exponentially stable, uniformly in the trajectories $y(t)$, hence, (after proving completeness of the x_1-subsystem) the estimation error equation (2.81) can be rewritten as a linear time-varying system $\dot{x}_2 = A_o(t)x_2$ where $A_o(t) := \bar{A}_o(y(t))$.

Finally, since it is not desirable to implement the controller (2.79) using state measurements, we use

$$\tau = -J^\top(y)b - K_d\hat{\nu} - J^\top(y)K_p\hat{\eta}\,. \tag{2.82}$$

In summary we have that the overall controller-observer-boat closed-loop system (2.78), (2.80) and (2.82) has the following desired cascaded structure:

$$\begin{aligned}\Sigma_1 &: \dot{x}_1 = A_c(x_1)x_1 + g(x_1)x_2\\ \Sigma_2 &: \dot{x}_2 = A_o(t)x_2\end{aligned}$$

where $g(x_1)x_2 = J^\top(y)\bar{b} - K_d\bar{\nu} - J^\top(y)K_p\bar{\eta}$ where $(\bar{\cdot}) := (\cdot) - (\hat{\cdot})$, which is uniformly bounded in x_1 since $J(\cdot)$ is uniformly bounded and $y := h(x_1)$.

Therefore, GAS for the closed-loop system can be concluded invoking any of the three theorems of Section 2.3.5, based on the stabilization property of the *state*-feedback (2.79) and the filtering properties of the observer (2.80). An immediate interesting consequence is that both, the observer and the controller can be tuned *separately*.

For further details and experimental results, see [27].

We also invite the reader to consult [25] for other simple cascaded-based controllers for ships. Specifically, one must remark the mathematical simplicity of controllers obtained using the cascades approach in contrast to the complexity of some backstepping designs. This has been nicely put in perspective in [25, Appendix A] where the 2782 terms of a backstepping controller are written explicitly[9].

2.4.2 Pressure Stabilization of a Turbo-Diesel Engine

Model and Problem Formulation

To further illustrate the utility of our main results we consider next the set-point control problem for the simplified emission VGT-EGR diesel engine depicted in Figure 2.4.

The simplified model structure consists of two dynamic equations derived by differentiation of ideal gas flow ([34]); they describe the intake pressure dynamics p_1 and the exhaust pressure p_2 dynamics under the assumption of time-independent temperatures. The third equation describes the dynamics of the compressor power P_c:

$$\dot{p}_1 = k_1(w_c + w_{egr} - k_{1e}p_1) \tag{2.83}$$

$$\dot{p}_2 = k_2(k_{1e}p_1 + w_f - w_{egr} - w_{turb}) \tag{2.84}$$

$$\dot{P}_c = \frac{1}{\tau_c}(-P_c + K_t(1 - p_2^{-\mu})w_{turb}) \tag{2.85}$$

where the fuel flow rate w_f and k_1, k_2, k_{1e}, K_t, τ_c are assumed to be positive constants. The control inputs are the back flow rate to the intake manifold w_{egr} and the flow rate through the turbine w_{turb}. The outputs to be controlled are the EGR flow rate w_{egr} and the compressor flow rate $w_c = K_c\frac{P_c}{p_1^\mu - 1}$, where the constants $K_c > 0$ and $0 < \mu < 1$. The compressor flow rate, intake and exhaust pressures are supposed to be the measurable outputs of the system, i.e. $z = col(p_1, p_2, w_c)$.

[9] No, there is no typographical error: *two thousand seven hundred and eighty two.*

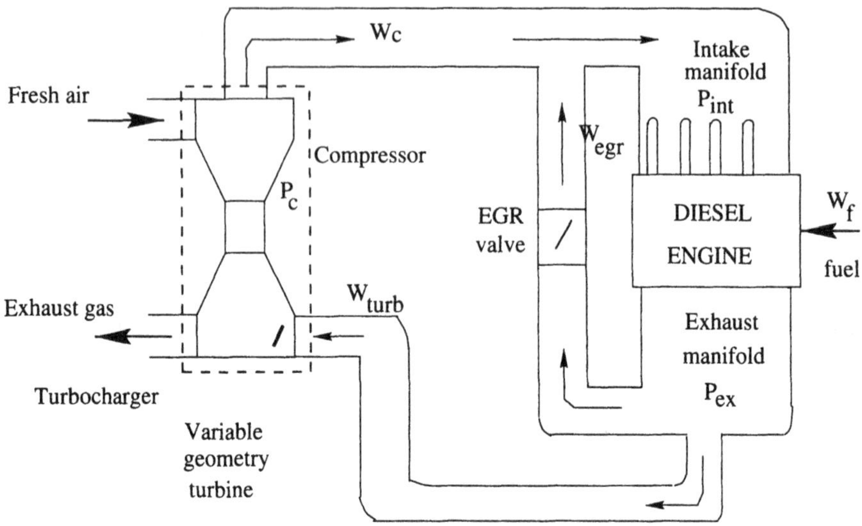

Fig. 2.4. Turbo charged VGT-EGR diesel engine.

From practical considerations it's reasonable to assume that $p_1, p_2 > 1$ (cf. [34]). As a matter of fact, the region of possible initial conditions $p_1(0) > 1$, $p_2(0) > 1$, $P_c(0) > 0$ is always known in practice, hence one can always choose the controller gains to ensure that $p_1, p_2 > 1$ for all $t \geq 0$ and thus to relax our practical assumption.

Under these conditions, the control objective is to assure asymptotic stabilization of the desired setpoint $y_d = \text{col}(w_{c,d}, w_{egr,d})$ for the controlled output $y = \text{col}(w_c, w_{egr})$.

Controller Design

The approach undertaken here consists on performing a suitable change of coordinates and designing decoupling control laws in order to put the controlled system into a cascaded form. Then, instead of looking for a Lyapunov function for the overall system we investigate the stability properties of both subsystems separately and exploit the structure of the interconnection, we do this by verifying the conditions of Theorem 2 which allow us to claim uniform global asymptotic stability.

We emphasize that, as shown in [34], fixing the outputs to their desired values $w_{c,d}$, $w_{egr,d}$ corresponds to setting the following desired equilibrium position of the diesel engine

$$p_{1*} = \frac{1}{k_{1e}}(w_{c,d} + w_{egr,d}) \tag{2.86}$$

$$p_{2*} = \left(1 - \frac{w_{c,d}}{K_c K_t(w_{c,d} + w_f)}(p_{1*}^{\mu} - 1)\right)^{-\frac{1}{\mu}} \tag{2.87}$$

$$P_c^* = \frac{w_{c,d}}{K_c}(p_{1*}^{\mu} - 1). \tag{2.88}$$

In other words, the stabilization problem of the output y to y_d reduces to the problem of stabilising the operating point p_{1*}, p_{2*}, P_{c*}. To that end, let us introduce the following change of variables

$$\begin{aligned} \tilde{p}_1 &= p_1 - p_{1*} \\ \tilde{p}_2 &= p_2 - p_{2*} \\ \tilde{p}_c &= P_c - P_c^* \end{aligned} \tag{2.89}$$

$$\begin{aligned} w_{egr} &= w_{egr,d} + u_1 \\ w_{turb} &= w_{c,d} + w_f + u_2 \end{aligned} \tag{2.90}$$

which will appear more suitable for our analysis. In these coordinates, and using (2.86)–(2.88), the system (2.83) –(2.85) with the new measurable output $z' = \text{col}[\tilde{p}_1 - p_{1*}, \tilde{p}_2 - p_{2*}, w_c - w_{c,d}]$ takes the form

$$\dot{\tilde{p}}_1 = k_1(z_3' - k_{1e}\tilde{p}_1 + u_1) \tag{2.91}$$

$$\dot{\tilde{p}}_2 = k_2(k_{1e}\tilde{p}_1 - u_1 - u_2) \tag{2.92}$$

$$\begin{aligned} \dot{\tilde{p}}_c = \frac{1}{\tau_c}\Big[&-\tilde{p}_c + K_t\left(p_{2*}^{-\mu} - (\tilde{p}_2 + p_{2*})^{-\mu}\right)(w_{c,d} + w_f) + \\ &K_t\left(1 - (\tilde{p}_2 + p_{2*})^{-\mu}\right)u_2\Big]. \end{aligned} \tag{2.93}$$

In order to apply the cascades-based approach, notice that system (2.91)–(2.92) has the required cascaded form with Σ_1 being equation (2.93) and Σ_2 being the pressure subsystem represented by equations (2.91) and (2.92).

Let us first consider the subsystem Σ_2 and let the control inputs be

$$u_1 = -z_3' - \gamma_1\tilde{p}_1 - \gamma_2\tilde{p}_2 \tag{2.94}$$

$$u_2 = z_3' + \gamma_1\tilde{p}_1 + \gamma_2'\tilde{p}_2 \tag{2.95}$$

where $\gamma_1, \gamma_2, \gamma_2'$ are arbitrary constants with the property $\gamma_2 < \gamma_2'$. Using the Lyapunov function candidate $V(\tilde{p}_1, \tilde{p}_2) = \frac{1}{2}\tilde{p}_1^2 + \frac{c}{2}\tilde{p}_2^2$ with $c = \frac{\gamma_2}{k_{1e}k_2}$ one can show that the closed loop system Σ_2 with (2.94,2.95) is globally exponentially stable uniformly in $\tilde{p}_c$.

To this point, it is important to remark that this closed-loop system actually depends on the variable $\tilde{p}_c$ due to the presence of z_3' in the control inputs.

However, the *uniform* character of the stability property established above implies that for any initial conditions, the signal $w_c(t)$ "inside" z_3' simply adds a time-varying character to the closed-loop system Σ_2 with (2.94)–(2.95) and hence it becomes non-autonomous. This allows us to consider these feedback interconnected systems as a *cascade* of an autonomous and a non autonomous nonlinear system.

Having established the stability property of system Σ_2 we proceed to investigate the properties of Σ_1 in closed loop with u_2. Substituting u_2 defined by (2.95) in (2.93) and after some lengthy but straightforward calculations involving the identity

$$1 - p_{2*}^{-\mu} = \frac{1}{K_c K_t} \frac{w_{c,d}}{w_{c,d} + w_f} (p_{1*}^{\mu} - 1),$$

we obtain that

$$\begin{aligned}\dot{\tilde{p}}_c = \underbrace{-\frac{1}{\tau_c}\frac{w_f}{w_{c,d}+w_f}\tilde{p}_c}_{f(x_1)} + \frac{1}{\tau_c}\Bigg[& \frac{w_{c,d}}{w_{c,d}+w_f}\frac{[\tilde{p}_c + P_{c*}][p_{1*}^{\mu} - (\tilde{p}_1 + p_{1*})^{\mu}]}{(\tilde{p}_1 + p_{1*})^{\mu} - 1} + \\ & K_t\left(1 - (\tilde{p}_2 + p_{2*})^{-\mu}\right)(\gamma_1 \tilde{p}_1 + \gamma_2' \tilde{p}_2) + \\ & K_t(w_{c,d} + w_f + z_3')\frac{(\tilde{p}_2 + p_{2*})^{\mu} - p_{2*}^{\mu}}{p_{2*}^{\mu}(\tilde{p}_2 + p_{2*})^{\mu}}\Bigg]\end{aligned}$$

which in compact form can be written as

$$\dot{x}_1 = f(x_1) + g(x_1, x_2)$$

where we recall that $x_1 = \tilde{p}_c$, $x_2 = \mathrm{col}(\tilde{p}_1, \tilde{p}_2)$. Notice that $g(x_1, x_2) \equiv 0$ if $x_2 = 0$ and $\dot{x}_1 = f(x_1)$ is GES with a quadratic Lyapunov function satisfying Assumption 3.

Since $(\tilde{p}_1 + p_{1*}) > 1$, $(\tilde{p}_2 + p_{2*}) > 1$ for all $t \geq 0$ and $0 < \mu < 1$ one can show that $g(x_1, x_2)$ is continuously differentiable and moreover notice that it grows linearly in x_1 (i.e., $\tilde{p}_c$) hence it satisfies the bound (2.56). Since $x_2 = 0$ is GES, $x_2(t)$ satisfies (2.57) with $\alpha(s) := \alpha s$, $\alpha > 0$. Thus all the conditions of Theorem 2.1 and Proposition 2.3 are satisfied and therefore the overall system is UGES.

2.4.3 Nonholonomic Systems

In recent years the control of nonholonomic dynamic systems has received considerable attention, in particular the stabilization problem. One of the reasons for this is that no smooth stabilizing state-feedback control law exists for these systems, since Brockett's necessary condition for smooth asymptotic stabilisability is not met [4]. For an overview we refer to the survey paper [22] and references cited therein. In contrast to the stabilization problem, the tracking control problem for nonholonomic control systems has received little attention. In [7, 18, 33, 36, 37] tracking control schemes have been proposed based on linearisation of the corresponding error model. In [2, 50] the feedback design issue was addressed via a dynamic feedback linearisation approach. All these papers solve the local tracking problem for some classes of nonholonomic systems. Some global tracking results are presented in [52, 17, 14].

Recently, the results in [17] have been extended to arbitrary chained form nonholonomic systems [16]. The proposed backstepping-based recursive design turned out to be useful for simplified dynamic models of such chained form systems (see [17, 16]). However, it is clear that the technique used in these references does not exploit the physical structure behind the model, and then the controllers may become quite complicated and computationally demanding when computed in original coordinates.

The purpose of this section is to show that the nonlinear controllers proposed in [17] can be simplified into *linear* controller for both the kinematic model and an 'integrated' simplified dynamic model of the mobile robot. Our approach is based on cascaded systems. As a result, instead of exponential stability for small initial errors as in [17], the controllers proposed here yield exponential stability for initial errors in a ball of arbitrary radius.

For a more detailed study on tracking control of nonholonomic systems based on the stability theorems for cascades presented here, see [25]. Indeed, the material presented in this section was originally reported in [40] and later in [25].

Model and Problem-Formulation

A kinematic model of a wheeled mobile robot with two degrees of freedom is given by the following equations

$$\begin{aligned} \dot{x} &= v\cos\theta \\ \dot{y} &= v\sin\theta \\ \dot{\theta} &= \omega \end{aligned} \tag{2.96}$$

where the forward velocity v and the angular velocity ω are considered as inputs, (x, y) is the center of the rear axis of the vehicle, and θ is the angle between heading direction and x-axis.

Consider the problem of tracking a reference robot as done in the well known article [18]. That is, given a mobile robot modeled by (2.96) it is desired that it follows a fictitious reference robot (cf. Figure 2.5) whose kinematic model is given by

$$\begin{aligned} \dot{x}_r &= v_r\cos\theta_r \\ \dot{y}_r &= v_r\sin\theta_r \\ \dot{\theta}_r &= \omega_r\,. \end{aligned}$$

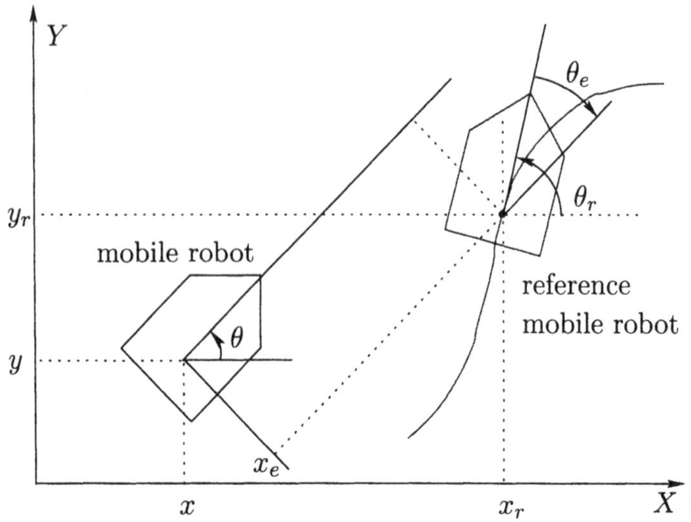

Fig. 2.5. Mobile robot: tracking problem.

Following [18] we define the error coordinates

$$\begin{bmatrix} x_e \\ y_e \\ \theta_e \end{bmatrix} = \begin{bmatrix} \cos\theta & \sin\theta & 0 \\ -\sin\theta & \cos\theta & 0 \\ 0 & 0 & 1 \end{bmatrix} \begin{bmatrix} x_r - x \\ y_r - y \\ \theta_r - \theta \end{bmatrix} .$$

It is easy to verify that in these new coordinates the error dynamics becomes

$$\begin{aligned} \dot{x}_e &= \omega y_e - v + v_r \cos\theta_e \\ \dot{y}_e &= -\omega x_e + v_r \sin\theta_e \\ \dot{\theta}_e &= \omega_r - \omega \,. \end{aligned} \tag{2.97}$$

Our aim is to find appropriate velocity control laws v and ω of the form

$$\begin{aligned} v &= v(t, x_e, y_e, \theta_e) \\ \omega &= \omega(t, x_e, y_e, \theta_e) \end{aligned} \tag{2.98}$$

such that the origin of the closed-loop system (2.97), (2.98) is USGES and UGAS.

Controller Design

Consider the error dynamics (2.97):

$$\dot{x}_e = \omega y_e - v + v_r \cos\theta_e \tag{2.99}$$

$$\dot{y}_e = -\omega x_e + v_r \sin\theta_e \tag{2.100}$$

$$\dot{\theta}_e = \omega_r - \omega \tag{2.101}$$

From a purely control theory viewpoint the problem is now to stabilize this system at the origin with the inputs v and ω.

To obtain simple (mathematically) controllerswe subdivide the tracking control problem into two simpler and 'independent' problems: for instance, positioning and orientation.

Firstly, we can easily stabilize the change rate of the mobile robot's orientation, that is, the linear equation (2.101), by using the linear controller

$$\omega = \omega_r + c_1\theta_e \tag{2.102}$$

which yields GES for θ_e, provided $c_1 > 0$.

If we now replace θ_e by 0 in (2.99), (2.100) we obtain

$$\begin{aligned} \dot{x}_e &= \omega_r y_e - v + v_r \\ \dot{y}_e &= -\omega_r x_e \end{aligned} \tag{2.103}$$

where we used (2.102). Concerning the positioning of the robot, if we choose the linear controller

$$v = v_r + c_2 x_e \tag{2.104}$$

where $c_2 > 0$, we obtain for the closed-loop system (2.103,2.104):

$$\begin{bmatrix} \dot{x}_e \\ \dot{y}_e \end{bmatrix} = \begin{bmatrix} -c_2 & \omega_r(t) \\ -\omega_r(t) & 0 \end{bmatrix} \begin{bmatrix} x_e \\ y_e \end{bmatrix} \tag{2.105}$$

which, as it is well known in the literature of adaptive control (see e.g. [3, 11]) is asymptotically stable if $\omega_r(t)$ is persistently exciting (PE), i.e., if there exist T, $\mu > 0$ such that $\omega_r(t)$ satisfies

$$\int_t^{t+T} \omega_r(\tau)^2 d\tau \geq \mu \quad \forall t \geq 0$$

and $c_2 > 0$. The following proposition makes this result rigorous.

Proposition 2.5. *Consider the system (2.97) in closed-loop with the controller*

$$\begin{aligned} v &= v_r + c_2 x_e \\ \omega &= \omega_r + c_1\theta_e \end{aligned} \tag{2.106}$$

where $c_1 > 0$, $c_2 > 0$. If $\omega_r(t)$, $\dot{\omega}_r(t)$, and $v_r(t)$ are bounded and ω_r is PE then, the closed-loop system (2.97), (2.106) is USGES and UGAS.

Proof. Observing that

$$\sin\theta_e = \theta_e \int_0^1 \cos(s\theta_e)ds \qquad \text{and} \qquad 1 - \cos\theta_e = \theta_e \int_0^1 \sin(s\theta_e)ds$$

we can write the closed-loop system (2.97), (2.106) as

$$\begin{bmatrix} \dot{x}_e \\ \dot{y}_e \end{bmatrix} = \begin{bmatrix} -c_2 & \omega_r(t) \\ -\omega_r(t) & 0 \end{bmatrix} \begin{bmatrix} x_e \\ y_e \end{bmatrix} + \begin{bmatrix} v_r \int_0^1 \sin(s\theta_e)ds + c_1 y_e \\ v_r \int_0^1 \cos(s\theta_e)ds - c_1 x_e \end{bmatrix} \theta_e \tag{2.107}$$
$$\dot{\theta}_e = -c_1\theta_e$$

which is of the form (2.35) with $x_1 := (x_e, y_e)^\top$, $x_2 := \theta_e$, $f_2(t, x_2) = -c_1\theta_e$,

$$f_1(t, x_1) = \begin{bmatrix} -c_2 & \omega_r(t) \\ -\omega_r(t) & 0 \end{bmatrix} \begin{bmatrix} x_e \\ y_e \end{bmatrix}, \quad g(t, x) = \begin{bmatrix} v_r \int_0^1 \sin(s\theta_e)ds + c_1 y_e \\ v_r \int_0^1 \cos(s\theta_e)ds - c_1 x_e \end{bmatrix}$$

To be able to apply Theorem 2.1 we need to verify the following three conditions

- The assumption on Σ_1: due to the assumptions on $\omega_r(t)$ we have that $\dot{x} = f_1(t, x_1)$ is GES and therefore UGAS. The assumptions on $V(t, x_1)$ hold for $V(t, x_1) = |x_1|^2$.
- The assumption on the interconnection term: since $|v_r(t)| \leq v_r^{max}$ for all $t \geq 0$ we have:
$$|g(t, x)| \leq v_r^{max}\sqrt{2} + c_1 |x_1| .$$
- The assumption on Σ_2 follows from GES of (2.101) in closed-loop with (2.102).

Therefore, we can conclude UGAS from Theorem 2.1. Since both Σ_1 and Σ_2 are GES, Proposition 2.3 gives the desired result. ∎

Remark 2.7. It is interesting to notice the link between the tracking condition that the reference trajectory should not converge to a point and the well known persistence-of-excitation condition in adaptive control theory. More precisely, we could think of (2.105) as a controlled system with state x_e, parameter estimation error y_e and regressor, the *reference* trajectory ω_r.

Remark 2.8. The cascaded decomposition used above is not unique. One may find other ways to subdivide the original system, for which different control laws may be found. Notice that the structure that we have used has an interesting physical interpretation: roughly speaking we have proved that the positioning and the orientation of the cart can be controlled *independently* of each other.

A Simplified Dynamic Model

Let us consider now the dynamic extension of (2.97) as studied in [17], i.e.,

$$\begin{aligned} \dot{x}_e &= \omega y_e - v + v_r \cos\theta_e \\ \dot{y}_e &= -\omega x_e + v_r \sin\theta_e \\ \dot{\theta}_e &= \omega_r - \omega \\ \dot{v} &= u_1 \\ \dot{\omega} &= u_2 \end{aligned} \tag{2.108}$$

where u_1 and u_2 are regarded as torques or generalized force variables for the two-degree-of-freedom mobile robot.

Our aim is to find a control law $u = (u_1, u_2)^T$ of the form

$$\begin{aligned} u_1 &= u_1(t, x_e, y_e, \theta_e, v, \omega) \\ u_2 &= u_2(t, x_e, y_e, \theta_e, v, \omega) \end{aligned} \tag{2.109}$$

such that the origin of the closed-loop system (2.108), (2.109) is USGES and UGAS. To solve this problem we start by defining

$$\begin{aligned} v_e &= v - v_r \\ \omega_e &= \omega - \omega_r \end{aligned}$$

which leads to

$$\begin{bmatrix} \dot{x}_e \\ \dot{v}_e \\ \dot{y}_e \end{bmatrix} = \begin{bmatrix} 0 & -1 & \omega_r(t) \\ 0 & 0 & 0 \\ -\omega_r(t) & 0 & 0 \end{bmatrix} \begin{bmatrix} x_e \\ v_e \\ y_e \end{bmatrix} + \begin{bmatrix} 0 \\ 1 \\ 0 \end{bmatrix} (u_1 - \dot{v}_r) + \begin{bmatrix} v_r \int_0^1 \sin(s\theta_e)ds & y_e \\ 0 & 0 \\ v_r \int_0^1 \cos(s\theta_e)ds & -x_e \end{bmatrix} \begin{bmatrix} \theta_e \\ \omega_e \end{bmatrix} \tag{2.110}$$

$$\begin{bmatrix} \dot{\theta}_e \\ \dot{\omega}_e \end{bmatrix} = \begin{bmatrix} 0 & -1 \\ 0 & 0 \end{bmatrix} \begin{bmatrix} \theta_e \\ \omega_e \end{bmatrix} + \begin{bmatrix} 0 \\ 1 \end{bmatrix} (u_2 - \dot{\omega}_r) \tag{2.111}$$

It is easy, again, to recognize a cascaded structure similar to the one in the previous example. We only need to find u_1 and u_2 such that the systems

$$\begin{bmatrix} \dot{x}_e \\ \dot{v}_e \\ \dot{y}_e \end{bmatrix} = \begin{bmatrix} 0 & -1 & \omega_r(t) \\ 0 & 0 & 0 \\ -\omega_r(t) & 0 & 0 \end{bmatrix} \begin{bmatrix} x_e \\ v_e \\ y_e \end{bmatrix} + \begin{bmatrix} 0 \\ 1 \\ 0 \end{bmatrix} u_1$$

and

$$\begin{bmatrix} \dot{\theta}_e \\ \dot{\omega}_e \end{bmatrix} = \begin{bmatrix} 0 & -1 \\ 0 & 0 \end{bmatrix} \begin{bmatrix} \theta_e \\ \omega_e \end{bmatrix} + \begin{bmatrix} 0 \\ 1 \end{bmatrix} u_2$$

are USGES and UGAS.

Proposition 2.6. *Consider the system (2.108) in closed-loop with the controller*

$$u_1 = \dot{v}_r + c_3 x_e - c_4 v_e \tag{2.112a}$$

$$u_2 = \dot{\omega}_r + c_5 \theta_e - c_6 \omega_e \tag{2.112b}$$

where $c_3 > 0$, $c_4 > 0$, $c_5 > 0$, $c_6 > 0$. *If* $\omega_r(t)$, $\dot{\omega}_r(t)$ *and* $v_r(t)$ *are bounded and* ω_r *is PE then, the closed-loop system (2.108), (2.112) is USGES and UGAS.*

Proof. The closed-loop system (2.108), (2.112) can be written as

$$\begin{bmatrix} \dot{x}_e \\ \dot{v}_e \\ \dot{y}_e \end{bmatrix} = \begin{bmatrix} 0 & -1 & \omega_r(t) \\ c_3 & -c_4 & 0 \\ -\omega_r(t) & 0 & 0 \end{bmatrix} \begin{bmatrix} x_e \\ v_e \\ y_e \end{bmatrix} + \begin{bmatrix} v_r \int_0^1 \sin(s\theta_e) ds & y_e \\ 0 & 0 \\ v_r \int_0^1 \cos(s\theta_e) ds & -x_e \end{bmatrix} \begin{bmatrix} \theta_e \\ \omega_e \end{bmatrix}$$

$$\begin{bmatrix} \dot{\theta}_e \\ \dot{\omega}_e \end{bmatrix} = \begin{bmatrix} 0 & -1 \\ c_5 & -c_6 \end{bmatrix} \begin{bmatrix} \theta_e \\ \omega_e \end{bmatrix}$$

which is of the form (2.35). We again have to verify the three assumptions of Theorem 2.1:

- The assumption on Σ_1: This system is GES (and therefore UGAS) under the assumptions imposed on $\omega_r(t)$ and c_2. The assumption on $V(t, x_1)$ holds with $V(t, x_1) := c_3 v_e^2 + x_e^2 + y_e^2$.
- The assumption on the interconnection term: since $|v_r(t)| \leq v_r^{max}$ for all $t \geq 0$ we have:
$$|g(t, x, y)| \leq v_r^{max}\sqrt{2} + |x| .$$
- The assumption on Σ_2 follows from GES of Σ_2. Note that the latter holds for any c_5 and $c_6 > 0$.

Therefore we can conclude UGAS from Theorem 2.1. Since both Σ_1 and Σ_2 are GES, Proposition 2.3 gives the desired result. ■

2.5 Conclusions

Motivated by practical problems such as global tracking of time-varying trajectories we have presented some results on the stability analysis problem of UGAS nonlinear non-autonomous systems in cascade. The theorems presented in this chapter establish sufficient conditions to ensure uniform global asymptotic stability of cascaded nonlinear systems. The most fundamental result is probably that, for UGAS nonlinear time-varying systems in cascade, it is sufficient and necessary that the solutions of the cascaded system remain uniformly globally bounded. Other conditions have been established so as to verify this fundamental property. Such conditions have been formulated mainly in terms of growth rates of converse Lyapunov functions and the terms that define the interconnection of the cascade.

We have also illustrated the technique of cascaded-based control through different applications.

References

1. M. Arcak, D. Angeli, and E. Sontag, "A unifying integral ISS framework for stability of nonlinear cascades," *SIAM J. on Contr. and Opt.*, vol. 40, pp. 888–1904, 2002.
2. G. Bastin, B. Brogliato, G. Campion, C. Canudas, M. Khalil, B. d'Andra Novel, A. de Luca, R. Lozano, R. Ortega, C. Samson, B. Siciliano, and P. Tomei (1996) *Theory of Robot Control.* Texts in Applied Mathematics. Springer-Verlag.
3. S. Boyd and S. Sastry (1989) *Adaptive control: stability, convergence, and robustness.* Prentice Hall, USA.
4. R. Brockett (1983) Asymptotic stability and feedback stabilization. In R. S. Millman R. W. Brocket and H. J. Sussmann, editors, *Differential geometric control theory*, pages 181–191. Birkhäuser.
5. C.I. Byrnes and A. Isidori (1989) New results and examples in nonlinear feedback stabilization. *Syst. & Contr. Letters*, 12:437–442.
6. M. Corless and L. Gilelmo (1998) New converse Lyapunov theorems and related results on exponential stability. *Math. of Cont. Sign. and Syst.*, 11(1):79–100.
7. R. Fierro and F.L. Lewis (1995) Control of a nonholonomic mobile robot: backstepping kinematics into dynamics. In *Proc. 34th. IEEE Conf. Decision Contr.*, pages 3805–3810, New Orleans, LA.
8. T. I. Fossen (1994) *Guidance and control of ocean vehicles.* John Wiley & Sons Ltd.
9. T. I. Fossen and J. P. Strand (1999) Passive nonlinear observer design for ships using Lyapunov methods: Full-scale experiments with a supply vessel. *Automatica*, 35(1):3–16.
10. W. Hahn (1967) *Stability of motion.* Springer-Verlag, New York.
11. P. Ioannou and J. Sun (1996) *Robust adaptive control.* Prentice Hall, New Jersey, USA.
12. M. Janković, R. Sepulchre, and P. V. Kokotović (1996) Constructive Lyapunov stabilization of non linear cascaded systems. *IEEE Trans. on Automat. Contr.*, 41:1723–1736.
13. R.H. Middleton J.H. Braslavsky (1996) Global and semi-global stabilizability in certain cascade nonlinear systems. *IEEE Trans. on Automat. Contr.*, 41:876–880.
14. Z.-P. Jiang, E. Lefeber, and H. Nijmeijer. (1998) Stabilization and tracking of a nonholonomic mobile robot with saturating actuators. In *Control'98*, Portugal.
15. Z. P. Jiang and I. Mareels (1997) A small gain control method for nonlinear cascaded systems with dynamic uncertainties. *IEEE Trans. on Automat. Contr.*, 42(3):1–17.
16. Z.-P. Jiang and H. Nijmeijer (1997) Backstepping based tracking control of nonholonomic chained systems. In *Proc. 4th. European Contr. Conf.*, Brussels, Belgium. Paper no. 672.
17. Z.-P. Jiang and H. Nijmeijer (1997) Tracking control of mobile robots: A case study in backstepping. *Automatica*, 33(7):1393–1399.
18. Y. Kanayama, Y. Kimura, F. Miyazaki, and T. Naguchi (1990) A stable traking control scheme for an autonomous vehicle. In *Proc. IEEE Conf. Robotics Automat.*, pages 384–389.
19. H. Khalil (1996) *Nonlinear systems.* Macmillan Publishing Co., 2nd ed., New York.

20. H. Khalil (2002) *Nonlinear systems.* Prentice Hall, 3rd ed., New York.
21. D.E. Koditschek (1987) Adaptive techniques for mechanical systems. In *Proceedings 5th Yale Workshop on Adaptive Systems*, pages 259–265, New Haven, CT.
22. I. Kolmanovsky and H. McClamroch (1995) Developments in nonholonomic control problems. *Control systems magazine*, pages 20–36.
23. M. Krstić, I. Kanellakopoulos, and P. Kokotović (1995) *Nonlinear and Adaptive control design.* John Wiley & Sons, Inc., New York.
24. J. Kurzweil (1956) On the inversion of Ljapunov's second theorem on stability of motion. *Amer. Math. Soc. Translations*, 24:19–77.
25. A. A. J. Lefeber (2000) *Tracking control of nonlinear mechanical systems.* PhD thesis, University of Twente, Enschede, The Netherlands.
26. Y. Lin, E. D. Sontag, and Y. Wang (1996) A smooth converse Lyapunov theorem for robust stability. *SIAM J. on Contr. and Opt.*, 34:124–160.
27. A. Loría T. I. Fossen, and E. Panteley (2000) A separation principle for dynamic positioning of ships: theoretical and experimental results. *IEEE Trans. Contr. Syst. Technol.*, 8(2):332–344.
28. A. Loría H. Nijmeijer, and O. Egeland (1998) Controlled synchronization of two pendula: cascaded structure approach. In *Proc. American Control Conference*, Philadelphia, PA.
29. I. J. Malkin (1958) Theory of stability of motion. Technical report, U.S. Atomic energy commission.
30. F. Mazenc and L. Praly (1996) Adding integrators, saturated controls and global asymptotic stabilization of feedforward systems. *IEEE Trans. on Automat. Contr.*, 41(11):1559–1579.
31. F. Mazenc and L. Praly (1997) Asymptotic tracking of a state reference for systems with a feedforward structure. In *Proc. 4th. European Contr. Conf.*, Brussels, Belgium. paper no. 954. To appear in *Automatica.*
32. F. Mazenc, R. Sepulchre, and M. Janković (1997) Lyapunov functions for stable cascades and applications to global stabilization. In *Proc. 36th. IEEE Conf. Decision Contr.*
33. A. Micaelli and C. Samson (1993) Trajectory tracking for unicycle-type and two-steering-wheels mobile robots. Technical Report 2097, INRIA.
34. P. Moraal, M. Van Nieuwstadt, and M. Jankovich (1997) Robust geometric control: An automotive application. In *Proc. 4th. European Contr. Conf.*, Brussels.
35. A. P. Morgan and K. S. Narendra (1977) On the stability of nonautonomous differential equations $\dot{x} = [A+B(t)]x$ with skew-symmetric matrix $B(t)$. *SIAM J. on Contr. and Opt.*, 15(1):163–176.
36. R.M. Murray, G. Walsh, and S.S. Sastry (1992) Stabilization and tracking for nonholonomic control systems using time-varying state feedback. In M. Fliess, editor, *IFAC Nonlinear control systems design*, pages 109–114, Bordeaux.
37. W. Oelen and J. van Amerongen (1994) Robust tracking control of two-degrees-of-freedom mobile robots. *Control Engineering Practice*, pages 333–340.
38. R. Ortega (1991) Passivity properties for stabilization of nonlinear cascaded systems. *Automatica*, 29:423–424.
39. R. Ortega, A. Loría P. J. Nicklasson, and H. Sira-Ramírez (1998) *Passivity-based Control of Euler-Lagrange Systems: Mechanical, Electrical and Electromechanical Applications.* Comunications and Control Engineering. Springer Verlag, London. ISBN 1-85233-016-3.

40. E. Panteley, E. Lefeber, A. Loría and H. Nijmeijer (1998) Exponential tracking of a mobile car using a cascaded approach. In *IFAC Workshop on Motion Control*, pages 221–226, Grenoble, France.
41. E. Panteley and A. Loría(1997) Global uniform asymptotic stability of cascaded non autonomous nonlinear systems. In *Proc. 4th. European Contr. Conf.*, Louvain-La-Neuve, Belgium. Paper no. 259.
42. E. Panteley and A. Loría(1998) On global uniform asymptotic stability of non linear time-varying non autonomous systems in cascade. *Syst. & Contr. Letters*, 33(2):131–138.
43. E. Panteley, A. Loría and A. Sokolov (1999) Global uniform asymptotic stability of nonlinear nonautonomous systems: Application to a turbo-diesel engine. *European J. of Contr.*, 5:107–115.
44. E. Panteley, A. Loría and A. Teel (1999) UGAS of NLTV systems: Applications to adaptive control. Technical Report 99-160, Lab. d'Automatique de Grenoble, UMR 5528, CNRS, France. Related publications: see *IEEE Trans. Aut. Contr.* 46(12), *Syst. Contr. Lett. 2002.*
45. E. Panteley and R. Ortega (1997) Cascaded control of feedback interconnected systems: Application to robots with AC drives. *Automatica*, 33(11):1935–1947.
46. E. Panteley, R. Ortega, and P. Aquino (1997) Cascaded Control of Feedback Interconnected Systems: Application to Robots with AC Drives. In *Proc. 4th. European Contr. Conf.*, Louvain-La-Neuve, Belgium.
47. L. Praly (2002) An introduction to lyapunov designs of global asymptotic stabilizers. Technical report, Banach Center Summer School on Mathematical control theory, Poland.
48. L. Praly and Y. Wang (1996) Stabilization in spite of matched unmodelled dynamics and an equivalent definition of input-to-state stability. *Math. of Cont. Sign. and Syst.*, 9:1–33.
49. N. Rouche and J. Mawhin (1980) *Ordinary differential equations II: Stability and periodical solutions.* Pitman publishing Ltd., London.
50. C. Rui and N.H. McClamroch (1995) Stabilization and asymptotic path tracking of a rolling disk. In *Proc. 34th. IEEE Conf. Decision Contr.*, pages 4294–4299, New Orleans, LA.
51. A. Saberi, P. V. Kokotović, and H. J. Sussmann (1990) Global stabilization of partially linear systems. *SIAM J. Contr. and Optimization*, 28:1491–1503.
52. C. Samson and K. Ait-Abderrahim (1991) Feedback control of a nonholonomic wheeled cart in cartesian space. In *Proc. IEEE Conf. Robotics Automat.*, pages 1136–1141, Sacramento.
53. G. Sansone and R. Conti (1964) *Nonlinear Differential equations.* Pergamon Press, Oxford, England.
54. S. Sastry and M. Bodson (1989) *Adaptive control: Stability, convergence and robustness.* Prentice Hall Intl.
55. P. Seibert and R. Suárez (1990) Global stabilization of nonlinear cascaded systems. *Syst. & Contr. Letters*, 14:347–352.
56. R. Sepulchre and M. Arcak abd A. R. Teel (2002) Trading the stability of finite zeros for global stabilization of nonlinear cascade systems. *IEEE Trans. on Automat. Contr.*, 47(3):521–525.
57. R. Sepulchre, M. Janković, and P. Kokotović (1997) *Constructive nonlinear control.* Springer Verlag.
58. E. Sontag (1989) Smooth stabilization implies coprime factorization. *IEEE Trans. on Automat. Contr.*, 34(4):435–443.

59. E. D. Sontag (1989) Remarks on stabilization and Input-to-State stability. In *Proc. 28th. IEEE Conf. Decision Contr.*, pages 1376–1378, Tampa, Fl.
60. E. D. Sontag (1995) On the input-to-state stability property. *European J. Control*, 1:24–36.
61. E. D. Sontag, "Comments on integral variants of ISS," *Syst. & Contr. Letters*, vol. 34, pp. 93–100, 1998.
62. E.D. Sontag and Y. Wang (1995) On characterizations of the input-to-state stability property. *Syst. & Contr. Letters*, 24:351–359.
63. H. J. Sussman and P. V. Kokotović (1991) The peaking phenomenon and the global stabilization of nonlinear systems. *IEEE Trans. on Automat. Contr.*, 36(4):424–439.
64. A. Teel. Converse Lyapunov function theorems: statements and constructions. Notes from *Advanced nonlinear systems theory.* Course no. ECE 594D, University of California, Santa Barbara, CA, USA, Summer 1998.
65. J. Tsinias (1989) Sufficient Lyapunov-like conditions for stabilization. *Math. Control Signals Systems*, pages 343–357.
66. M. Vidyasagar (1993) *Nonlinear systems analysis.* Prentice Hall, New Jersey.

3

Control of Mechanical Systems from Aerospace Engineering

Bernard Bonnard, Mohamed Jabeur, and Gabriel Janin

Département de Mathématiques, Laboratoire Analyse Appliquée et Optimisation B.P 47870, 21078 Dijon Cedex, France. E-mail: bbonnard@u-bourgogne.fr, mjabeur@u-bourgogne.fr, gabriel.janin@ensta.org.

The objective of this chapter is to describe geometric methods applied to aerospace systems: attitude control of a rigid spacecraft, orbital transfer, shuttle re-entry. We characterize the controllability of a rigid spacecraft controlled by one gas jet and the set of orbits reached in orbital transfer, depending upon the orientation of the thrust. Then we construct controls in orbital transfer and attitude control, using stabilization techniques and path planning. The optimal control is analyzed in orbital transfer, the cost being the time and in the shuttle re-entry, where the cost is the total amount of thermal flux, taking into account the state constraints. We present an analysis of extremals, solutions of the maximum principle and second-order sufficient optimality conditions.

3.1 Introduction

The aim of this chapter is to present recent advances of non linear control applied and developed for systems from aerospace engineering. Our presentation is founded on three problems studied with aerospace agencies (ESTEC - CNES) which are:

- Control the attitude (or orientation) of a rigid spacecraft governed by gas jets.
- Orbital transfer: a satellite is on an elliptic orbit and has to be transferred to an other elliptic orbit, e.g. geostationary, with or without rendezvous. Here the control is the thrust and we have a standard controlled Kepler equation.

F. Lamnabhi-Lagarrigue et al. (Eds.): Adv. Top. in Cntrl. Sys. Theory, LNCIS 311, pp. 65–113, 2005

• Shuttle re-entry: in this problem we are in the atmospheric part of the trajectory, the thrust is turned off and the shuttle behaves as a glider, the control being the lift force. There is also a drag force. The model is Kepler equation, with aerodynamic forces.

The studied problems are not independent because:

• The attitude control can be used to orientate the thrust in the orbital transfer.

• In the re-entry problem, a preliminary part of the trajectory is an orbital transfer.

The analyzed problems in this chapter are:

1) *Controllability problem.* Given a system of the form $\dot{q}(t) = f(q(t), u(t))$, where $q \in M$, $u \in U$. We note $A(q_0, t)$ the accessibility set in time t and $A(q_0)$ the accessibility set. Preliminary questions are the controllability problems: $A(q_0, t) = M$ or $A(q_0) = M$ for $q_0 \in M$.

Our aim is to present techniques to decide about controllability using Lie algebraic conditions, which are applicable to our problems. In particular we shall derive necessary and sufficient controllability conditions in the following cases:

• In the attitude control problem we shall characterize how many gas jets we need to control the orientation and what the positions of the gas jets on the satellite are.

• In the orbital transfer, the thrust F can be decomposed with respect to a frame e.g. the tangential/normal frame: $F = u_t F_t + u_n F_n + u_c F_c$, where $F_t = \frac{\dot{q}}{|\dot{q}|}$, F_n perpendicular to F_t in the q, $\dot{q}$ plane and F_c normal to this plane. The controllability problem is to characterize the accessibility set if the control is oriented along one of the F_t, F_n or F_c direction only.

The second step is:

2) *Construction of a control.* After having characterized the controllability property of the system, a problem is to construct a control to steer the system to the desired state. We shall present techniques we use in our problems.

2.a) *Stabilization technique.* Construct a feedback $q \mapsto u(q)$ to steer q_0 to the final state q_f asymptotically. This technique will be presented in the orbital transfer, it can also be used in the attitude control problem, for instance to control the angular velocity.

2.b) *Path planning using geometric techniques.* The method is to fix a path joining q_0 to the desired state q_f and to compute a control to track this path. Such a technique required a good knowledge of the admissible trajectories of the system. It can be used in the attitude problem where the transfer can be realized using rotations around the principal axis of the body, and in the orbital transfer.

3) *Optimal control.* In our problems a control law can be designed minimizing a cost:

• *Attitude control.* A cost can be the time or the fuel consumption.

• *Orbital transfer.* In recent research projects we consider a low propulsion and it can take several months to steer the system to a geostationary orbit, the problem is then to minimize the time.

• *Shuttle re-entry.* An important problem is to minimize the total amount of thermal flux.

To analyze the problems, we use the maximum principle. If the cost is $\int_0^T f^0(q,u)dt$, we consider the extended system: $\dot{q} = f(q,u)$, $\dot{q_0} = f^0(q,u)$ written shortly: $\dot{\tilde{q}} = \tilde{f}(\tilde{q},u)$. The optimal solutions are extremals, solutions of the Hamiltonian equations,

$$\dot{\tilde{q}} = \frac{\partial \tilde{H}}{\partial \tilde{p}} \quad , \quad \dot{\tilde{p}} = -\frac{\partial \tilde{H}}{\partial \tilde{q}} ,$$
$$\tilde{H}(\tilde{q},\tilde{p},u) = \max_{v \in U} \tilde{H}(\tilde{q},\tilde{p},v),$$

where $\tilde{H} = \langle \tilde{p}, \tilde{f}(\tilde{q},u) \rangle$ is the Hamiltonian, $\tilde{p}$ is the adjoint vector and $\langle , \rangle$ denotes the scalar product. In order to compute the optimal solution one needs to analyze the extremal flow. This analysis is complicated and we shall describe some of the tools coming from singularity analysis to achieve this task.

The optimal control problem will be analyzed in details in the re-entry problem where we have also to take into account state constraints, on the thermal flux and the normal acceleration. An analysis of the solutions of the maximum principle and direct computations will allow to describe the structure of the control, each optimal solution being a concatenation of bang and boundary arcs. Using this geometric characterization, we can use numerical methods e.g. shooting methods to compute only the switching times to realize the transfer.

The synthesis of the optimal law will be also analyzed in the orbital transfer, where the structure of extremals solutions is much simpler. Also we use this problem to present second-order sufficient conditions based on the concept of conjugate points which can be interpreted and computed using the extremal flow.

3.2 Mathematical Models

In this section, we describe the systems representing the attitude control problem, the orbital transfer and the shuttle re-entry.

3.2.1 The Attitude Control Problem

Let G be the center of mass of the rigid body, $k = (e_1, e_2, e_3)$ be an inertial frame with origin G. The attitude or orientation of the body is characterized by a frame $K(t) = (E_1(t), E_2(t), E_3(t))$ attached to the body. Let $R(t)$ be the matrix defined by $R(t) = (r_{ij})$, $1 \leq i, j \leq 3$, $r_{ij}(t) = \langle E_i(t), e_j \rangle$, $\langle .,. \rangle$ standard scalar product, $R(t) \in SO(3)$, the group of direct rotations. Each vectors of $\mathbb{R}^3$ can be measured in k or K, with respective notations v and V and we have $v = {}^tR\,V$. We choose the moving frame so that E_1, E_2, E_3 are principal axis and the system is described by

$$\frac{dR(t)}{dt} = S(\Omega(t))R(t)$$

$$S(\Omega) = \begin{pmatrix} 0 & \Omega_3 & -\Omega_2 \\ -\Omega_3 & 0 & \Omega_1 \\ \Omega_2 & -\Omega_1 & 0 \end{pmatrix}, \tag{3.1}$$

$$\begin{aligned} I_1 \frac{d\Omega_1(t)}{dt} &= (I_2 - I_3)\Omega_2(t)\Omega_3(t) + F_1(t) \\ I_2 \frac{d\Omega_2(t)}{dt} &= (I_3 - I_1)\Omega_3(t)\Omega_1(t) + F_2(t) \\ I_3 \frac{d\Omega_3(t)}{dt} &= (I_1 - I_2)\Omega_1(t)\Omega_2(t) + F_3(t). \end{aligned} \tag{3.2}$$

Equation (3.1) is equivalent to the equation $\frac{dq}{dt} = \omega \wedge q$ defining the angular velocity, the vector Ω whose components are Ω_i, representing the angular velocity measured in the moving frame. The parameters $I_1 \geq I_2 \geq I_3 > 0$ are the principal moments of inertia and they will be assumed distinct. The vector $F(t)$ with components $F_i(t)$ represents the applied torque measured in the body.

The equation (3.2) is called Euler equations and can be written:

$$\frac{dM(t)}{dt} = M(t) \wedge \Omega(t) + F(t),$$

where m is the momentum of the rigid body, $m = {}^tR\,M$.

In our system, the attitude is controlled using gas jets opposite thrusters. The applied torque for one pair is:

$$F(t) = u(t)(I_1 b^1, I_2 b^2, I_3 b^3),$$

where (b^1, b^2, b^3) are constants representing the position on the thruster and the control $u(t)$, $t \in [0, T]$ is a piecewise constant mapping with values in $\{-M, 0, M\}$. For m pairs of gas jets, we get a system of the form:

$$\begin{aligned}\frac{dR(t)}{dt} &= S(\Omega(t))R(t)\\ \frac{d\Omega(t)}{dt} &= Q(\Omega(t)) + \sum_{k=1}^{m} u_k(t)b_k\,,\end{aligned} \tag{3.3}$$

where $Q(\Omega_1, \Omega_2, \Omega_3) = (a_1\Omega_2\Omega_3, a_2\Omega_1\Omega_3, a_3\Omega_1\Omega_2)$, $a_1 = (I_2 - I_3)I_1^{-1}$, $a_2 = (I_3 - I_1)I_2^{-1}$, $a_3 = (I_1 - I_2)I_3^{-1}$, $a_1, a_3 > 0$, $a_2 < 0$ and each input $u_k(t)$ is a piecewise constant mapping taking its values in {-1,0,+1}.

3.2.2 Orbital Transfer

Model

We note m the mass and F the thrust of the engine. The equations, using coordinates, are:

$$\ddot{q} = -\mu\frac{q}{|q|^3} + \frac{F}{m}\,, \tag{3.4}$$

where q is the position of the satellite measured in a frame I, J, K, whose origin is the center of the Earth and μ is the gravitation constant, the free motion where $F = 0$ being the Kepler equation. The thrust is bounded: $|F| \leq F_{max}$ and the equations describing the mass variation is:

$$\dot{m}(t) = -\frac{|F|}{v_e}\,. \tag{3.5}$$

The thrust can be decomposed into a cartesian frame, or a moving frame. In particular, if $q \wedge \dot{q} \neq 0$, we can use a tangential/normal frame: $F = u_tF_t + u_nF_n + u_cF_c$, F_t being collinear to $\dot{q}$, F_n normal to F_t in the osculating plane $Span\{q, \dot{q}\}$ and F_c is orthogonal to this plane.

First Integrals and Orbit Elements

Proposition 3.1. *Let* $\ddot{q} = -\mu\frac{q}{|q|^3}$ *be Kepler equation, the following vectors are first integrals*

1. $c = q \wedge \dot{q}$ *(momentum)*
2. $L = -\mu\frac{q}{|q|} + \dot{q} \wedge c$ *(Laplace integral).*

Moreover, the energy: $H(q, \dot{q}) = \frac{1}{2}\dot{q}^2 - \frac{\mu}{|q|}$ *is preserved and the following equations hold:* $L.c = 0$, $L^2 = \mu^2 + 2Hc^2$.

Proposition 3.2. *For Kepler equation, if the momentum $c = 0$, q and $\dot{q}$ are collinear and the motion is on a line. If $c \neq 0$, we have:*

1. *If $L = 0$, the motion is circular*
2. *If $L \neq 0$ and $H < 0$, the trajectory is an ellipse given by*

$$|q| = \frac{c^2}{\mu + |L| \cos(\theta - \theta_0)} \tag{3.6}$$

where θ_0 is the angle of the pericenter.

Next we introduce the orbit elements and Lagrange equations which are adapted for low thrust propulsion, each trajectory being a perturbed ellipse.

Definition 3.1. *The domain $\Sigma_e = \{(q, \dot{q}); H < 0, c \neq 0\}$ is filled by elliptic orbits and to each (c, L) corresponds a unique oriented ellipse. Σ_e is called the elliptic domain.*

We identify I, J plane to the equatorial plane, each point $(q, \dot{q})$ in the elliptic domain is represented by standard orbit elements:

Ω: longitude of the ascending node,

ω: angle of the pericenter,

i: inclination of the osculating plane,

a: semi-major axis of the ellipse,

$|e|$: eccentricity.

If e is the eccentricity vector collinear to L with modulus $|e|$, let $\tilde{\omega}$ be the angle between I and e. We set: $e_1 = |e| \cos(\tilde{\omega})$, $e_2 = |e| \sin(\tilde{\omega})$. The line of nodes is represented by $h_1 = \tan(\frac{i}{2}) \cos \Omega$, $h_2 = \tan(\frac{i}{2}) \sin \Omega$.

If we use the orbit elements and decompose the thrust in the tangential/normal frame, the system (3.4) becomes:

$$
\begin{aligned}
\frac{da}{dt} &= 2\sqrt{\frac{a^3}{\mu}} \frac{B}{A} u_t \\
\frac{de_1}{dt} &= \sqrt{\frac{a}{\mu}} \frac{A}{D} \left(\frac{2(e_1 + \cos(l))D}{B} \right) u_t \\
&- \sqrt{\frac{a}{\mu}} \frac{A}{D} \left(\frac{2e_1 e_2 \cos(l) - \sin(l)(e_1^2 - e_2^2) + 2e_2 + \sin(l)}{B} \right) u_n \\
&- \sqrt{\frac{a}{\mu}} \frac{A}{D} e_2 (h_1 \sin(l) - h_2 \cos(l)) u_c
\end{aligned}
$$

$$\begin{aligned}
\frac{de_2}{dt} &= \sqrt{\frac{a}{\mu}\frac{A}{D}}\left(\frac{2(e_1+\sin(l))D}{B}\right)u_t \\
&+\sqrt{\frac{a}{\mu}\frac{A}{D}}\left(\frac{2e_1e_2\sin(l)+\cos(l)(e_1^2-e_2^2)+2e_1+\cos(l)}{B}\right)u_n \\
&+\sqrt{\frac{a}{\mu}\frac{A}{D}}e_1(h_1\sin(l)-h_2\cos(l))u_c \\
\frac{dh_1}{dt} &= \frac{1}{2}\sqrt{\frac{a}{\mu}\frac{A}{D}}(1+h_1^2+h_2^2)\cos(l)u_c \\
\frac{dh_2}{dt} &= \frac{1}{2}\sqrt{\frac{a}{\mu}\frac{A}{D}}(1+h_1^2+h_2^2)\sin(l)u_c \\
\frac{dl}{dt} &= \sqrt{\frac{\mu}{a^3}\frac{D^2}{A^3}}+\sqrt{\frac{a}{\mu}\frac{A}{D}}(h_1\sin(l)-h_2\cos(l))u_c\,,
\end{aligned} \tag{3.7}$$

with

$$\begin{aligned}
A &= \sqrt{1-e_1^2-e_2^2} \\
B &= \sqrt{1+2e_1\cos(l)+2e_2\sin(l)+e_1^2+e_2^2} \\
D &= 1+e_1\cos(l)+e_2\sin(l)\,.
\end{aligned}$$

The respective distances of the apocenter and pericenter are

$$\begin{aligned}
r_a &= a(1+|e|)\,, \\
r_p &= a(1-|e|)
\end{aligned}$$

and a geostationary orbit corresponds to $|e|=0$, $|h|=0$.

3.2.3 Shuttle Re-entry

Model

Let O be the center of the planet, $K=NS$ be the axis of rotation of the Earth and $\vec{\Omega}$ be the angular velocity oriented along K and with modulus Ω. We note $E=(e_1,e_2,e_3)$, $e_3=K$ an inertial frame with center O. The reference frame is the quasi-inertial frame $R=(I,J,K)$ with origin O, rotating around K with angular velocity Ω and I is chosen to intersect Greenwich meridian. Let r_T denote the Earth radius and let G be the mass center of the shuttle. The spherical coordinates of G are denoted (r,l,L), $h=r-r_T$ is the altitude, l the longitude and L the latitude. We denote $R_1'=(e_r,e_l,e_L)$ a moving frame with center G, where e_r is the local vertical and (e_l,e_L) is the local horizontal plane, e_L pointing to the north.

Let $t \mapsto (x(t), y(t), z(t))$ be a trajectory of G measured in the frame attached to the planet and let $\vec{v}$ be the relative velocity. We parameterize $\vec{v}$ by the modulus and the two angles:

- the path inclination γ, angle of $\vec{v}$ with respect to the horizontal plane
- the azimuth angle χ which is the angle of the projection of $\vec{v}$ in the horizontal plane with respect to e_L.

We denote (i, j, k) a frame defined by $i = \dfrac{\vec{v}}{v}$, j is the unitary vector in the plane (i, e_r) perpendicular to i and oriented by $j.e_r > 0$ and let $k = i \wedge j$.

The system is written in the coordinates $(r, v, \gamma, l, L, \chi)$. The actions on the vehicle are the gravitational force $\vec{P} = m\vec{g}$ and the aerodynamic force consisting of:

- A drag force $\vec{D} = (\frac{1}{2}\rho S C_D v^2) i$ opposite to $\vec{v}$
- A lift force: $\vec{L} = (\frac{1}{2}\rho S C_L v^2)(\cos(\mu) j + \sin(\mu) k)$ perpendicular to $\vec{v}$,

where μ is the angle of bank, $\rho = \rho(r)$ is the air density, S is the reference area, C_D, C_L are respectively the lift and drag coefficients, depending on the angle of attack of the vehicle and the Mach number. Concerning the air density, we take an exponential model: $\rho(r) = \rho_0 \exp(-\frac{r - r_T}{h_s})$.

The system being represented in a non inertial frame, the spacecraft is submitted to a Coriolis force $O(\Omega)$ and a centripetal force $O(\Omega^2)$.

During the atmospheric arc, the shuttle behaves as a glider, the physical control being the lift force, whose orientation is represented by the angle of bank which can be adjusted.

The equations of the system are

$$\begin{aligned}
\frac{dr}{dt} &= v \sin(\gamma) \\
\frac{dv}{dt} &= -g \sin(\gamma) - \frac{1}{2}\rho \frac{SC_D}{m} v^2 + O(\Omega^2) \\
\frac{d\gamma}{dt} &= \left(-\frac{g}{v} + \frac{v}{r}\right) \cos(\gamma) + \frac{1}{2}\rho \frac{SC_L}{m} v \cos(\mu) + O(\Omega) \\
\frac{dL}{dt} &= \frac{v}{r} \cos(\gamma) \cos(\chi) \\
\frac{dl}{dt} &= \frac{v}{r} \frac{\cos(\gamma) \sin(\chi)}{\cos(L)} \\
\frac{d\chi}{dt} &= \frac{v}{r} \cos(\gamma) \tan(L) \sin(\chi) + \frac{1}{2}\rho \frac{SC_L}{m} \frac{v}{\cos(\gamma)} \sin(\mu) + O(\Omega) .
\end{aligned} \tag{3.8}$$

The control is the angle of bank μ. The first system represents the lateral motion. If we neglect the Earth rotation, it is decoupled with respect to the second system which represents the longitudinal motion, $\sin\mu$ allows the shuttle to turn left or right.

Optimal Control Problem

The problem is to steer the system from an initial manifold M_0 to a terminal manifold M_1, imposed by CNES research project. The transfer time t_f is free. There are state constraints of the form $c_i(q) \leq 0$ for $i = 1, 2, 3$ which consist of:

- A constraint on the thermal flux:
$$\varphi = C_q\sqrt{\rho}v^3 \leq \varphi^{max} .$$
- A constraint on the normal acceleration:
$$\gamma_n = \gamma_{n_0}\rho v^2 \leq \gamma_n^{max} .$$
- A constraint on the dynamic pressure:
$$\frac{1}{2}\rho v^2 \leq P^{max} .$$

The optimal control problem is to minimize the total amount of thermal flux:

$$J(\mu) = \int_0^{t_f} C_q\sqrt{\rho}v^3 dt .$$

If we introduce the new time parameter $ds = \varphi\, dt$, our optimal control problem is reduced to a time minimum problem.

3.3 Controllability and Poisson Stability

In order to characterize the controllability properties in the attitude control and the orbital transfer we present a general theorem which can be applied to every system where the free motion is Poisson stable.

3.3.1 Poisson Stability

Definition 3.2. *Let $\dot{q} = X(q)$ be a (smooth) differential equation on M. We note $q(t, q_0)$ the solution with $q(0) = q_0$. We assume it is defined on $\mathbb{R}$. The point q_0 is Poisson stable if for each V neighborhood of q_0, for each $T > 0$ there exist $t_1, t_2 \geq T$ such that $q(t_1, q_0), q(-t_2, q_0) \in V$. The vector field X is said to be Poisson stable if almost every point is Poisson stable.*

Theorem 3.1 (Poincaré theorem). *If X is a conservative vector field and if every trajectory is bounded, then X is Poisson stable.*

Corollary 3.1. *Let $\vec{H}$ be a Hamiltonian vector field, if every trajectory is bounded, then $\vec{H}$ is Poisson stable.*

Example 3.1. Consider the free motion in the orientation of a rigid body. Let $H(R,\Omega) = \frac{1}{2}(I_1\Omega_1^2 + I_2\Omega_2^2 + I_3\Omega_3^2)$ be the kinetic energy. The free motion corresponds to Euler-Lagrange equations and induces a Hamiltonian system. The trajectories are bounded because the attitude $R \in SO(3)$ and the angular velocity is bounded, H being is a first integral.

Example 3.2. Consider Kepler equation: $\ddot{q} = -\mu \frac{q}{|q|^3}$, $H = \frac{1}{2}\dot{q}^2 - \frac{\mu}{|q|}$ be the energy. If we restrict our equation to the elliptic domain Σ_e where the trajectories are ellipses, the vector field is Poisson stable. It corresponds to bounded trajectories.

3.3.2 General Results About Controllability

Notations

- $V(M)$: set of (smooth) vector fields on M.
- If $f : M \mapsto \mathbb{R}$ smooth, $L_X f = df(X) = \frac{\partial f}{\partial q}.X(q)$: Lie derivative.
- Lie bracket: $X, Y \in V(M), [X,Y] = L_Y \circ L_X - L_X \circ L_Y = \frac{\partial X}{\partial q}(q)Y(q) - \frac{\partial Y}{\partial q}(q)X(q)$ in local coordinates
- If $\dot{q} = X(q)$, we note $q(t, q_0)$ the maximal solution with $q(0) = q_0$. We set: $(\exp tX)(q_0) = q(t, q_0)$, $\exp tX$ is a local diffeomorphism on M, X is complete if $\exp tX$ is defined $\forall t \in \mathbb{R}$.

Definition 3.3. *A polysystem D is a family $\{X_i, i \in I\}$ of vector fields. We note $D : q \mapsto SpanD(q)$ the associated distribution; D is involutive if $[X_i, X_j] \in D, \forall X_i, X_j \in D$. We note $D_{A.L.}$ the Lie algebra generated by D computed by: $D_1 = SpanD$, $D_2 = Span\{D_1 + [D_1, D_1]\}$,..., $D_k = Span\{D_{k-1} + [D_1, D_{k-1}]\}$, $D_{A.L.} = \bigcup_{k \geq 1} D_k$.*

Definition 3.4. *Let $\dot{q} = f(q,u)$ be a system on M. Let U be the control domain and we can restrict our controls to the set $\mathcal{U}$ of piecewise mappings with values in U. Let $q(t, q_0, u)$ be the solution with $q(0) = q_0$ associated to u. We note:*

- $A^+(q_0, t) = \bigcup_{u(.)} q(t, q_0, u)$: *accessibility set in time t*, $A^+(q_0) = \bigcup_{t \geq 0} A^+(q_0, t)$

the accessibility set

- $A^-(q_0,t)$: *set of points which can be steered to* q_0 *in time* t, $A^-(q_0) = \bigcup_{t\geq 0} A^-(q_0,t)$.

The system is controllable in time t *if* $A^+(q_0,t) = M, \forall q_0$, *and controllable if* $A^+(q_0) = M, \forall q_0$. *The system is locally controllable at* q_0 *if* $\forall t \geq 0, A^+(q_0,t)$ *and* $A^-(q_0,t)$ *are neighborhood of* q_0.

We introduce the polysystem $D = \{f(q,u); u \in U\}$ and the set $S_T(D) = \{\exp t_1X_1 \circ ... \circ \exp t_kX_k; X_i \in D, k \geq 0, t_i > 0, t_1 + ... + t_k = T\}$ and the pseudo semi-group $S(D) = \bigcup_{T\geq 0} S_T(D)$; $G(D)$ is the pseudo-group $\{\exp t_1X_1 \circ ... \circ \exp t_kX_k; X_i \in D, k \in \mathbb{N}, t_i \in \mathbb{R}\}$.

Lemma 3.1. *We have:* $A^+(q_0,T) = S_T(D)(q_0)$, $A^+(q_0) = S(D)(q_0)$.

Definition 3.5. *Let* D *be a polysystem,* D *is said to be controllable if* $S(D)(q_0) = M$, $\forall q_0$ *and weakly controllable if the orbit* $O(q_0) = G(D)(q_0) = M, \forall q_0$. *The polysystem is said to satisfy the rank condition if* $D_{A.L.}(q_0) = T_{q_0}M, \forall q_0$.

The following two results are standard ones.

Theorem 3.2 (Chow). *Let* D *be a polysystem on a connected manifold* M. *If it satisfies the rank condition, then* D *is weakly controllable.*

Proposition 3.3. *Assume* D *satisfies the rank condition. Then for each* $q_0 \in M$, *for each* V *neighborhood of* q_0, *there exist* U^+, U^- *open sets respectively contained in* $V \bigcap A^+(q_0)$, $V \bigcap A^-(q_0)$.

Theorem 3.3 (Lobry). *Let* D *be a polysystem on a connected manifold* M. *Assume:*

1. $D_{A.L.}(q_0) = T_{q_0}M, \forall q_0$,
2. $\forall X_i \in D, X_i$ *is Poisson stable.*

Then D *is controllable.*

Proof. Let $q_0, q_1 \in M$, one must construct a trajectory joining q_0 to q_1. Using the previous proposition, we take U^- and U^+ contained respectively in $A^-(q_0)$ and $A^+(q_1)$. Let $q_0' \in U^-$ and $q_1' \in U^+$, the problem is equivalent to steer q_0' to q_1'. From Chow's theorem, we can join q_0' to q_1' using concatenation of integrals curves or vector fields in D, with positive or negative time t. Each curve with $t < 0$ is replaced by a curve of the same vector field, with positive time using Poisson stability (replacing a trajectory if necessary by a neighboring trajectory). This proves the assertion. ∎

This result can be improved to get controllability conditions with only one Poisson stable vector field.

3.3.3 Controllability and Enlargement Technique (Jurdjević-Kupka)

We restrict our analysis to polysystems satisfying the rank condition. Then D is controllable if and only if $\forall q_0, \overline{S(D)(q_0)} = M$.

Definition 3.6. *Let D, D' be two polysystems. They are called equivalent if $\forall q_0, \overline{S(D)(q_0)} = \overline{S(D')(q_0)}$. The union of all polysystems equivalent to D is called the saturated, with notation $satD$.*

Proposition 3.4. *Let D be a polysystem.*

1. *If $X \in D$ and X is Poisson stable, then $-X \in D$.*
2. *The convex hull $convD \subset satD$.*

Proof. For condition 1, use the proof of theorem (3.3).
For 2, if $X \in D$, then $\lambda X \in satD, \forall \lambda > 0$. If $X, Y \in D$, from Baker-Campbell-Hausdorff formula we have:
$$\prod_{n\ times} (\exp \frac{t}{n} X) \circ (\exp \frac{t}{n} Y) = \exp t(X+Y) + o(\frac{1}{n}).$$
Taking the limit when n goes to $+\infty$, we have $X + Y \in satD$. ■

Theorem 3.4. *Let $\dot{q} = F_0(q) + \sum_{i=1}^{p} u_i F_i(q)$ be a system on a connected manifold M, with $u_i \in \{-1, +1\}$. Assume:*
1. *F_0 is Poisson stable*
2. *$\{F_0, ..., F_p\}_{A.L.}(q_0) = T_{q_0}M, \forall q_0$.*

Then the polysystem is controllable. Moreover, condition 2. is also necessary in the analytic case.

Proof. Let $D = \{F_0(q) + \sum_{i=1}^{p} u_i F_i(q),\ u_i \in \{-1, +1\}\}$. We check:
- $D_{A.L.} = \{F_0, \cdots, F_p\}_{A.L.}$
- $F_0 = \frac{1}{2}[(F_0 + F_1) + (F_0 - F_1)] \in satD$
- $-F_0 \in satD$
- $\pm F_i \in satD$, for $i = 1, \cdots, p$.

Hence the symmetric polysystem $\{\pm F_i; i = 0, ..., p\} \in satD$ and the controllability result follows from Chow's theorem. The necessary condition in the analytic case follows from Nagano-Sussmann theorem. ■

In the next section, we apply this result to characterize controllability computing Lie brackets.

3.3.4 Application to the Attitude Problem

Theorem 3.5. *Consider the system (3.3) describing the attitude control problem, with $m = 1$, i.e. a single pair of opposite gas jets. Then the system is controllable except when two b_1^i are zero or $\sqrt{a_3}b_1^1 = \pm\sqrt{a_1}b_1^3$.*

Sketch of proof. The free motion is Poisson stable and the system is controllable if and only if the rank condition is satisfied.

First of all, consider the subsystem describing the evolution of the angular velocity: $\dot{\Omega} = Q(\Omega) + u_1 b_1$. We must have $\{Q, b_1\}_{A.L.}$ of rank 3 at each point.

Since Q is homogeneous, this condition is satisfied if and only if the Lie algebra of constant vector fields in $\{Q, b_1\}_{A.L.}$ is of rank 3. A computation shows that this Lie subalgebra is generated by: $f_1 = b_1$, $f_2 = [[Q, f_1], f_1]$, $f_3 = [[Q, f_1], f_2]$, $f_4 = [[Q, f_1], f_3]$ and $f_5 = [[Q, f_2], f_2]$.

The conditions follow.

Geometric Interpretation

Consider Euler equation with no applied torque, describing the evolution of the angular velocity. The system can be integrated using the conservation of the energy and of the momentum. If we fix the energy to $H = 1$, we can represent the trajectories. The system has three pairs of singular points corresponding to the axis E_i, the rigid body describing stationary rotations. Singular points corresponding to the major and the minor axis are centers and for the intermediate axis we have saddles. Each trajectory is periodic except separatrices connecting opposite saddle points.

If the applied torque is oriented along one axis, we control only the corresponding stationary rotation. This corresponds to the first condition. The condition $\sqrt{a_3}b_1^1 = \pm\sqrt{a_1}b_1^3$ means that the torque is oriented in one of the plane filled by separatrices.

3.3.5 Application to the Orbital Transfer

We restrict our analysis to the case when the mass is constant, the system being given by equation (3.4).

First we make Lie bracket computations using cartesian coordinates and the thrust being decomposed in the tangential/normal frame. Computations give us:

Proposition 3.5. *Let $x = (q, \dot{q})$ with $q \wedge \dot{q} \neq 0$, then:*

1. $\{F_0, F_t, F_c, F_n\}_{A.L.}(x) = \mathbb{R}^6$ *and coincides with*
 $Span\{F_0(x), F_t(x), F_c(x), F_n(x), [F_0, F_c](x), [F_0, F_n](x)\}$.

2. *The dimension of* $\{F_0, F_t\}_{A.L.}(x)$ *is four.*
3. *The dimension of* $\{F_0, F_n\}_{A.L.}(x)$ *is three.*
4. *The dimension of* $\{F_0, F_c\}_{A.L.}(x)$ *is four if* $L(0) \neq 0$ *and three if* $L(0) = 0$.

Also the orbit corresponding to each direction can be characterized using the orbital coordinates and system (3.7).

Proposition 3.6. *Consider the single input system:* $\dot{x} = F_0(x) + uG(x)$*, where* G *is* F_t*,* F_c *or* F_n *and* x *is restricted to the elliptic domain, then:*

1. *If* $G = F_t$*, then the orbit is the 2D-elliptic domain.*
2. *If* $G = F_n$*, then the orbit is of dimension three and is the intersection of the 2D-elliptic domain with* $a = a(0)$ *constant.*
3. *If* $G = F_c$*, the orbit is of dimension four if* $L(0) \neq 0$*, and of dimension three if* $L(0) = 0$ *and is given by* $a = a(0)$, $|e| = |e(0)|$.

Theorem 3.6. *Consider the single input system:* $\dot{x} = F_0(x) + uG(x)$*, where* G *is* F_t*,* F_c *or* F_n *and* x *is restricted to the elliptic domain, then each point in the orbit is accessible.*

3.4 Constructive Methods

The aim of this section is to present two methods which can be applied to compute a control. The first one is a stabilization technique and the second is a path planning method, a final section will be optimal control which is also a method to design controls.

3.4.1 Stabilization Techniques

Generalities

Definition 3.7. *Let* $\dot{q} = X(q)$ *be a smooth differential equation on a open set* $U \subset \mathbb{R}^n$*,* $q_0 \in U$ *be an equilibrium point of* X*. We say:*

- q_0 *is stable if* $\forall \epsilon > 0$, $\exists \eta > 0$, $|q_1 - q_0| \leq \eta \Rightarrow |q(t, q_1) - q_0| \leq \epsilon \ \forall t \geq 0$.
- *The attraction basin is* $D(q_0) = \{q_1\,;\, q(t, q_1) \to q_0 \text{ when } t \to +\infty\}$.
- q_0 *is exponentially stable if* q_0 *is stable and* $D(q_0)$ *is neighborhood of* q_0.
- q_0 *is GAS (globally asymptotically stable) if* $D(q_0) = U$.

Definition 3.8. *Let* $V : U \to \mathbb{R}$ *be a smooth function,* V *is a Lyapunov function if locally* $V > 0$ *for* $q \neq q_0$ *and* $\dot{V} = L_X V \leq 0$*,* V *is strict if* $\dot{V} < 0$ *for* $q \neq q_0$.

Theorem 3.7 (Lyapunov). *If there exists a Lyapunov function, q_0 is stable and if V is strict, q_0 is asymptotically stable.*

Lyapunov functions are important tools to check stability, moreover it allows to get estimates of the attraction basin. In many applications q_0 is exponentially stable but we can only construct easily Lyapunov functions which are not strict. In order to get stability result we must use properties of the ω-limit set of q_0 and the following result.

Theorem 3.8 (LaSalle-global formulation). *Let $\dot{q} = X(q)$ be a differential equation on $\mathbb{R}^n$, $X(0) = 0$. Assume there exists a function V such that: $V > 0$ for $q \neq 0$, $\dot{V} \leq 0$ and $V(q) \to +\infty$ when $|q| \to +\infty$. Let M be the largest invariant set contained in $E = \{q\,;\,\dot{V}(q) = 0\}$. Then all the solutions are bounded and converge to M when $t \to +\infty$.*

Jurdjević-Quinn Theorem

The aim of this section is to present a result to construct standard stabilizing feedback adapted to mechanical systems. We present only the single-input formulation, the extension to the multi-inputs case being straightforward.

Theorem 3.9. *Consider a system on $\mathbb{R}^n$ of the form: $\dot{q} = X(q) + uY(q)$, $X(0) = 0$. Assume the following:*

1. $\exists V : \mathbb{R}^n \to \mathbb{R}, V > 0$ *on* $\mathbb{R}^n - \{0\}, V(q) \to +\infty$ *when* $|q| \to +\infty$ *such that*
   a. $\dfrac{\partial V}{\partial q} \neq 0$ *if* $q \neq 0$,
   b. V *is a first integral of* X: $L_X V = 0$.
2. $F(q) = Span\{X(q), Y(q), [X,Y](q), ..., ad^k X(Y)(q), ...\} = \mathbb{R}^n$ *for* $q \neq 0$ *where* $adX(Y) = [X,Y]$, $ad^k X(Y) = [X, ad^{k-1}X(Y)]$.

Then the canonical feedback $\hat{u}(q) = -L_Y V(q)$ stabilizes globally and asymptotically the origin.

Proof. Plugging $\hat{u}(q)$ in the system, we get an ordinary differential equation: $\dot{\hat{q}} = X(\hat{q}) + \hat{u}Y(\hat{q})$. We have: $\dot{V}(\hat{q}) = L_{X+\hat{u}Y}(V) = L_X V + \hat{u}L_Y V = -(L_Y V(\hat{q}))^2 \leq 0$. Using LaSalle theorem, $\hat{q}(t) \to M$, when $t \to +\infty$ and M is the largest invariant set in $\dot{V} = 0$.

We can evaluate this set. Indeed, since M is invariant, if $\hat{q}(0) \in M$ then $\hat{q}(t) \in M$. Moreover, on M, $\hat{u}(\hat{q}) = 0$ and $\hat{q}$ is solution of the free motion $\dot{\hat{q}} = X(\hat{q})$.

Hence deriving $\dot{V}(\hat{q}) = (L_Y V)(\hat{q}) = 0$, we get:

$$\frac{d}{dt}L_Y V(\hat{q}) = L_X L_Y V = 0.$$

Moreover $L_{[X,Y]}V = L_Y L_X V - L_X L_Y V$ and $L_X V = 0$, hence we get the condition:

$$L_{[X,Y]}V = 0.$$

Iterating the derivation, one gets:
$L_X V = L_Y V = L_{[X,Y]}V = ... = L_{ad^k X(Y)}(V) = 0.$

Hence $\frac{\partial V}{\partial q} \perp F = Span\{X, Y, ..., ad^k X(Y), ...\}$ and $dimF = n$ for $q \neq 0$. Since $\frac{\partial V}{\partial q} \neq 0$ except at 0, we get $M = \{0\}$. ■

Application to Euler Equation

Consider Euler's equation (3.2) describing the evolution of the angular velocity: $\dot{\Omega} = Q(\Omega) + Bu$, B being a constant matrix. Along a motion, the energy H and the modulus m of the momentum are positive first integrals and 0 can be stabilized if the Lie algebraic condition is satisfied.

But more generally, consider a system (Q, B) defined on $\mathbb{R}^n$, where Q is quadratic and B constant, called a quadratic system, we can apply the following algorithm:

Step 1. Find a quadratic feedback $u = \alpha(\Omega) + \beta v$ where α is quadratic, β constant such that for the system $(Q + B\alpha, B\beta)$ the free motion has a quadratic positive first integral.

Step 2. Check the algebraic condition (which is not feedback invariant).

Example 3.3. Consider the 2D-case. Assume that the system satisfies the rank condition. Using linear change of coordinates and quadratic feedback, we have, in the single-input case, 3 classes whose normal forms are:

$$\begin{cases} \dot{x} = x^2 + y^2 \\ \dot{y} = u \end{cases} \begin{cases} \dot{x} = x^2 - y^2 \\ \dot{y} = u \end{cases} \begin{cases} \dot{x} = y^2 \\ \dot{y} = u\,. \end{cases}$$

We check that the only classes which have a positive definite first integral are the two classes,

$$\begin{cases} \dot{x} = x^2 - y^2 \\ \dot{y} = u \end{cases} \begin{cases} \dot{x} = y^2 \\ \dot{y} = u\,, \end{cases}$$

and among those two classes, only the first one can be stabilized using our technique.

The same computation can be applied for Euler equation, with two inputs.

Application to the Orbital Transfer

We present here the results of [5] to transfer locally the system from a given elliptic orbit to a final orbit.

From section 3.2.2, each oriented elliptic orbit is represented by (c, L) where c is the momentum and L is the Laplace vector, c being orthogonal to L. The evolution of (c, L) is given by

$$\dot{c} = q \wedge \frac{F}{m}$$
$$\dot{L} = F \wedge c + \dot{q} \wedge (q \wedge \frac{F}{m}).$$

If the target is the orbit parameterized by (c_T, L_T), we introduce:

$$V = \frac{1}{2}(|c - c_T|^2 + |L - L_T|^2)$$

where $|.|$ is the usual distance. Let $\Delta L = (L - L_T)$, $\Delta c = (c - c_T)$. Differentiating, we get

$$\dot{V} = \frac{F}{m} W \text{ and } W = \Delta c \wedge q + c \wedge \Delta L + (\Delta L \wedge \dot{q}) \wedge q$$

and to get $\dot{V} \leq 0$, we apply the standard feedback $\frac{F}{m} = -k(q, \dot{q})w$ where $k > 0$ is arbitrary.

We choose l small enough such that $B_l = \{(c, L)\,;\, \frac{1}{2}(|c-c_T|^2+|L-L_T|^2) \leq l\}$ is contained in the elliptic domain. From LaSalle theorem, each trajectory is converging to the largest invariant set contained in $W = 0$.

Computations detailed in [5] give us that it is reduced to (c_T, L_T). Hence we obtain the local result.

To get a global result, in the elliptic domain Σ_e: $c \neq 0$, $L < \mu$ we can proceed as follows.

A first possibility is to modify the Lyapunov function V in order to get a proper function on Σ_e satisfying $V \to +\infty$ on the boundary of the domain, and this can be achieved because Σ_e is a simple domain.

Another possibility is to use path planning to transfer the initial orbit (c_i, L_i) to the terminal orbit (c_f, L_f), and then apply our local stabilization result along the path to transfer the system using intermediate points $(c_1, L_1) = (c_i, L_i)$, (c_2, L_2),..., $(c_N, L_N) = (c_f, L_f)$.

Hence, next we make a heuristic introduction to path planning in our problems.

3.4.2 Path Planning

Standard methods to do path planning is to approximate trajectories of the system using Baker-Campbell-Hausdorff formula or nilpotent approximations. This approach will be used in optimal control to compute approximating small time solutions. Here, we shall restrict our analysis to geometric methods.

Orbital Transfer

The problem is to transfer the system to a final geostationary orbit. The terminal conditions using orbit elements are: $|e| = 0$ and $|h| = 0$ where the modulus of h measures the inclination with respect to the equatorial plane. Decomposing the thrust into the tangential/normal frame, we have from proposition (3.6):

- If $u_c = 0$, we have a 2-D second order system with 2 inputs u_t, u_n.
- If $u_t, u_n = 0$, we have a 3-D system but we have no action on the semi-axis $|a|$ and the eccentricity $|e|$.

Hence we can construct trajectories in two steps:

Step 1. 2-D problem using u_t, u_n to reshape the parameter of the ellipse.

Step 2. Use u_c to control the inclination to get $|h| = 0$ at the end.

This approach combined with the local stabilization algorithm allows to construct a control in the orbital transfer.

Attitude Control

A first method is to change the attitude using three successive rotations. Introduce respectively the matrixes A_1, A_2, A_3 given by:

$$\begin{pmatrix} 0 & 0 & 0 \\ 0 & 0 & 1 \\ 0 & -1 & 0 \end{pmatrix}, \quad \begin{pmatrix} 0 & 0 & -1 \\ 0 & 0 & 0 \\ 1 & 0 & 0 \end{pmatrix}, \quad \begin{pmatrix} 0 & 1 & 0 \\ -1 & 0 & 0 \\ 0 & 0 & 0 \end{pmatrix}$$

corresponding to rotations around the principal axis E_1, E_2, E_3, rotations around E_1, E_3 being stable and unstable around E_2. A standard (constructive) result is:

Lemma 3.2. *Let $R \in SO(3)$, then there exists χ, θ, $\psi \in [0, 2\pi]$ called the Euler angles such that:* $R = (\exp \chi A_3)(\exp \theta A_1)(\exp \psi A_3)$

Application to the attitude control. Assume that the system is controlled by two pairs of opposite gas jets oriented along E_1, E_2; the equation describing the control of the angular velocity are:

$$\begin{aligned}\dot{\Omega}_1 &= a_1\Omega_2\Omega_3 + u_1\\ \dot{\Omega}_2 &= a_2\Omega_1\Omega_3\\ \dot{\Omega}_3 &= a_3\Omega_1\Omega_2 + u_3\,.\end{aligned}$$

Since Euler equation is controllable, we can assume that we have to transfer between stationary states with zero angular velocity: $(R_0, 0)$, $(R_1, 0)$. From the previous lemma, there exist Euler angles such that:

$$R_1 = (\exp \chi A_3)(\exp \theta A_1)(\exp \Phi A_3)R_0\,.$$

To track this path, we proceed as follows. First of all, construct a control to transfer $(R_0, 0)$ to $(\exp \Phi A_3, 0)$. For this, observe that if $u_1 = 0$, the solutions starting from $(R_0, 0)$ are solutions of:

$$\dot{R} = (\Omega_3(t)A_3)R(t) \text{ and } \dot{\Omega}_3 = u_3(t)\,. \tag{3.9}$$

Hence setting $R(t) = (\exp \Phi(t)A_3)R_0$, then Φ is solution of: $\ddot{\Phi} = u_3$, $\Phi(0) = 0$. Therefore, it is sufficient to transfer $(0,0)$ to $(\Phi, 0)$ for our second order system, using for instance a standard time optimal control. Iterating the process, we can transfer $(R_0, 0)$ to $(R_1, 0)$.

Geodesic Paths

In the previous algorithm, a path of $SO(3)$ is realized using three successive rotations, which are stationary rotations of the body, and they are stable. Only two inputs along E_1, E_3 are necessary but if we want a small rotation around E_2, we have $\Phi = \chi = \frac{\pi}{2}$ and we have two large rotations. We can use an other algorithm founded on the local property of the angular velocity at 0, trivially satisfied with the two inputs. We have:

Lemma 3.3. *The free motion corresponds to geodesic motion for the metric given by the kinetic energy. The metric is complete, and for each $R_0, R_1 \in SO(3)$, there exists an initial angular velocity Ω_0 to transfer R_0 to R_1, the terminal velocity being noted Ω_1.*

This gives us the following algorithm to transfer $(R_0, 0)$ to $(R_1, 0)$. Using the local controllability at 0 at Euler equation, we transfer $(R_0, 0)$ to a point close to $(R_0, \lambda\Omega_0)$, $\lambda > 0$ small. Then applying a zero control, we transfer $(R_0, \lambda\Omega_0)$ to a point $(R_1, \lambda'\Omega_1)$, using previous lemma, where λ' is small. Once again using local controllability of the angular velocity, we can transfer $(R_1, \lambda'\Omega_1)$ to a point close to $(R_1, 0)$. Physically, a small control is used to steer the system of a geodesic joining R_0 to R_1, at the beginning, using local controllability at 0 of Euler equation.

Shuttle Re-entry

In the shuttle re-entry, a practical method to steer the shuttle is to use the so-called Harpold and Graves strategy. The basis of the method is the following. The cost is the thermal flux:

$$J(\mu) = \int_0^{t_f} C_q \sqrt{\rho} v^3 dt$$

and approximating the system by the drag: $\dot{v} = -d$, the cost can be written:

$$J(\mu) = \kappa \int_{v_f}^{v_0} \frac{v^2}{\sqrt{d}} dv \text{ with } \kappa > 0$$

and the optimal policy consists in maximizing the drag term d during the flight. Taking into account the state constraints, the strategy of tracking the boundary of the domain in the following order: thermal flux $\to$ normal acceleration $\to$ dynamic pressure.

3.5 Optimal Control

3.5.1 Geometric Framework

We consider a smooth system of the form: $\dot{q} = f(q, u)$ and a cost to be minimized $\min \int_0^T f^0(q, u)dt$, where the integral is smooth and the time T can be fixed or not. The set of admissible controls is the set $\mathcal{U}$ of bounded measurable mappings with values in a control domain U. We introduce the cost extended system: $\dot{\tilde{q}} = \tilde{f}(\tilde{q}, u)$ defined with $\tilde{q} = (q, q^0)$ by:

$$\dot{q} = f(q, u) \ , \ \dot{q}^0 = f^0(q, u) \ , \ q^0(0) = 0,$$

and $q^0(T)$ represents the cost. Hence if u is optimal, the extremity $\tilde{q}(T, q_0, u)$ of the trajectory has to belong to the boundary of the accessibility set. Therefore we need necessary conditions satisfied by trajectories in the boundary of the accessibility set.

The simplest result is obtained if the control domain is open: this is the weak maximum principle that we present first.

3.5.2 Weak Maximum Principle

Definition 3.9. *Let $\dot{q} = f(q, u)$, $u(t) \in U$ be a control system. The extremity mapping is $E : u \in \mathcal{U} \mapsto q(T, q_0, u)$ where $q(0) = q_0$ and T are fixed.*

Proposition 3.7. *Let u be a control defined on $[0,T]$ such that the corresponding trajectory $q(t)$ is defined on $[0,T]$. The extremity mapping is C^∞ near u for the L^∞–norm topology and the Fréchet derivative is:*

$$E_u^{'}(v) = \int_0^T \Phi(T)\Phi^{-1}(s)B(s)v(s)ds,$$

where Φ is the matrix solution of $\dot{\Phi} = A\Phi$, $\Phi(0) = Id$, $A(t) = \frac{\partial f}{\partial q}(q(t),u(t))$, $B(t) = \frac{\partial f}{\partial u}(q(t),u(t))$.

Sketch of proof. Consider a L^∞ variation of u denoted δu and $q+\delta q$ be the response to $u+\delta u$ with $q(0)+\delta q(0) = q(0)$. We have:

$$\begin{aligned}\frac{d}{dt}(q+\delta q) &= f(q+\delta q, u+\delta u)\\ &= f(q,u) + \frac{\partial f}{\partial q}(q,u)\delta q + \frac{\partial f}{\partial u}(q,u)\delta u + o(\delta q, \delta u).\end{aligned}$$

If we write $\delta q = \delta_1 q + \delta_2 q + \cdots$, where $\delta_1 q$ is linear in δu, $\delta_2 q$ quadratic, etc, we obtain:

$$\frac{d}{dt}\delta_1 q = A(t)\delta_1 q + B(t)\delta u.$$

Integrating with $\delta_1 q(0) = 0$ we get:

$$\delta_1 q(T) = \int_0^T \Phi(T)\Phi^{-1}(s)B(s)\delta u(s)ds,$$

and $E_u^{'}(\delta u) = \delta_1 q(T)$.

Definition 3.10. *The control is said to be singular on $[0,T]$ if u is a singular point of the extremity mapping, i.e., $E^{'}$ is not of full rank at u. Otherwise u is called regular.*

Proposition 3.8. *Assume u singular on $[0,T]$ with corresponding trajectory q. Then there exists $p(.) \in \mathbb{R}^n - \{0\}$ absolutely continuous such that the following equations are satisfied for a.e. t, for the triplet (q,p,u):*

$$\dot{q} = \frac{\partial H}{\partial p}, \quad \dot{p} = -\frac{\partial H}{\partial q}, \quad \frac{\partial H}{\partial u} = 0,$$

where $H(q,p,u) = \langle p, f(q,u)\rangle$ is the Hamiltonian.

Proof. Since $E^{'}$ is not of full rank at u, there exists $\bar{p} \in \mathbb{R}^n - \{0\}$ such that $\langle \bar{p}, E_u^{'}(v)\rangle = 0$. We set $p(t) = \bar{p}\Phi(T)\Phi^{-1}(t)$, writing $\bar{p}$, p as line vectors.

Hence: $\dot{p} = -pA(t)$ and $p(T) = \bar{p}$.
Since $\langle p, E_u^{'}(v)\rangle = 0 \ \forall v \in L^\infty$ we get:

$$\langle p(t), B(t)\rangle = 0 \ \ a.e.$$

Introducing H we obtain our conditions in Hamiltonian form. ∎

Proposition 3.9. *Consider a control system $\dot{q} = f(q,u)$, $u(t) \in \mathbb{R}^m$. If $q(T, q_0, u)$ belongs to the boundary of $A(q_0, T)$ then the control and the corresponding trajectory are singular.*

Proof. Assume u regular, then $Im\, E_u^{'} = \mathbb{R}^n$ and $q(T, q_0, u)$ belongs to the interior of the accessibility set. This proves the result. ∎

Remark 3.1. If the time is not fixed, we can make the same reasoning for the time extended extremity mapping and we get also the condition $H = 0$ for a solution in the boundary of the accessibility set.

Remark 3.2. If U is open, the same result holds for u such that $|u|_{L^\infty} \in U$.

Connection with Jurdjević-Quinn theorem. Consider a system of the form: $\dot{q} = X + uY$. Assume $u = 0$ singular. Then we get $\langle p(t), Y(q(t))\rangle = 0$. Differentiating with respect to t and using $u = 0$, we have: $\langle p(t), ad^k X(Y)(q(t))\rangle = 0$, $k \geq 0$. Moreover if $H = 0$ then $\langle p(t), X(q(t))\rangle = 0$. Therefore if $Span\{X, ad^k X(Y);\ k = 0, \cdots, +\infty\}$ is of full rank, $u = 0$ is a regular point of the time extended extremity mapping and $q(T, q_0, 0)$ belongs to the interior of the accessibility set for each $T > 0$.

3.5.3 Maximum Principle

Theorem 3.10 (Maximum Principle). *Consider the problem of* $\min \int_0^T f^0$ $(q,t)dt$ *for a system of the form $\dot{q} = f(q,u)$, $u \in U$ with boundary condition $q(0) \in M_0$, $q(T) \in M_1$. The maximum principle tells us that if $u(.)$ is an optimal control on $[0,T]$ with response $q(.)$ then there exists an absolutely continuous vector function $p(t) \in \mathbb{R}^n$ such that the following equations are satisfied a.e:*

$$\dot{q} = \frac{\partial \tilde{H}}{\partial p} \quad , \quad \dot{p} = -\frac{\partial \tilde{H}}{\partial q},$$
$$\tilde{H}(q,p,u) = \max_{v \in U} \tilde{H}(q,p,v) = M(q,p), \tag{3.10}$$

where $\tilde{H} = \langle p(t), f(q,u)\rangle + p_0 f^0(q,u)$, with p_0 constant ≤ 0 and $(p, p_0) \neq 0$.

The vector p satisfies the transversality conditions:

$$p(0) \perp T_{q(0)}M_0 \quad , \quad p(T) \perp T_{q(T)}M_1. \tag{3.11}$$

Moreover M is a constant and this constant is zero if the transfer time T is free.

Definition 3.11. *An extremal is a triplet (q,p,u) solution of the equations (3.10). It is called a BC−extremal if it satisfies the transversality conditions (3.11) which can be written $(q(0),p(0)) \in M_0^\perp$, $(q(T),p(T)) \in M_1^\perp$ with $N^\perp = \{(q,p) \in T^*N,\ q \in N,\ p \perp T_qN\}$.*

A first step in the analysis of an optimal problem via the maximum principle is to compute the extremals. We present an important example.

3.5.4 Extremals in SR-Geometry

Let $q_0 \in O \subset \mathbb{R}^n$, where O is an open set, and $\{F_1, \cdots, F_m\}$ are m independent vector fields near q_0. consider the system:

$$\dot{q}(t) = \sum_{i=1}^{m} u_i(t) F_i(q(t)),$$

and the problem of minimizing the energy of a curve $q(.)$ tangent to the distribution $D(q) = Span\{F_1(q), \cdots, F_m(q)\}$ defined by:

$$E(q) = \int_0^T \sum_{i=1}^{m} u_i^2(t)dt.$$

We introduce the Hamiltonian:

$$H(q,p,u) = \langle p, \sum_{i=1}^{m} u_i F_i(q) \rangle + p_0 \sum_{i=1}^{m} u_i^2,$$

where p_0 is a constant which can be normalized to 0 or $-\frac{1}{2}$. Hence we distinguish two cases.

Normal case. We assume $p_0 = -\frac{1}{2}$. The extremals controls are given by the relations $\frac{\partial H}{\partial u} = 0$ and we get: $u_i = \langle p, F_i(q) \rangle$. To compute easily the normal extremals it is convenient to use the following coordinates on T^*O. On O we complete $\{F_1, \cdots, F_m\}$ to form a frame $\{F_1, \cdots, F_n\}$. We set

$$P_i = \langle p, F_i(q) \rangle \quad , \quad P = (P_1, \cdots, P_n),$$

and we use on T^*O the coordinates (q,P). Since $u_i = P_i$ plugging in H we get the Hamilton function: $H_n = \frac{1}{2}\sum_{i=1}^{m} P_i^2$.
The normal extremals are solutions of:

$$\dot{q} = \sum_{i=1}^{m} P_i F_i(q),$$

$$\dot{P}_i = \{P_i, H_n\} = \sum_{j=1}^{m} \{P_i, P_j\} P_j,$$

where $\{\ ,\ \}$ is the Poisson bracket and here we have:

$$\{P_i, P_j\} = \langle p, [F_i, F_j](q)\rangle,$$

and since F_i form a frame we can write:

$$[F_i, F_j](q) = \sum_{k=1}^{n} c_{ij}^k(q) F_k(q)$$

where the c_{ij}^k are smooth functions.

Abnormal extremals. They correspond to $p_0 = 0$ and are in fact singular trajectories of the system. They satisfy the constraints:

$$\langle p(t), F_i(q(t))\rangle = 0, \qquad i = 1, \cdots, m.$$

A general algorithm exists to compute such trajectories in the generic case, see [2].

3.5.5 SR-Systems with Drift

In the previous problem, minimizing the energy of a curve is equivalent to minimize the length $l(q) = \int_0^T (\sum_{i=1}^{m} u_i^2)^{\frac{1}{2}} dt$ and parameterizing by arc-length: $\sum_{i=1}^{m} u_i^2 = 1$, we get a time optimal control problem, for a symmetric system.
More generally, we can define:

Definition 3.12. *We call SR-problem with drift the time optimal control problem for a system of the form:*

$$\frac{dq(t)}{dt} = F_0(q(t)) + \sum_{i=1}^{m} u_i(t) F_i(q(t))$$

where $q(t) \in \mathbb{R}^n$ and the control $u = (u_1, ..., u_m)$ is bounded by: $\sum_{i=1}^{m} u_i^2 \leq 1$.

Generic Computations of Extremals

Let P_i be the Hamiltonian: $\langle p, F_i(q)\rangle$, $i = 0, ..., m$ and let Σ be the surface called the switching surface defined by $P_i = 0$ for $i = 1, ..., m$. The Hamiltonian associated to the problem reduces to: $H = P_0 + \sum_{i=1}^{m} u_i P_i$ and the maximization of H with $|u| \leq 1$ gives us, outside the surface Σ, the relation:

$$u_i = \frac{P_i}{\sqrt{\sum_{i=1}^{m} P_i^2}}$$

and plugging u_i into H defines the Hamiltonian function $H_0 = P_0 + (\sum_{i=1}^{m} P_i^2)^{\frac{1}{2}}$. The associated extremals are called of order 0. From the maximum principle, optimal extremals are contained in $H_0 \geq 0$. Those contained in $H_0 = 0$ are called exceptional.

Proposition 3.10. *The extremals of order* 0 *are smooth, the associated controls are on the boundary* $|u| = 1$ *of the control domain, such extremals correspond to a singularity of the extremity mapping:* $u \mapsto q(T, q_0, u)$ *for the* L^∞*-norm topology, where* u *is restricted to the unit sphere* S^{m-1}.

Broken and Singular Extremals

In order to construct all the extremals, we must analyze the behaviors of extremals of order 0 near the switching surface. In particular, we can connect two arcs of order 0 at a point of Σ if we respect the necessary optimality condition:

$$p(t_i+) = p(t_i-) \text{ and } H_0(t_i+) = H_0(t_i-)$$

where t_i is the time to reach Σ.

Other extremals are singular extremals, satisfying $P_i = 0$, $i = 1, ..., m$ and contained in Σ. They correspond to singularity of the extremity mapping with $u(t) \in \mathbb{R}^m$.

In order to analyze the behavior of extremals near Σ, we proceed as follows. Let $z(t) = (q(t), p(t))$ be an extremal. The curves $t \mapsto P_i(z(t))$ are a.c. and, differentiating, we get:

$$\dot{P}_i = \{P_i, P_0\} + \sum_{j \neq i} u_j \{P_i, P_j\} \text{ for } i = 1, ..., m. \tag{3.12}$$

If we denote by D the distribution $Span\{F_1(q), ..., F_m(q)\}$, the following is clear:

Lemma 3.4. *We can connect every extremal of order* 0 *converging to* $z_0 = (q_0, *) \in \Sigma$ *with every extremal of order* 0*, starting from* z_0*, the Hamiltonian being* P_0 *at* z_0*. If* $[D, D](q_0) \subset D(q_0)$*, then the coordinates* P_i*,* $i = 1, ..., m$ *are* C^1*.*

Π-Singularity and Its Nilpotent Model

The aim of this section is to do an analysis of the first singularity encountered when making junctions between extremals of order 0, this analysis being useful in orbital transfer and shuttle re-entry. We limit our analysis to the case $m = 2$, the generalization being straightforward. The system is: $\dot{q} = F_0(q) + u_1F_1(q) + u_2F_2(q)$, $H = P_0 + u_1P_1 + u_2P_2$ and the extremal controls are:

$$u_1 = \frac{P_1}{\sqrt{P_1^2 + P_2^2}} \, , \, u_2 = \frac{P_2}{\sqrt{P_1^2 + P_2^2}} .$$

They correspond to singularities of the extremity mapping with $u_1^2 + u_2^2 = 1$ and we can parameterize by: $u_1 = \cos\mu$, $u_2 = \sin\mu$. The system (3.12) takes the form:

$$\begin{aligned} \dot{P}_1 &= \{P_1, P_0\} + u_2\{P_1, P_2\} \\ \dot{P}_2 &= \{P_2, P_0\} - u_1\{P_1, P_2\} \end{aligned}$$

and we make a polar blowing up:

$$P_1 = r\cos\theta \, , \, P_2 = r\sin\theta$$

we get:

$$\begin{aligned} \dot{\theta} &= \frac{1}{r}[\{P_1, P_2\} + \sin\theta\{P_1, P_0\} - \cos\theta\{P_2, P_0\}] \\ \dot{r} &= \cos\theta\{P_1, P_0\} + \sin\theta\{P_2, P_0\} \end{aligned} \tag{3.13}$$

In order to evaluate the solutions for small t, we can make a nilpotent approximation. We choose vector fields F_0, F_1, F_3 such that all Lie brackets of length ≥ 3 are 0. Differentiating, we get:

$$\frac{d}{dt}\{P_1, P_2\} = \frac{d}{dt}\{P_1, P_0\} = \frac{d}{dt}\{P_2, P_0\} = 0 .$$

Hence for a given extremal, we can set:

$$\{P_1, P_2\} = b \, , \, \{P_1, P_0\} = a_1 \, , \, \{P_2, P_0\} = a_2$$

where a_1, a_2, b are constants. Now (3.13) can be integrated using the time parameterization: $ds = \dfrac{dt}{r}$. Trajectories crossing Σ with a well-defined slope are obtained solving $\dot{\theta} = 0$. We introduce:

Definition 3.13. *A point $z_0 \in \Sigma$ is called of order one if at least one of the Lie bracket $\{P_1, P_0\}$ or $\{P_2, P_0\}$ is non zero at z_0.*

Let z_0 be such a point. Up to a rotation in the $\{F_1, F_2\}$ frame, we can impose: $\{P_2, P_0\}(z_0) = 0$. The relation $\dot{\theta} = 0$ is then $\{P_1, P_2\} + \sin\theta\{P_1, P_0\} = 0$ where $\{P_1, P_0\} \neq 0$. Hence we get using the approximation: $b + a_1 \sin\theta = 0$, with $a_1 \neq 0$. The equation has two roots $\theta_0 < \theta_1$ on $[0, 2\pi]$ if and only if $|\frac{b}{a_1}| < 1$ and $\theta_0 = \theta_1 = \pi$ if and only if $b = 0$.

This last condition is satisfied in the involutive case $[D, D] \subset D$. Moreover if $\theta_0 \neq \theta_1$, then $\cos\theta_0$ and $\cos\theta_1$ have opposite signs. Then one extremal of order 0 is reaching Σ and one is starting from Σ.

Definition 3.14. *A generic nilpotent model associated to a point of order one is a six dimensional system of independent vector fields F_0, F_1, F_2, $[F_0, F_1]$, $[F_0, F_2]$, $[F_1, F_2]$, all the Lie brackets of order ≥ 3 being 0.*

Proposition 3.11. *In the generic nilpotent model, the extremals project onto:*

$$\dot{\theta} = \frac{1}{r}[b + a_1 \sin\theta - a_2 \cos\theta]$$
$$\dot{r} = a_1 \cos\theta + a_2 \sin\theta$$

where a_1, a_2, b are constant parameters defined by: $b = [P_1, P_2]$, $a_1 = [P_1, P_0]$, $a_2 = [P_2, P_0]$. The involutive case is $[D, D] \subset D$ and is of dimension 5, and $b = 0$ in the previous equations.

Definition 3.15. *In the involutive case $[D, D] \subset D$, when crossing Σ at a point of order one, the control rotates instantaneously of an angle π and the corresponding singularity is called a π-singularity.*

Applications to SR-Problems in Dimension 4

Consider a system of the form: $\frac{dq}{dt} = F_0(q) + \sum_{i=1}^{2} u_i F_i(q)$, where $q(t) \in \mathbb{R}^4$ and $[D, D] \subset D$. We assume that the system is regular in the following sense: for each $q \in \mathbb{R}^4$, the rank of $\{F_1(q), F_2(q), [F_1, F_0](q), [F_2, F_0](q)\}$ is 4. Hence there exists a vector $\lambda(q) = (\lambda_1(q), \lambda_2(q))$ such that:

$$F_0(q) = \lambda_1(q)[F_1, F_0](q) + \lambda_2(q)[F_2, F_0](q) \mod (D).$$

Each adjoint vector p such that $P_1 = P_2 = 0$ is defined by a vector $a = (a_1, a_2)$ where $a_1 = \{P_1, P_0\}$, $a_2 = \{P_2, P_0\}$.

Proposition 3.12. *In the regular case, the only discontinuities of an extremal control correspond to π-singularities, where the control rotates instantaneously of π in the directions defined by $\langle\lambda, a\rangle \geq 0$. In the non exceptional case, $\langle\lambda, a\rangle \neq 0$, the extremal crosses the switching surface in a single direction.*

Proof. In the regular case, the conditions $P_1 = P_2 = \{P_1, P_0\} = \{P_2, P_0\} = 0$ imply $p = 0$ and hence the only singularities correspond to π-singularities. When crossing Σ, we have $P_1 = P_2 = 0$ and $H = P_0$. Hence:

$$P_0 = \lambda_1(q)\{P_1, P_0\} + \lambda_2(q)\{P_2, P_0\}$$

and $H \geq 0$ imposes $\langle\lambda, a\rangle \geq 0$. Time maximal trajectories satisfy $H \leq 0$ and the exceptional case corresponds to $\langle\lambda, a\rangle = 0$.

When crossing Σ, the extremal is solution of:

$$\dot{r} = a_1 \cos\theta + a_2 \sin\theta$$

where θ is solution of $\tan\theta = \dfrac{a_2}{a_1}$.

Hence except in the exceptional case where we can change p into $-p$, changing (a_1, a_2) into $(-a_1, -a_2)$, the orientation of the trajectory crossing Σ with the slope θ is imposed by $H \geq 0$. ■

Computations on the nilpotent model. Take $q = (q_1, q_2, q_3, q_4)$, $F_1 = \dfrac{\partial}{\partial q_1}$, $F_2 = \dfrac{\partial}{\partial q_2}$ and $F_0 = (1+q_1)\dfrac{\partial}{\partial q_3} + q_2\dfrac{\partial}{\partial q_4}$. Hence $[F_0, F_1] = \dfrac{\partial}{\partial q_3}$, $[F_0, F_2] = \dfrac{\partial}{\partial q_4}$. All Lie brackets with length ≥ 3 being zero. We have:

$$F_0 = (1 + q_1)[F_0, F_1] + q_2\frac{\partial}{\partial q_4}.$$

Hence, for $q = 0$ we get

$$F_0 = -[F_1, F_0],$$

and using our previous notations: $\lambda(0) = (-1, 0)$. If $p = (p_1, p_2, p_3, p_4)$, the condition $\langle\lambda, a\rangle \geq 0$ gives $p_3 \geq 0$ and $p_3 = 0$ in the exceptional case. Introducing the planes $H_1 = (q_1, q_3)$ and $H_2 = (q_2, q_4)$, the system is decomposed into:

$$\begin{cases} \dot{q}_3 = 1 + q_1 \\ \dot{q}_1 = \quad u_1 \end{cases} \text{and} \quad \begin{cases} \dot{q}_4 = \ q_2 \\ \dot{q}_2 = u_2 \, . \end{cases}$$

Optimal synthesis. It can be easily computed using our representation. For instance, from 0, in each plane H_1, H_2 the synthesis is the following:

Plane H_1. A time minimal (resp. maximal) trajectory is an arc with $u_1 = 1$ followed by an arc with $u_1 = -1$ denoted $\gamma_+\gamma_-$ (resp. $\gamma_-\gamma_+$), u being defined by $u_2 = 0$, $u_1 = sign P_1$ and $H \geq 0$.

Plane H_2. Optimal controls are defined by $u_1 = 0$, $u_2 = signP_2$, 0 corresponding to an exceptional direction which is locally controllable. An optimal policy is of the form $\gamma_+\gamma_-$ or $\gamma_-\gamma_+$.

In particular, we proved:

Proposition 3.13. *There exist extremal trajectories that have a π-singularity and are optimal.*

Application to the 2D-orbital transfer. Assuming the mass constant, the system can be written: $m\ddot{q} = K(q) + u_1 F_t + u_2 F_n$, where K is Kepler equation and the thrust is decomposed in the tangential /normal frame. We restrict the system to the elliptic domain. The regularity condition is satisfied and we have:

Proposition 3.14. *For every compact M of the elliptic domain, the only extremals are a finite concatenation of extremals of order 0, each switching corresponding to a π-singularity and the number of switchings is uniformly bounded on M.*

3.5.6 Extremals for Single-Input Affine Systems

In this section, we consider the time optimal control problem for a system of the form: $\dot{q} = X + uY$, $|u| \leq 1$ where $q(t) \in \mathbb{R}^n$. We introduce the Hamilton functions $P_X = \langle p, X\rangle$ and $P_Y = \langle p, Y\rangle$.

Singular Extremals

Proposition 3.15. *Singular extremals are contained in $P_Y = \{P_Y, P_X\} = 0$ and generic singular extremals are solutions of: $\dot{z}(t) = \mathbf{H}_s(z(t))$, where $H_s = \langle p, X + u_s Y\rangle$, the singular extremal control being given by:*

$$u_s(q,p) = -\frac{\{\{P_Y, P_X\}, P_X\}(q,p)}{\{\{P_Y, P_X\}, P_Y\}(q,p)}.$$

Proof. From the maximum principle, singular controls correspond to $\frac{\partial H}{\partial u} = 0$, and hence $\langle p(t), Y(q(t))\rangle = 0$, $\forall t$.

Differentiating twice and using Hamiltonian formalism, we get:

$$\langle p, [Y, X]\rangle = \{P_Y, P_X\} = 0$$

$$\{\{P_Y, P_X\}, P_X\} + u\{\{P_Y, P_X\}, P_Y\} = 0$$

and the singular control is computed using the previous equation. ∎

Regular Extremals

From the maximum principle, a regular extremal is defined by $u(t) = sign\langle p(t), Y(q(t))\rangle$ a.e.. To analyze the switching sequence we introduce the switching mapping $\Phi : t \mapsto P_Y(z(t))$ evaluated along the reference regular extremal. It is denoted Φ^+, Φ^- respectively if the corresponding extremal control is $+1$ or -1. If z is smooth, we have:

Lemma 3.5. *The first two derivatives of the switching mapping are:*

$$\dot{\Phi}(t) = \{P_Y, P_X\}(z(t))$$
$$\ddot{\Phi}(t) = \{\{P_Y, P_X\}, P_X\}(z(t)) + u(t)\{\{P_Y, P_X\}, P_Y\}(z(t)).$$

Next we describe the generic classification of regular extremals near the switching surface that we shall need in the sequel.

Normal Switching Points

Let Σ_1 be the switching surface $P_Y = 0$ and Σ_2 be the surface $P_Y = \{P_Y, P_X\} = 0$. Let $z_0 = (q_0, p_0)$ and assume $Y(q_0) \neq 0$, $z_0 \in \Sigma_1$, $z_0 \notin \Sigma_2$. The point z_0 is called a normal switching point. From the previous lemma, we have:

Lemma 3.6. *Let t_0 be the switching time defined by $z^+(t_0) = z^-(t_0) = z_0$, z^+, z^- being extremals with respectively $u = +1, -1$. Then, the following holds:*

$$\dot{\Phi}^+(t_0) = \dot{\Phi}^-(t_0) = \{P_Y, P_X\}(z_0)$$

and the extremal passing through z_0 is of the form:

$$z = \gamma_+\gamma_- \text{ if } \{P_Y, P_X\}(z_0) < 0$$

$$z = \gamma_-\gamma_+ \text{ if } \{P_Y, P_X\}(z_0) > 0.$$

The Fold Case

Let $z_0 \in \Sigma_2$, assuming $Y(q_0) \neq 0$ and Σ_2 smooth surface of codimension 2. If $\mathbf{H}_+$, $\mathbf{H}_-$ are Hamiltonian vector fields associated to $\langle p, X \pm Y\rangle$, then $\Sigma_1 = \{z; H_+ = H_-\}$ and at $z_0 \in \Sigma_2$, both vector fields $\mathbf{H}_+$ and $\mathbf{H}_-$ are tangent to Σ_1. We set:

$$\lambda_\pm = \{\{P_Y, P_X\}, P_X\}(z_0) \pm \{\{P_Y, P_X\}, P_Y\}(z_0)$$

and we assume that both $\lambda_\pm \neq 0$. Then the contact of $\mathbf{H}_+$, $\mathbf{H}_-$ with Σ_1 is of order 2 and the point is called a fold. We distinguish three cases:

a. $\lambda_+\lambda_- > 0$: parabolic case
b. $\lambda_+ > 0$, $\lambda_- < 0$: hyperbolic case
c. $\lambda_+ < 0$, $\lambda_- > 0$: elliptic case.

We have the following:

Proposition 3.16. *Let z_0 be a fold point, then there exists a neighborhood V of z_0 such that:*

1. *In the hyperbolic case, each extremal trajectory has at most two switchings and is of the form $\gamma_\pm\gamma_s\gamma_\pm$ where γ_+, γ_- are bang arcs and γ_s a singular arc.*
2. *In the parabolic case, each extremal arc is bang-bang with at most two switchings and has one of the form: $\gamma_+\gamma_-\gamma_+$ or $\gamma_-\gamma_+\gamma_-$.*
3. *In the elliptic case, each extremal arc is bang-bang, but with no uniform bound on the number of switchings.*

3.5.7 Second-Order Conditions

Definition 3.16. *Consider an optimal control problem with fixed extremities, and let $z(t) = (q(t), p(t))$ be an extremal trajectory defined on $[0, T]$. We call first conjugate time t_{1c} the first time when the trajectory ceases to be optimal for the C^0-topology on the set of curves q and the corresponding point $q(t_{1c})$ is called the first conjugate point. If we fix $q(0) = q_0$, the set of conjugate points is the conjugate locus denoted $C(q_0)$. The cut point is the point where the trajectory ceases to be globally optimal and the set of such points forms the cut-locus denoted $L(q_0)$.*

Second-Order Condition in the Classical Calculus of Variations

Consider the problem: $\min \int_0^T L(q, \dot{q}, t)dt$ where $q(t) \in \mathbb{R}$, T is fixed and $q(0) = q_0$, $q(T) = q_1$. We set $\dot{q} = u$, $u(t) \in \mathbb{R}$ and $H = pu - L(q, u, t)$. From $\frac{\partial H}{\partial u} = 0$, we get: $p = \frac{\partial L}{\partial \dot{q}}$ which corresponds to Legendre transformation. Moreover $\dot{p} = -\frac{\partial H}{\partial q} = \frac{\partial L}{\partial q}$, hence we get Euler-Lagrange equation: $\frac{d}{dt}\frac{\partial L}{\partial \dot{q}} = \frac{\partial L}{\partial q}$.

Let q be a reference trajectory and let δq be a variation. Computing the cost variation:

$$\int_0^T [L(q + \delta q, \dot{q} + \delta\dot{q}, t) - L(q, \dot{q})]dt = \int_0^T \frac{\partial L}{\partial q}\delta q + \frac{\partial L}{\partial \dot{q}}\delta\dot{q} + \circ(\delta q, \delta\dot{q})$$

and with $\delta q(0) = \delta q(T)$ we get:

$$\int_0^T \left(\frac{\partial L}{\partial q} - \frac{d}{dt}\frac{\partial L}{\partial \dot{q}}\right)\delta q dt + \circ(\delta q, \delta \dot{q}) .$$

If q is an extremal, the integral is 0. Expanding at order 2, we get:

$$\int_0^T \left(\frac{1}{2}\frac{\partial^2 L}{\partial q^2}\delta q^2 + \frac{1}{2}\frac{\partial^2 L}{\partial \dot{q}^2}\delta \dot{q}^2 + \frac{\partial^2 L}{\partial q \partial \dot{q}}\delta q \delta \dot{q}\right)dt$$

which corresponds to the second-order derivative. Using $\delta q(0) = \delta q(T) = 0$ it can be written after an integration by part as:

$$\delta^2 C = \int_0^T (P(t)\delta \dot{q}^2(t) + Q(t)\delta q^2(t))dt$$

where

$$P = \frac{1}{2}\frac{\partial^2 L}{\partial \dot{q}^2}$$
$$Q = \frac{1}{2}\frac{\partial^2 L}{\partial q^2} - \frac{d}{dt}\frac{\partial^2 L}{\partial q \partial \dot{q}} .$$

Integrating by parts, we obtain:

$$\delta^2 C = \int_0^T (Q(t)\delta q - \frac{d}{dt}P(t)\delta \dot{q})\delta q dt$$

and introducing the linear operator:

$$D : h \mapsto Qh - \frac{d}{dt}P\dot{h}$$

the second-order derivative is:

$$\delta^2 C = (D\delta q, \delta q)$$

where $(,)$ is the scalar product on $L^2[0, T]$.

The equation $Dh = 0$ is called Jacobi equation and is derived from Euler equation by taking the variational equation. It is also Euler-Lagrange equation associated to the so-called accessory problem:

$$\min \int_0^T (P\dot{h}^2 + Qh^2)dt$$

among the set of curves satisfying the conditions $h(0) = h(T) = 0$.

In the standard theory, we call conjugate time t a $0 < t \leq T$ such that there exists a non trivial solution h of $Dh = 0$, with $h(0) = h(t) = 0$. And we have:

Proposition 3.17. *Assume the strict Legendre condition $P > 0$, then beyond the first conjugate time, the quadratic form $\int_0^T (P\dot{h}^2 + Qh^2)dt$, $h(0) = h(t) = 0$, takes strictly negative values and the reference extremal is no more C^1-optimal.*

The condition $\delta^2 C \geq 0$ is a second-order C^1-necessary condition. An important C^0-sufficient condition is obtained using the concept of central field defined as follows. Let $\mathcal{F} = (t, q(t)) \in \mathbb{R} \times \mathbb{R}$ be the set of extremal solutions on the time extended space state with initial condition $q(0) = q_0$. If $t < t_{1c}$, the first conjugate time, the reference extremal can be imbedded in the central field $\mathcal{F}$ which forms, for $t > 0$ a tubular neighborhood, on which the extremals are not overlapping. We have

Proposition 3.18. *On $\mathcal{F}$ the Hilbert-Cartan form $\omega = pdq - Hdt$ where $p = \dfrac{\partial L}{\partial \dot{q}}$, $H = p\dot{q} - L$ is closed. Moreover, the reference extremal is C^0-optimal for the problem with fixed extremities, with respect to every curve contained in the tubular neighborhood covered by $\mathcal{F}$.*

Hence, from this important result, in the standard calculus of variations case, the construction of a central field gives us an estimate of C^0-optimality of the reference optimal.

Next, we shall generalize the concept of central field to get second-order optimality conditions for control systems. The important concept in Hamiltonian formalism is the concept of Lagrangian manifold.

Lagrangian Manifold and Jacobi Equation

Definition 3.17. *Let M be a n-dimensional manifold, T^*M the cotangent bundle, $\omega = pdq$ be the standard Liouville form. Let $L \subset T^*M$ be a regular submanifold. We say that L is isotropic if $d\omega$ restricted to TL is zero, and if L is of maximal dimension n, L is called Lagrangian.*

Proposition 3.19. *Let L be a Lagrangian submanifold. If the standard projection $\Pi : (q, p) \mapsto q$ is regular on L, then there exist canonical coordinates (Q, P) preserving Π such that L is given locally by a graph of the form $(Q, P = \dfrac{\partial S}{\partial Q}(Q))$ and S is called the generating function of L.*

Definition 3.18. *Let M be a Lagrangian submanifold of T^*M. A tangent vector $v \neq 0$ to L is said vertical if $d\Pi(v) = 0$. We call caustic the set of points q such that there exists at least one vertical vector.*

Definition 3.19. *Let $\mathbf{H}(z)$ be a smooth Hamiltonian vector field associated to an optimal control problem and let $z(t) = (q(t), p(t))$, $t \in [0, T]$ be a reference curve. The variational equation:*

$$\frac{d\delta z}{dt} = \frac{\partial \mathbf{H}}{\partial z}(z(t))\delta z(t)$$

is called Jacobi equation. We call Jacobi field $J(t) = (\delta q(t), \delta p(t))$ *a non trivial solution of Jacobi equation;* J *is vertical at time* t *if* $\delta q(t) = 0$. *We say that* t_c *is a geometric conjugate time if there exits a Jacobi field* $J(t)$ *with* $\delta q(0) = \delta q(t_c) = 0$ *and the point* $q(t_c)$ *is said geometrically conjugate to* $q(0)$.

Proposition 3.20. *Let* $z(t) = (q(t), p(t))$, $t \in [0, T]$ *be the reference extremal and let* $z(t, p_1) = (q(t, p_1), p(t, p_1))$ *be the solution of* $\mathbf{H}$ *starting from* $t = 0$ *of* $q_0 = q(0)$ *fixed and corresponding to* p_1. *Let* L_0 *be the fiber* $T^*_{q_0}M$ *and* L_t *be the image of* L_0 *by* $\exp t\mathbf{H}$. *Then* L_t *is a Lagrangian manifold and* t_c *is geometrically conjugate if and only if* $(t, p_1) \mapsto q(t, p_1)$ *is not an immersion at* $p_0 = p(0)$ *for* $t = t_c$, *i.e.* (L_{t_c}, Π) *is singular.*

Proof. Let $\dot{x} = X(x)$ be a differential equation and $\{\exp tX\}$ be the corresponding local parameter group. If $\epsilon \mapsto \alpha(\epsilon)$ is a curve with $\alpha(0) = x_0$, then the image $\beta(\epsilon) = \exp tX(\alpha(\epsilon))$ is a curve with $\beta(0) = \exp tX(x_0) = y_0$. If $v = \dot{\alpha}(0)$, $w = \dot{\beta}(0)$ are the tangent vectors, then from standard calculus, w is the image of v by $d\exp tX$ and $w = \delta x(t)$ where $\dot{\delta x} = \frac{\partial X}{\partial x}(\exp tX(x_0))\,\delta x$, $\delta x(0) = v$. In our case, we apply this result with curves $\alpha(\epsilon)$ in the fiber $T^*_{q_0}M$, which imposed $\delta q(0) = 0$. Observe also that w can be computed in the analytic case using the ad-formula:

$$d\exp tX(Y(x_0)) = \sum_{k \geq 0} \frac{t^k}{k!} ad^k X(Y)(y_0)$$

t small enough and Y is a vector field with $Y(x_0) = v$. ■

Conjugate Points in The SR-case

Consider a SR-problem:

$$\dot{q} = \sum_{i=1}^{m} u_i F_i(q) \quad , \quad \min l(q),$$

where $l(q) = \int_0^T (\sum_{i=1}^m u_i^2)^{\frac{1}{2}} dt$ is the length of a curve tangent to $D(q) = Span\{F_1(q), \cdots, F_m(q)\}$, F_i being chosen orthonormous. Our problem is called parametric, because the length does not depend upon the parametrization and for instance we can parameterize by arc length, imposing: $\sum_{i=1}^m u_i^2 = 1$, hence our problem is: minimize time. In order to fit in a framework where the time T is fixed, we use Maupertuis principle, minimizing the length being equivalent to minimizing the energy $E(q) = \int_0^T (\sum_{i=1}^m u_i^2)dt$. From our previous computation in section 3.5.4 extremals in the normal case are associated to $H = \sum_{i=1}^m u_i P_i - \frac{1}{2}\sum_{i=1}^m u_i^2$, $P_i = \langle p, F_i \rangle$, and a normal control is $u_i = P_i$; extremals solutions are associated to the Hamilton function $H_n = \frac{1}{2}\sum_{i=1}^m P_i^2$.

Algorithm. In order to compute geometric conjugate points in the normal case we proceed as follows: we note $J_i(t) = (\delta q_i(t), \delta p_i(t))$ a solution of Jacobi equation with $\delta q_i(0) = 0$, $\delta p_i(0) = e_i$ canonical basis. Along a reference extremal we evaluate the corresponding determinant:

$$D = \det |\delta q_1(t), \cdots, \delta q_n(t)|,$$

a geometric conjugate time is defined by $D(t_c) = 0$.

Now we observe the following important reduction. An extremal can be computed using the level energy: $H_n = \frac{1}{2}$, which corresponds to parametrize by arc-length.

In particular:

Lemma 3.7. *Geometric first conjugate points are contained in the caustic, projection on the q-space of* $\exp \mathbf{H}_n(L_0)$ *where* $L_0 = T^*_{q_0}M$.

Definition 3.20. *Let* $q(t, q_0, p_0)$, $p(t, q_0, p_0)$ *be the solution of* $\mathbf{H}_n$ *starting at* $t = 0$ *from* (q_0, p_0). *We fix* q_0, *the exponential mapping is* $\exp_{q_0} : p_0 \mapsto q(t, q_0, p_0)$ *where we can restrict* p_0 *to* $H_n = \frac{1}{2}$.

We have:

Lemma 3.8. *The solutions of* $\mathbf{H}_n$ *satisfy the homogeneity relation:*

$$q(t, q_0, \lambda p_0) = q(\lambda t, q_0, p_0)\ ,\ p(t, q_0, \lambda p_0) = \lambda p(\lambda t, q_0, p_0).$$

Hence the Jacobi field corresponding to a variation: $\alpha(\epsilon) = (q_0, p_0 + \epsilon p_0)$ *and denoted* $J(t)$ *satisfies* $\pi(J(t))$ *collinear to* $\dot{q}$. *Therefore* t_c *is conjugate if and only if the rank of the matrix* $R(t) = |\delta q_1(t), \cdots, \delta q_{n-1}(t)|$ *is* $< n-1$ *where* $(\delta q_i, \delta p_i)$ *are Jacobi fields with* $\delta q_i(0) = 0$ *and* $\delta p_i(0)$ *form a basis of* $p_0 \delta p = 0$. *Moreover* t_c *is conjugate if* $\exp_{q_0}$ *is not an immersion at* $t = t_c$, $p = p_0$.

Remark 3.3. In SR-geometry, the computation of geometric conjugate points amounts to test the rank of $(n-1)$ vectors. The same will be applied in time optimal problems.

Conjugate Points in SR-Case with Drift

We consider a time optimal control problem for a system of the form:

$$\dot{q} = F_0 + \sum_{i=1}^{m} u_i F_i \ , \quad |u| \leq 1.$$

From section 3.5.5, we set $P_i = \langle p, F_i(q) \rangle$, $i = 0, \cdots, m$ and the extremals of order 0 defined outside Σ : $P_i = 0$, $i = 1, \cdots, m$ are solutions corresponding to the Hamiltonian function $H_0 = P_0 + (\sum_{i=1}^{m} P_i^2)^{\frac{1}{2}}$.
If (q_0, p_0) is an initial point, we note $z(t) = (q(t, q_0, p_0), p(t, q_0, p_0))$ the associated extremal.

By homogeneity, we can restrict p_0 to the projective space $\mathbb{P}^{n-1}$. Similarly to the SR-case, we introduce the exponential mapping:

$$\exp_{q_0} : \ (t, p_0) \mapsto q(t, q_0, p_0).$$

The Jacobi equation corresponding to an extremal of order 0 is:

$$\frac{d}{dt}(\delta z(t)) = \frac{\partial \mathbf{H}_0}{\partial z}(z(t))\delta z(t).$$

We introduce the $(n-1)$ Jacobi fields $J_i(t) = (\delta q_i(t), \delta p_i(t))$ solution of the above equation and with initial condition $\delta q_i(0) = 0$, $\delta p_i(0) = e_i$ where e_i : $i = 1, \cdots, n-1$ is a basis of the set of δp with $p_0 \delta p = 0$.

A geometric conjugate time t_c is a t such that rank of the matrix $C(t) = (\delta q_1(t), \cdots, \delta q_{n-1}(t))$ is strictly less than $n-1$.

Let $z(t) = (q(t), p(t))$ be an extremal of order 0, it corresponds to a singularity of the extremity mapping, where the control u is restricted to $|u| = 1$.

We make the following assumptions:

Assumption 10 *The codimension of the singularity is one.*

Assumption 11 *The extremal is not exceptional, i.e. $H_0 > 0$.*

From [2] we have:

Proposition 3.21. *Under our previous assumptions, if $t < t_{1c}$ first geometric conjugate point, the reference extremal is optimal for neighboring controls in the L^∞-norm and no more optimal if $t > t_{1c}$.*

Application. This algorithm can be applied to the orbital transfer to evaluate the conjugate locus, using numerical simulations.

3.5.8 Optimal Controls with State Constraints

In many applications, we have an optimal control problem: $\dot{q} = f(q,u)$, $\min \int_0^T f^0(q,u)dt$ with state constraints of the form $c(q) \leq 0$. This is the case for the shuttle re-entry problem, where we have two active constraints, on the thermal flux and the normal acceleration. Our aim is to present tools which can be used to analyze such problems. A first method is to apply necessary conditions from maximum principle. They are not straightforward and we shall restrict our presentation to the conditions obtained by Weierstrass for Riemannian problems in the plane with obstacles. A second method is the one used to analyze the re-entry problem which is the following. First we analyze the small time optimal control for the system without taking into account the state constraints using the maximum principle and second-order conditions. Secondly we construct a model to take into account the constraints.

Weierstrass Necessary Optimality Conditions

We consider the problem of minimizing the length of a curve $q(.)$ of the form $\int_0^T L(q,\dot{q})dt$, avoiding a domain D of the plane, with smooth boundary.

In order to analyze the problem we must:

1) Determine if a boundary arc is optimal.

2) Determine what the conditions for an optimal arc are when entering, departing or reflecting on the boundary.

To get necessary optimality conditions we make variations of a reference trajectory and we use a standard formula to estimate the length variation.

The variations introduced by Weierstrass are the following (see Fig. 3.1).

To determine if a reference boundary arc is optimal he introduced at a point P of the interior of the previous arc a vector u normal to boundary. He compares the length of the reference boundary arc to trajectories of the domain with same extremities and passing through u. We get:

Proposition 3.22. *A necessary optimality condition for a boundary arc is* $\frac{1}{r} \geq \frac{1}{\tilde{r}}$ *at each point* P *of the boundary where* $\frac{1}{\tilde{r}}$ *is the curvature of the boundary arc and* $\frac{1}{r}$ *the curvature of the extremal curve tangent to* P *to the boundary.*

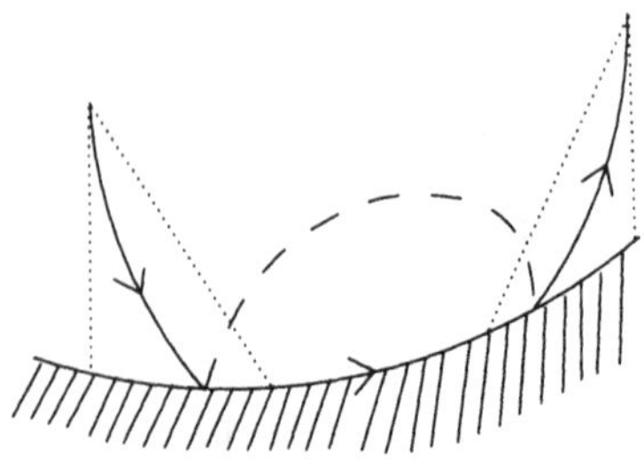

Fig. 3.1. Weierstrass variations.

Next we present a necessary condition satisfied by an optimal arc connecting to or departing from the boundary. They correspond to variations of the point entering the boundary and departing.

Proposition 3.23. *A necessary optimality condition for connecting or departing the boundary is to be tangent to the boundary.*

Finally, when reflecting on the boundary on a point making variations of the point, we obtain the following condition.

Proposition 3.24. *Assume that the metric is given by* $L = \sqrt{\dot{x}^2 + \dot{y}^2}$ *with* $q = (x, y)$. *Then when reflecting the boundary the optimal straight lines must have equal angles with the tangent to the boundary.*

Small Time Minimal Syntheses for Planar Affine Systems with State Constraints

Generalities. We consider a system of the form: $\dot{q} = X + uY$, $|u| \leq 1$, $q = (x, y) \in \mathbb{R}^2$ with state constraints. We denote by $\omega = pdq$ the clock form defined on the set where X, Y are independent by: $\omega(X) = 1$ and $\omega(Y) = 0$. The singular trajectories are located on the set $S:\ \det(Y, [Y, X]) = 0$ and the singular control u_s is solution of:

$$\langle p, [[Y, X], X] \rangle + u_s \langle p, [[Y, X], Y] \rangle = 0,$$

the two form $d\omega$ being 0 on S. A boundary arc γ_b is defined by $c(\gamma_b) = 0$ and differentiating we get: $\dot{c}(\gamma_b) = L_X c + u_b L_Y c = 0$. If $L_Y c \neq 0$ along the boundary arc, it is called of order one and the boundary control is given by $u_b = -\dfrac{L_X c}{L_Y c}$ and it has to be admissible $|u_b| \leq 1$.

We take $q_0 \in \{c = 0\}$, identified to 0. The problem is to determine the local optimality status of a boundary arc $t \mapsto \gamma_b(t)$ corresponding to a control u_b and to describe the time minimal synthesis near q_0. The first step is to construct a normal form, assuming the constraint of order one.

Lemma 3.9. *Assume:*
1) $X(q_0), Y(q_0)$ *linearly independent*
2) *The constraint is of order* 1, *i.e.* $L_Y c(q_0) \neq 0$.

Then changing if necessary u *into* $-u$, *there exists a local diffeomorphism preserving* q_0 *identified to* 0 *such that the constrained system is:*

$$\dot{x} = 1 + ya(q) \ , \ \dot{y} = b(q) + u \ , \ y \leq 0.$$

Proof. Using a local change of coordinates preserving 0, we can identify Y to $\frac{\partial}{\partial y}$ and the boundary arc to $\gamma_b = t \mapsto (t, 0)$. The admissible space is $y \leq 0$ or $y \geq 0$. Changing if necessary u into $-u$ it can be identified to $y \leq 0$. ■

The generic case. In this case, we make additional assumptions:
1. $Y(q_0), [X, Y](q_0)$ are linearly independent.
2. The boundary arc is admissible and not saturating at 0.

Under these assumptions we construct a local model, which corresponds to a nilpotent approximation to evaluate the small time policy. Indeed we have $a(0) \neq 0$, $|b(0)| < 1$ and we can take $a = a(0)$, $b = b(0)$, the model being:

$$\dot{x} = 1 + ya,$$
$$\dot{y} = b + u, \ \ y \leq 0,$$

the clock form is: $\omega = \dfrac{dx}{1+ay}$ and $d\omega = \dfrac{a}{(1+ay)^2}\, dx \wedge dy$.

Next we compute the local syntheses. First consider the unconstrained case. For small time, each point can be reached by an arc $\gamma_+\gamma_-$ or $\gamma_-\gamma_+$. If $a > 0$, $d\omega > 0$ and each time minimal policy is of the form $\gamma_+\gamma_-$, $\gamma_-\gamma_+$ being time maximal, and the opposite if $a < 0$, this can be obtained by considering the clock form or direct computations, because observe that if $a > 0$, if $y > 0$ we increase the speed along the $x - axis$ and decrease if $y < 0$.

For the constrained case, the same reasoning shows that the boundary arc is optimal if and only if $a > 0$, more precisely:

Lemma 3.10. *Under our assumptions, we have:*
1. *For the unconstrained problem, if* $a > 0$ *an arc* $\gamma_+\gamma_-$ *is time minimal and an arc* $\gamma_-\gamma_+$ *is time maximal and conversely if* $a < 0$.
2. *For the constrained problem, a boundary arc is optimal if and only if* $a > 0$ *and in this case, each optimal trajectory is of the form* $\gamma_+\gamma_b\gamma_-$. *If* $a < 0$ *each optimal arc is of the form* $\gamma_-\gamma_+$.

Small Time Minimal Syntheses for Systems in Dimension 3 with State Constraints

Preliminaries. We consider a system of the form: $\dot{q} = X + uY$, $|u| \leq 1$, $c(q) \leq 0$ with $q = (x, y, z) \in \mathbb{R}^3$. A boundary arc γ_b is defined by $c(\gamma_b) = 0$

and differentiating we get:

$$\dot{c}(\gamma_b) = (L_X c + L_Y c)(\gamma_b) = 0.$$

The arc is of order one if $L_Y c \neq 0$ and in this case the generic small time optimal synthesis is as in the planar case. Hence in the sequel we shall restrict our analysis to the case where $L_Y c = 0$ holds identically, that is Y is tangent to each hypersurface $c =$ *constant*. It is also the situation encountered in the re-entry problem.

In this case, along a boundary arc we have $L_X c = 0$ and differentiating we get:

$$L_X^2 c + u_b L_Y L_X c = 0.$$

If $L_Y L_X c \neq 0$ along the boundary, the constraint is called of order 2 and a boundary control is given by:

$$u_b = -\frac{L_X^2 c}{L_Y L_X c}.$$

In order to describe the small time optimal synthesis, we proceed as in the planar case, considering first the unconstrained case.

Assumption 12 *Let $q_0 \in \mathbb{R}^3$ and we assume that $X, Y, [X,Y]$ are linearly independent at q_0.*

In this case we have a standard result.

Lemma 3.11. *If $X, Y, [X,Y]$ form a frame at q_0, the small time accessibility set: $A^+(q_0) = \bigcup_{t\ small} q(t, q_0, u)$ is homeomorphic to a closed convex cone, where the boundary is formed by the surfaces S_1, S_2, where S_1 is the union of trajectories of the form $\gamma_-\gamma_+$ and S_2 is the union of trajectories of the form $\gamma_+\gamma_-$. Each point in the interior of the cone is reached by a unique trajectory with two switchings of the form $\gamma_-\gamma_+\gamma_-$ and a unique trajectory of the form $\gamma_+\gamma_-\gamma_+$.*

To construct optimal trajectories we must analyze the boundary of the small time accessibility set for the time extended system. We proceed as follows. Differentiating twice the switching function $\Phi = \langle p(t), Y(q(t))\rangle$ we get:

$$\dot{\Phi}(t) = \langle p(t), [Y, X](q(t))\rangle,$$
$$\ddot{\Phi}(t) = \langle p(t), [[Y, X], X + uY](q(t))\rangle.$$

If $\langle p(t), [[Y,X],Y](q(t))\rangle$ is not vanishing we can solve $\ddot{\Phi}(t) = 0$ to compute the singular control:

$$u_s = -\frac{\langle p, [[Y,X],X](q)\rangle}{\langle p, [[Y,X],Y](q)\rangle}.$$

If Y and $[X,Y]$ are independent, p can be eliminated by homogeneity and u_s computed as a feedback control. Introducing $D = \det(Y,[Y,X],[[Y,X],Y])$ and $D' = \det(Y,[Y,X],[[Y,X],X])$, we get $D'(q) + u_s D(q) = 0$. Hence in dimension 3 through each generic point there is a singular direction. Moreover, as in the planar case, the Legendre-Clebsh condition allows to distinguish between slow and fast directions in the non exceptional case where $X, Y, [X,Y]$ are not collinear. We have two cases, see [7].

- Case 1 : If X and $[[X,Y],Y]$ are on the opposite side with respect to the plane generated by $Y,[X,Y]$, then the singular arc is locally time optimal if $u(t) \in \mathbb{R}$.
- Case 2 : On the opposite, if X and $[[X,Y],Y]$ are in the same side, the singular arc is locally time maximal.

In the two cases, the constraint $|u_s| \leq 1$ is not taken into account and the singular control can be strictly admissible if $|u_s| < 1$, saturating if $|u_s| = 1$ at q_0, or non admissible if $|u_s| > 1$. We have 3 generic cases. Assume $X, Y, [X,Y]$ not collinear and let p oriented with the convention of the maximum principle: $\langle p(t), X + uY\rangle \geq 0$. Let t be a switching time of a bang-bang extremal: $\Phi(t) = \langle p(t), Y(q(t))\rangle = 0$. It is called of order one if $\dot{\Phi}(t) = \langle p(t), [Y,X](q(t))\rangle \neq 0$ and of order two if $\dot{\Phi}(t) = 0$ but $\ddot{\Phi}(t) = \langle p(t), [[Y,X], X + uY](q(t))\rangle \neq 0$ for $u = \pm 1$. The classification of extremals near a point of order two is similar to the planar case. We have three cases:

- parabolic case: $\ddot{\Phi}_\pm$ have the same sign.
- hyperbolic case: $\ddot{\Phi}_+ > 0$ and $\ddot{\Phi}_- < 0$.
- elliptic case: $\ddot{\Phi}_+ < 0$ and $\ddot{\Phi}_- > 0$.

In both hyperbolic and parabolic cases, the local time optimal synthesis are obtained by using only the first-order conditions from the minimum principle and hence from extremality, together with Legendre-Clebsch condition in the hyperbolic case. More precisely we have:

Lemma 3.12. *In the hyperbolic or parabolic case, each extremal policy is locally time optimal. In the hyperbolic case each optimal policy is of the form* $\gamma_\pm\gamma_s\gamma_\pm$. *In the parabolic case, each optimal policy is bang-bang with at most two switchings.*

The set $B(q_0, T)$ describing the time minimal policy at fixed time T is homeomorphic to a closed disk, whose boundary is formed by extremities of arcs $\gamma_-\gamma_+$ and $\gamma_+\gamma_-$ with length T and the stratification in the hyperbolic case is represented on Fig. 3.2.

In the elliptic case, the situation is more complicated because there exists a cut-locus. The analysis is related to the following crucial result based on the concept of conjugate points defined in Section 3.5.7, adapted to the bang-bang case.

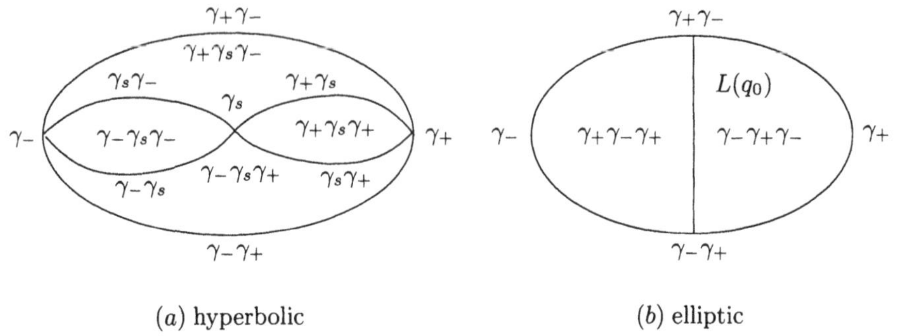

Fig. 3.2. Time optimal synthesis.

Lemma 3.13. *Consider a system* $\dot{q} = X + uY$, $|u| \leq 1$, $q(t) \in \mathbb{R}^3$. *Let* q_0 *be a point such that each of the two triplets* $Y, [X,Y], [X-Y,[X,Y]]$ *and* $Y,[X,Y],[X+Y,[X,Y]]$ *are linearly independent at* q_0. *Then near* q_0 *each bang-bang locally time optimal trajectory has at most two switchings.*

The local time optimal policy at fixed time is represented on the previous figures. There exists a cut locus $L(q_0)$ whose extremities are conjugate points on the boundary of the reachable set.

We shall now analyze the constrained case. If the order of the constraint is one, the situation is similar to the planar case analyzed in Section 3.5.8. Hence we shall assume that the constraint is of order 2. We restrict our analysis to the parabolic case, which corresponds to the space shuttle problem .

Geometric normal form in the constrained parabolic case and local synthesis. For the unconstrained problem the situation is clear in the parabolic case. Indeed $X, Y, [X,Y]$ form a frame near q_0 and writing:

$$[[Y,X], X \pm Y] = aX + bY + c[X,Y],$$

the synthesis depends only upon the sign of a at q_0. The small time reachable set is bounded by the surfaces formed by arcs $\gamma_-\gamma_+$ and $\gamma_+\gamma_-$. Each interior point can be reached by an arc $\gamma_-\gamma_+\gamma_-$ and an arc $\gamma_+\gamma_-\gamma_+$. If $a < 0$ the time minimal policy is $\gamma_-\gamma_+\gamma_-$ and the time maximal policy is $\gamma_+\gamma_-\gamma_+$ and the opposite if $a > 0$. To construct the optimal synthesis one can use a nilpotent model where all Lie brackets of length greater than 4 are 0. In particular the existence of singular direction is irrelevant in the analysis and a model where $[Y,[X,Y]]$ is zero can be taken. This situation is called the geometric model. A similar model is constructed next taking into account the constraints, which are assumed of order 2. Moreover we shall first suppose that is $Y.Xc \neq 0$ along γ_b and the boundary control is admissible and not saturating. We have the following:

Lemma 3.14. *Under our assumptions, a local geometric model in the parabolic case is:*

$$\dot{x} = a_1 x + a_3 z$$
$$\dot{y} = 1 + b_1 x + b_3 z$$
$$\dot{z} = (c + u) + c_1 x + c_2 y + c_3 z, \quad |u| \le 1$$

with $a_3 > 0$*, where the constraint is* $x \le 0$ *and the boundary arc is identified to* $\gamma_b : t \mapsto (0, t, 0)$*. Moreover we have* $[Y, X] = -a_3 \frac{\partial}{\partial x} - b_3 \frac{\partial}{\partial y}$, $[[Y, X], Y] = 0$, $[[Y, X], X] = (a_1 a_3 + a_3 c_3) \frac{\partial}{\partial x} + (a_3 b_1 + b_3 c_3) \frac{\partial}{\partial y} + (a_3 c_1 + b_3 c_2 + c_3^2) \frac{\partial}{\partial z}$*, and* $[[Y, X], X] = aX \; mod\{Y, [X, Y]\}$*, with* $a = a_3 b_1 - a_1 b_3 \neq 0$*. If the boundary arc is admissible and not saturating we have* $|c| < 1$*. Moreover* $a_3 = [X, Y]c$*.*

Proof. We give the details of the normalizations.
Normalization 1. Since $Y(0) \neq 0$, we identify locally Y to $\frac{\partial}{\partial z}$. The local diffeomorphisms $\varphi = (\varphi_1, \varphi_2, \varphi_3)$ preserving 0 and Y satisfy: $\frac{\partial \varphi_1}{\partial z} = \frac{\partial \varphi_2}{\partial z} = 0$ and $\frac{\partial \varphi_3}{\partial z} = 1$. Since the constraint is of order 2, $L_Y c = 0$ near 0 and Y is tangent to all surfaces $c = \alpha$, α small enough, hence $\frac{\partial c}{\partial z} = 0$.
Normalization 2. Since c is not depending on z, using a local diffeomorphism preserving 0 and $Y = \frac{\partial}{\partial z}$, we can identify the constraint to $c = x$. Then the system can be written: $\dot{x} = X_1(q)$, $\dot{y} = X_2(q)$, $\dot{z} = X_3(q) + u$, and $x \le 0$. The secondary constraint is $\dot{x} = 0$, and by assumption a boundary arc γ_b is contained in $x = \dot{x} = 0$ and passing through 0. In the parabolic case the affine approximation is sufficient for our analysis and the geometric model is:

$$\dot{x} = a_1 x + a_2 y + a_3 z$$
$$\dot{y} = b_0 + b_1 x + b_2 y + b_3 z$$
$$\dot{z} = c_0 + c_1 x + c_2 y + c_3 z + u,$$

where γ_b is approximated by the straight line: $x = 0$, $a_2 y + a_3 z = 0$.
Normalization 3. Finally we normalize the boundary as follows. In the plane $x = 0$, making a transformation of the form: $z' = \alpha y + z$, we can normalize the boundary arc to $x = z = 0$. Using a diffeomorphism $y' = \varphi(y)$, the boundary arc can be parameterized as $\gamma_b : t \mapsto (0, t, 0)$. The normal form follows, changing if necessary u to $-u$, and hence permuting the arcs γ_+ and γ_-. ■

Theorem 3.11. *Consider the time minimization problem for the system:* $\dot{q} = X(q) + uY(q)$, $q(t) \in \mathbb{R}^3$, $|u| \le 1$ *with the constraint* $c(q) \le 0$*. Let* $q_0 \in \{c = 0\}$ *and assume the following:*

1. *At* q_0, X, Y *and* $[X, Y]$ *form a frame and* $[[Y, X], X \pm Y](q_0) = aX(q_0) + bY(q_0) + c[X, Y](q_0)$*, with* $a < 0$*.*

2. *The constraint is of order 2 and both assumptions* $L_Y L_X c \neq 0$ *along* γ_b *and the boundary control is admissible and not saturating, are satisfied at* q_0.

Then the boundary arc through q_0 *is small time optimal if and only if the arc* γ_- *is contained in the non admissible domain* $c \geq 0$. *In this case the local time minimal synthesis with a boundary arc is of the form* $\gamma_- \gamma_+^T \gamma_b \gamma_+^T \gamma_-$, *where* γ_+^T *are arcs tangent to the boundary arc.*

Proof. The proof is straightforward and can be done using a simple reasoning visualized on the normal form. In this case $q_0 = 0$, the boundary arc is identified to $t \mapsto (0, t, 0)$ and due to $a_3 > 0$, arcs tangent to γ_b corresponding to $u = \pm 1$, are contained in $c \leq 0$ if $u = -1$ and in $c \geq 0$ if $u = +1$. Let B be a point of the boundary arc γ_b, for small enough $B = (0, y_0, 0)$. If $u = \pm 1$, we have the following approximations for arcs initiating from B :

$$x(t) = a_3(c_0 + c_2 y_0 + u)\frac{t^2}{2} + o(t^2)$$
$$z(t) = (c_0 + c_2 y_0 + u)t + o(t).$$

The projections in the plane (x, y) of the arcs $\gamma_-\gamma_+\gamma_-$ and $\gamma_+\gamma_-\gamma_+$ joining 0 to B are loops denoted $\tilde{\gamma}_-\tilde{\gamma}_+\tilde{\gamma}_-$ and $\tilde{\gamma}_+\tilde{\gamma}_-\tilde{\gamma}_+$. The loops $\tilde{\gamma}_-\tilde{\gamma}_+\tilde{\gamma}_-$ (*resp.* $\tilde{\gamma}_+\tilde{\gamma}_-\tilde{\gamma}_+$) are contained in $x \leq 0$ (*resp.* $x \geq 0$). We can now achieve the proof.
Taking the original system, if the arc $\gamma_-\gamma_+\gamma_-$ joining 0 to B which is the optimal policy for the unconstrained problem is contained in $c \leq 0$, it is time minimal for the constrained case and the boundary arc is not optimal. In the opposite, we can join 0 to B by an arc $\gamma_+\gamma_-\gamma_+$ in $c \leq 0$, but this arc is time maximal. Hence clearly the boundary arc γ_b is optimal. ■

Lie Bracket Properties of the Longitudinal Motion and Constraints

Consider the system $\dot{q} = X + uY$, $q = (x, y, z) \in \mathbb{R}^3$, where

$$X = \psi\left(v \sin\gamma \frac{\partial}{\partial r} - (g \sin\gamma - k\rho v^2)\frac{\partial}{\partial v} + \cos\gamma\left(-\frac{g}{v} + \frac{v}{r}\right)\frac{\partial}{\partial \gamma}\right),$$
$$Y = \psi k' \rho v \frac{\partial}{\partial \gamma}, \quad \psi = \frac{1}{\varphi},$$

$\varphi =$ thermal flux, which describes longitudinal motion when the rotation of the Earth is neglected ($\Omega = 0$) and $g = \dfrac{g_0}{r^2}$ is assumed to be constant. The following results, coming from computations, are crucial:

Lemma 3.15. *In the flight domain where* $\cos\gamma \neq 0$, *we have:*
1. $X, Y, [X, Y]$ *are linearly independent.*
2. $[[Y, X], Y] \in \mathit{Span}\{Y, [Y, X]\}$.
3. $[[Y, X], X](q) = a(q)X(q) + b(q)Y(q) + c(q)[X, Y](q)$ *with* $a < 0$.

Lemma 3.16. *Assuming C_D and C_L are constant, the constraints are of order 2.*

Application of classification to the space shuttle. The constraints are of order 2. In the part of the flight domain where the boundary arc is admissible and not saturating, the arc γ_- is violating the constraint along the boundary. Hence we proved:

Corollary 3.2. *Assume $\Omega = 0$ and consider the longitudinal motion in the re-entry problem. Then in the flight domain, a boundary arc is locally optimal and the small time optimal synthesis with fixed boundary conditions on (r, v, γ) is of the form $\gamma_-\gamma_+^T\gamma_b\gamma_+^T\gamma_-$.*

3.6 Indirect Numerical Methods in Optimal Control

Our approach to solve optimal control problems with the maximum principle is the following. A first step is to make a geometric analysis of the extremal flow. The most important consequence is to choose coordinates to represent the systems. This is well illustrated by the orbital transfer, with small propulsion. Using orbits elements which are parameters of the ellipses, we get slowly evoluating coordinates, decoupled from the fast variable corresponding to the longitude. Another aspect of the preliminary analysis is also to identify first integrals. After this analysis it remains to integrate the equations. Hence we need numerical integrations. Also the concept of conjugate point is simple, but it is necessary to use numerical integrations to compute the conjugate locus. The algorithm to handle our problems are mainly:

- Shooting or multiple-shooting techniques.
- Computations of conjugate points.

The remaining of this section will be devoted to the discussion of the problems in orbital transfer.

3.6.1 Shooting Techniques

Consider an optimal control problem: $\dot{q} = f(q, u)$, $q(t) \in \mathbb{R}^3$ $\min \int_0^T f^0(q, u)dt$ with $q(0) = 0$, $q(T) \in M_1$. By definition, $BC-$extremals are solutions of the maximum principle, satisfying the transversality conditions. In order to compute an optimal solution, we must solve a problem with boundary conditions.

Definition 3.21. *We introduce the shooting function $S : p \mapsto S(p)$ where $S(p) = (q(T,p) - q_1)$ with $q_1 \in M_1$ and $q(T,p)$ corresponds to a $BC-$extremal, with $q(0) = q_0$.*

First it is important to observe that in the maximum principle the number of equations equals to the number of unknown variables. Consider for instance, the time optimal control problem with fixed extremities. By homogeneity, we can replace p by λp, hence p has to be taken in the projective space $\mathbb{P}^{n-1}$ of dimension $(n-1)$ but the final time T is unknown.

In order to solve numerically an important question is the regularity of the shooting mapping. Behind smoothness is the concept of conjugate point which is explained using the example of orbital transfer. We proceed as follows:

Step 1. We restrict our analysis to extremal of order 0, evading switching due to $\pi-$singularity. Our extremals are solutions of a smooth Hamiltonian vector field $\mathbf{H_0}$ and before the first conjugate point, we can imbed each reference extremal in a central field: $\mathcal{F} : p \mapsto q(t,p)$, $q(0) = q_0$ in which p is uniquely defined by the boundary conditions and can be computed using the implicit function theorem.

Hence the restriction of the shooting function to the corresponding field is smooth, moreover it is an immersion, and in particular locally injective. But in general, it is not globally injective. An interesting model to understand the problem is the flat torus.

Flat torus. Consider a torus T^2 represented in the plane by a square $[0,1] \times [0,1]$, identifying opposite sides. It is endowed with usual metric of the plane, whose geodesics are straight lines. We have the following properties. First of all, the conjugate locus is empty. Now take a point q_0 of the torus which can be taken by invariance as the center of the square. Each geodesic line is optimal up to meeting the sides of the square and this gives us the cut locus, a cut point corresponding generically to the situation where two symmetric lines meet on the torus when reaching the sides, except for the lines reaching the corners where four symmetric lines meet. Then we have the following important properties.

Let q_1 be a point of the square, the minimizing geodesic is the line joining q_0 to q_1 but on the torus they are many geodesics joining q_0 to q_1 (see Fig. 3.3).

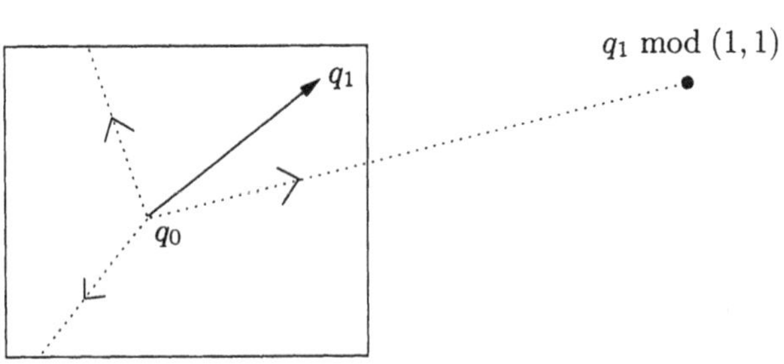

Fig. 3.3. Geodesics on a Flat torus.

In order to compute the minimizing geodesic one needs an estimate of the slope, which means that the localize our study for the central field around the given slope (micro-local analysis), or an estimate of the optimal length.

Shooting method and homotopy technique. When solving $S = 0$ a classical tool in the smooth case is a Newton type algorithm. One needs a preliminary estimate of the solution. To get such an estimate one can use an homotopy technique. This is illustrated in the orbital transfer. In this problem, there is a parameter λ corresponding to the modulus of the thrust and the shooting function restricted to extremals of order 0 takes the form $S(p, \lambda) = 0$, where S is smooth. The homotopy method is applied, starting from a large thrust to small propulsion. In order to solve this equation one can use iteratively Newton algorithm or solving the initial value problem:

$$\frac{\partial S}{\partial p}\dot{p}(s) + \frac{\partial S}{\partial \lambda}\dot{\lambda}(s) = 0,$$

where the set of solutions $S = 0$ is represented by a parametrized curve.

In general the method may fail to converge. For instance in the case of the torus we have several curves $p_i(s)$ solutions of the shooting equation. In the orbital transfer, in order to select the correct branch we use in fact an experimental fact, which gives the optimal time (see [4]).

Experimental fact. In orbital transfer, if t^* is the optimal time and λ the modulus of the maximal thrust we have: $t^*\lambda \simeq$ constant.

This result allows to compute the true minimizing curves because as observed numerically, like in the torus case, there are minimizing solutions and many other extremals solutions with bigger rotations numbers.

Numerical remarks in orbital transfer. The continuation method can be applied to the orbital transfer. We restrict our study to extremals of order 0. Nevertheless we observed experimentally that we are close to a $\pi-$singularity localized near a pericenter passage, for the given boundary conditions. To take into account this singularity, we use a multi-step method to integrate the differential equation.

Multiple shooting algorithm. In some case like the re-entry problem a shooting method is not sufficient and must be replaced by a multiple-shooting algorithm. This is connected to an instability property of Hamiltonian systems. We can not have stability in both state and costate variables because for any Hamiltonian matrix H if λ is an eigenvalue, $-\lambda$ is an eigenvalue.

In order to implement a multiple-shooting algorithm, we need to know the structure of the optimal policy. Applied to the re-entry problem we must a priori know the sequence of concatenating bang and boundary arcs. This is the result of the analysis. Only the switching times has to be computed, to estimate the optimal trajectory.

3.6.2 Second-Order Algorithms in Orbital Transfer

We restrict our analysis to the planar case, the $3D-$case being similar. First of all, assume that the mass is constant. We use the notations introduced in section 3.5.7., the system being written: $\dot{q} = F_0 + \sum_{i=1}^{2} u_i F_i$, $q(t) \in \mathbb{R}^4$, $u = (u_1, u_2)$, $|u| \leq 1$, and q are the orbits elements.

Let $z(t) = (q(t), p(t))$ be a reference extremal of order 0, solution of $\mathbf{H}_0$, with $H_0 = P_0 + \left(\sum_{i=1}^{2} P_i^2\right)^{\frac{1}{2}}$, $P_i = \langle p, F_i(q)\rangle$, $i = 0, 1, 2$. According to our general analysis, we must test the rank associated to the three Jacobi fields defined by the initial conditions $\delta q(0) = 0$ and δp such that $p(0)\delta p(0) = 0$.

Consider now the case where the mass variation is taken into account. We write the system as:

$$\dot{q} = F_0(q) + \frac{1}{m}\sum_{i=1}^{2} u_i F_i(q)$$
$$\dot{m} = -\delta|u| \quad , \quad |u| \leq 1.$$

The time minimum control corresponds to a maximum thrust $|u| = 1$. Therefore the mass is decreasing and minimizing time is equivalent to maximizing the final mass $m(t_f)$. If p_m denotes the dual variable of m, we have:

$$\dot{p}_m = \frac{1}{m^2}\sum_{i=1}^{2} u_i F_i.$$

The terminal mass $m(t_f)$ is free and hence from the transversality condition we have $p_m(t_f) = 0$. Therefore we must test a focal point which is generalization of the concept of conjugate point to a problem with terminal manifold. The algorithm is the following. We integrate backwards in time the variational equation with the initial conditions $\delta p_m = 0$, $\delta q = 0$, up to a first focal point where $\delta m = 0$, $\delta q = 0$.

Observe that since $\dot{\delta m} = 0$, then $\delta m \equiv 0$ and a focal point is also a conjugate point. Moreover if $\delta m = 0$ and if p denotes the vector dual to q, then the variational equation satisfied by δp is as in the constant mass case.

Finally the algorithm to test second-order condition for mass varying system is to compute the three Jacobi fields J_1, J_2, J_3 with initial conditions $\delta q(0) = 0$ and δp such that $p_0 \delta p(0) = 0$ for the time dependant system:

$$\dot{q} = F_0(q) + \frac{1}{m(t)}\sum_{i=1}^{2} u_i F_i(q)$$
$$\dot{p} = -p\left(\frac{\partial F_0}{\partial q}(q) + \frac{1}{m(t)}\sum_{i=1}^{2} u_i \frac{\partial F_i}{\partial q}(q)\right)$$

with $m(t) = m_0 - \delta t$.

References

1. O. Bolza (1961) *Lectures on the calculus of variations.* Chelsea Publishing Co., New York, 2nd edition.
2. B. Bonnard and M. Chyba (2003) *Singular trajectories and their role in control theory,* volume 40 of *Mathematiques and Applications.* Springer-Verlag, Berlin.
3. B. Bonnard, Faubourg L., and Trélat E. Mécanique céleste et contrôle des véhicules spatiaux. To appear.
4. J.B. Caillau (2000) *Contribution à l'étude du contrôle en temps minimal.* PhD thesis, Institut National Polytechnique de Toulouse.
5. D. E. Chang, D. F. Chichka, and J. E. Marsden (2002) Lyapunov-based transfer between elliptic keplerian orbits. *Discrete and Continuous Dynamical Systems-Series B,* 2(1):57–67.
6. V. Jurdjevic and J. P. Quinn (1978) Controllability and stability. *J. Differential Equations,* 28(3):381–389.
7. A. J. Krener and H. Schättler (1989) The structure of small-time reachable sets in low dimensions. *SIAM J. Control Optim.,* 27(1):120–147.
8. E. B. Lee and L. Markus (1986) *Foundations of optimal control theory.* Robert E. Krieger Publishing Co., Inc., Melbourne, FL, Second edition.
9. J. Stoer and R. Bulirsch (1980) *Introduction to numerical analysis.* Springer-Verlag, New York-Heidelberg. Translated from the Germany by R. Bartels, W. Gautschi and C. Witzgall.

4

Compositional Modelling of Distributed-Parameter Systems

Bernhard Maschke[1] and Arjan van der Schaft[2]

[1] Laboratoire d'Automatique et de Génie des Procédés, UCB Lyon 1 - UFR Génie Electrique et des Procédés - CNRS UMR 5007, CPE Lyon - Bâtiment 308 G, Université Claude Bernard Lyon-1, 43, bd du 11 Novembre 1918, F-69622 Villeurbanne cedex, France. maschke@lagep.univ-lyon1.fr

[2] Dept. of Applied Mathematics, University of Twente, P.O. Box 217, 7500 AE Enschede, The Netherlands. a.j.vanderschaft@math.utwente.nl

4.1 Introduction

The Hamiltonian formulation of distributed-parameter systems has been a challenging reserach area for quite some time. (A nice introduction, especially with respect to systems stemming from fluid dynamics, can be found in [26], where also a historical account is provided.) The identification of the underlying Hamiltonian structure of sets of p.d.e.s has been instrumental in proving all sorts of results on integrability, the existence of soliton solutions, stability, reduction, etc., and in *unifying* existing results, see e.g. [11], [24], [18], [17], [25], [14].

Recently, there has been also a surge of interest in the *design* and *control* of nonlinear distributed-parameter systems, motivated by various applications. At the same time, it is well-known from *finite-dimensional* nonlinear control systems [35], [32], [6], [21], [28], [27], [34] a Hamiltonian formulation is helpful in the control design, and the same is to be expected in the distributed-parameter case. However, in extending the theory as for instance exposed in [26] to distributed-parameter *control* systems a fundamental difficulty arises in the treatment of *boundary conditions*. Indeed, the treatment of infinite-dimensional Hamiltonian systems in the literature is mostly focussed on systems with infinite spatial domain, where the variables go to zero for the spatial variables tending to infinity, or on systems with boundary conditions such that the energy exchange through the boundary is *zero*. On the other hand, from a control and interconnection point of view it is quite essential to be able describe a distributed-parameter system with varying boundary conditions inducing *energy exchange through the boundary*, since in many applications the interaction with the environment (e.g. actuation or measurement) will ac-

F. Lamnabhi-Lagarrigue et al. (Eds.): Adv. Top. in Cntrl. Sys. Theory, LNCIS 311, pp. 115–154, 2005

tually take place through the boundary of the system. Clear examples are the telegraph equations (describing the dynamics of a transmission line), where the boundary of the system is described by the behavior of the voltages and currents at both ends of the transmission line, or a vibrating string (or, more generally, a flexible beam), where it is natural to consider the evolution of the forces and velocities at the ends of the string. Furthermore, in both examples it is obvious that in general the boundary exchange of power (voltage times current in the transmission line example, and force times velocity for the vibrating string) will be non-zero, and that in fact one would like to consider the voltages and currents or forces and velocities as additional *boundary variables* of the system, which can be interconnected to other systems. Also for numerical integration and *simulation* of complex distributed-parameter systems it is essential to be able to describe the complex system as the interconnection or coupling of its subsystems via their boundary variables; for example in the case of coupled fluid-solid dynamics.

From a mathematical point of view, it is not obvious how to incorporate non-zero energy flow through the boundary in the existing Hamiltonian framework for distributed-parameter systems. The problem is already illustrated by the Hamiltonian formulation of e.g. the Korteweg-de Vries equation (see e.g. [26]). Here for zero boundary conditions a *Poisson bracket* can be formulated with the use of the differential operator $\frac{d}{dx}$, since by integration by parts this operator is obviously skew-symmetric. However, for boundary conditions corresponding to non-zero energy flow the differential operator is not skew-symmetric anymore (since after integrating by parts the remainders are not zero).

In [37], see also [20], we proposed a framework to overcome this fundamental problem by using the notion of a *Dirac structure*. Dirac structures were originally introduced in [5],[7] as a geometric structure generalizing both *symplectic* and *Poisson* structures. Later on (see e.g. [35], [6], [19], [2]) it was realized that in the finite-dimensional case Dirac structures can be naturally employed to formalize Hamiltonian systems with *constraints* as *implicit* Hamiltonian systems. It turns out that in order to allow the inclusion of boundary variables in distributed-parameter systems the concept of Dirac structure again provides the right type of generalization with respect to the existing framework using Poisson structures.

The Dirac structure for distributed-parameter systems employed in this paper has a specific form by being defined on certain spaces of differential forms on the spatial domain of the system and its boundary, and making use of Stokes' theorem. Its construction emphasizes the geometrical content of the physical variables involved, by identifying them as differential k-forms, for appropriate k. This interpretation is rather well-known (see e.g. [12]) in the case of Maxwell's equations (and actually directly follows from Faraday's law and Ampère's law), but seems less well-known for the telegraph equations and the description of the Euler's equations for an ideal isentropic fluid.

From the systems and control point of view the approach taken in this paper can be seen as providing the extension of the port-Hamiltonian framework established for lumped-parameter systems in [35], [6], [27], [33], [35], [34], [3] to the distributed-parameter case. In the lumped-parameter case this Hamiltonian framework has been successfully employed in the consistent (modular) modeling and simulation of complex *interconnected* lumped-parameter physical systems, including (actuated) multi-body systems with kinematic constraints and electro-mechanical systems [35], [19], [6], [34], and in the design and *control* of such systems, exploiting the Hamiltonian and passivity structure in a crucial way [32], [21], [28], [27], [34]. Similar developments can be pursued in the distributed-parameter case; see already [30], [36] for developments in this direction. The remaining of the chapter is organized as follows. In Section 2 we give a general introduction to systems of conservation laws, together with the closure equations relating the conserved quantities to the flux variables. Furthermore, we show how this leads to infinite-dimensional power-continuous interconnection structures and the definition of Hamiltonian functions for energy storage. After this general introduction the main mathematical framework is given in Section 3 and 4, following [37]. In Section 3 it is shown how the notion of a power-continuous interconnection structure as discussed before can be formalized using the geometric concept of a Dirac structure, and in particular the Stokes-Dirac structure. In Section 4 it is shown how this leads to the Hamiltonian formulation of distributed-parameter systems with boundary energy flow, generalizing the notion of finite-dimensional port-Hamiltonian systems. In Section 5 (again following [37]) this is applied to Maxwell's equations on a bounded domain (Subsection 5.1), the telegraph equations for an ideal transmission line (Subsection 5.2), and the vibrating string (Subsection 5.3). Furthermore, by modifying the Stokes-Dirac structure with an additional term corresponding to three-dimensional convection, Euler's equations for an ideal isentropic fluid are studied in Section 6. Section 7 treats the basic notions of Casimir functions determined by the Stokes-Dirac structure. This can be seen as a starting point for control by interconnection of distributed-parameter port-Hamiltonian systems. Finally, Section 8 contains the conclusions.

4.2 Systems of Two Physical Domains in Canonical Interaction

The aim of this section is to introduce a class of infinite-dimensional physical systems and to show how they can be represented as port-Hamiltonian systems defined with respect to a special type of infnite-dimensional Dirac structure, called Stokes-Dirac structure. This will be done by formulating the distributed-parameter system as a system of conservation laws [10] [31], each describing the balance equation associated with some conserved physical quantity, coupled with a set of closure equations. These balance laws will define the Stokes-Dirac structure, while the closure equations will turn out to be equivalent with the definition of the Hamiltonian of the system.

4.2.1 Conservation Laws, Interdomain Coupling and Boundary Energy Flows: Motivational Examples

In this paragraph we shall introduce the main concepts of conservation law, interdomain coupling and boundary energy flow by means of three simple and classical examples of distributed-parameter systems.

The first example is the simplest one, and consists of only one conservation law on a one-dimensional spatial domain. With the aid of this simple example we shall introduce the notions of conservation law, balance equation, variational derivative, finally leading to the definition of a port-Hamiltonian system.

Example 4.1 (The inviscid Burger's equation).

The *viscous Burger's equation* is a scalar parabolic equation which represents the simplest model for a fluid flow (often used as a numerical test for the asymptotic theory of the Navier-Stokes equations) [31]. It is defined on a one-dimensional spatial domain (an interval) $Z = [a, b] \subset \mathbb{R}$, while its state variable is $\alpha(z,t) z \in Z,\ t \in I$, where I is an interval of $\mathbb{R}$ satisfying the partial differential equation

$$\frac{\partial \alpha}{\partial t} + \alpha \frac{\partial \alpha}{\partial z} - \nu \frac{\partial^2 \alpha}{\partial z^2} = 0 \tag{4.1}$$

In the following we shall consider the *inviscid* Burger's equations (corresponding to the case $\nu = 0$), which may be alternatively expressed by the following *conservation law*:

$$\frac{\partial \alpha}{\partial t} + \frac{\partial}{\partial z} \beta = 0 \tag{4.2}$$

where the state variable $\alpha(z,t)$ is called the *conserved quantity* and the function $\beta(z,t)$ is called *the flux variable* and is given by $\beta = \frac{\alpha^2}{2}$. Indeed, integrating the partial differential equation (4.2) on the interval Z, one obtains the following *balance equation*:

$$\frac{d}{dt} \int_a^b \alpha dz = \beta(a) - \beta(b) \tag{4.3}$$

Furthermore, according to the framework of Irreversible Thermodynamics [29], one may express the flux β as a function of the *generating force* which is the *variational derivative* (or, functional derivative,) of some generating functional $H(\alpha)$ of the state variable. This variational derivative plays the same role as the gradient of a function when considering functionals instead of functions. The variational derivative $\frac{\delta H}{\delta \alpha}$ of the functional $H(\alpha)$ is uniquely defined by the requirement:

$$H(\alpha + \epsilon\eta) = H(\alpha) + \epsilon \int_a^b \frac{\delta H}{\delta \alpha}\, \eta\, dz + O(\epsilon^2) \tag{4.4}$$

for any $\epsilon \in \mathbb{R}$ and any smooth function $\eta(z,t)$ such that $\alpha + \epsilon\eta$ satisfies the same boundary conditions as α [26]. For the inviscid Burger's equation it is easy to see that $\beta = \frac{\alpha^2}{2}$ can be expressed as $\beta = \frac{\delta H}{\delta \alpha}$, where

$$H(\alpha) = \int_a^b \frac{\alpha^3}{6} dz \tag{4.5}$$

Hence the inviscid Burger's equation may be also expressed as

$$\frac{\partial \alpha}{\partial t} = -\frac{\partial}{\partial z}\frac{\delta H}{\delta \alpha} \tag{4.6}$$

This defines an infinite-dimensional Hamiltonian system [26] with respect to the skew-symmetric operator $\frac{\partial}{\partial z}$ (defined on the functions with support strictly contained in the interval Z).

From this formulation one immediately derives that the Hamiltonian $H(\alpha)$ is *another* conserved quantity. Indeed, by integration by parts

$$\frac{d}{dt} H = \int_a^b \frac{\delta H}{\delta \alpha} . \frac{\partial \alpha}{\partial t} dz = \int_a^b \frac{\delta H}{\delta \alpha} . - \frac{\partial}{\partial z}\frac{\delta H}{\delta \alpha} dz = \left(\beta^2(a) - \beta^2(b)\right) \tag{4.7}$$

Here it is worth to notice that the time variation of the Hamiltonian functional is a *quadratic function of the flux variables* evaluated at the boundaries of the spatial domain Z.

The second example, the *p-system*, is a classical example that we shall use in order to introduce the concept of an infinite-dimensional *port-Hamiltonian system*. It corresponds to the case of two physical domains in interaction and consists of a system of *two* conservations laws.

Example 4.2 (The p-system). The p-system is a model for a *1-dimensional isentropic gas dynamics in Lagrangian coordinates.* The independent variable z belong to an interval $Z \subset \mathbb{R}$, It is defined with the following variables: the specific volume $v(z,t) \in \mathbb{R}^+$, the velocity $u(z,t)$ and the pressure functional $p(v)$ (which is for instance in the case of a polytropic isentropic ideal gas given by $p(v) = A\, v^{-\gamma}$ where $\gamma \geq 1$). The *p-system* is then defined by the following system of partial differential equations:

$$\begin{aligned} \frac{\partial v}{\partial t} - \frac{\partial u}{\partial z} &= 0 \\ \frac{\partial u}{\partial t} + \frac{\partial\, p(v)}{\partial z} &= 0 \end{aligned} \tag{4.8}$$

representing the conservation of mass and of momentum. By defining the state vector $\alpha(z,t) = \begin{pmatrix} \alpha_1 \\ \alpha_2 \end{pmatrix} = \begin{pmatrix} v \\ u \end{pmatrix}$ and the vector valued flux $\beta(z,t) = \begin{pmatrix} \beta_1 \\ \beta_2 \end{pmatrix} = \begin{pmatrix} -u \\ p(v) \end{pmatrix}$ the p-system is rewritten as

$$\frac{\partial \alpha}{\partial t} + \frac{\partial}{\partial z}\beta = 0 \tag{4.9}$$

Again, according to the framework of Irreversible Thermodynamics, the flux variables may be written as functions of the variational derivatives of some generating functionals. Consider the functional $H(\alpha) = \int_a^b \mathcal{H}(v,u)dz$ where $\mathcal{H}(v,u)$ denotes the energy density, which is given as the sum of the internal energy and the kinetic energy densities

$$\mathcal{H}(v,u) = \mathcal{U}(v) + \frac{u^2}{2}, \tag{4.10}$$

where $-\mathcal{U}(v)$ is a primitive function of the pressure. Note that the expression of the kinetic energy does not depend on the mass density which is assumed to be constant and for simplicity is set equal to 1. Hence no difference is made between the velocity and the momentum. The vector of fluxes β may now be expressed in terms of the generating forces as follows

$$\beta = \begin{pmatrix} -\frac{\delta H}{\delta u} \\ -\frac{\delta H}{\delta v} \end{pmatrix} = \begin{pmatrix} 0 & -1 \\ -1 & 0 \end{pmatrix} \begin{pmatrix} \frac{\delta H}{\delta v} \\ \frac{\delta H}{\delta u} \end{pmatrix} \tag{4.11}$$

The anti-diagonal matrix represents the canonical coupling between two physical domains: the kinetic and the potential (internal) domain (for lumped parameter systems this is discussed e.g. in [4]). The variational derivative of the total energy with respect to the state variable of one domain generates the flux variable for the other domain.

Combining the equations (4.9) and (4.11), the p-system may thus be written as the following Hamiltonian system:

$$\frac{\partial \alpha}{\partial t} = \begin{pmatrix} 0 & -\frac{\partial}{\partial z} \\ -\frac{\partial}{\partial z} & 0 \end{pmatrix} \begin{pmatrix} \frac{\delta H}{\delta \alpha_1} \\ \frac{\delta H}{\delta \alpha_2} \end{pmatrix} \tag{4.12}$$

From the Hamiltonian form of the system and using again integration by parts, one may derive that the total energy obeys the following *power balance equation*:

$$\frac{d}{dt}H = \beta_1(a)\,\beta_2(a) - \beta_1(b)\,\beta_2(b) \tag{4.13}$$

Notice again that the right-hand side of this power-balance equation is a quadratic function of the fluxes at the boundary of the spatial domain.

Remark 4.1. It is important to note that any non-linear wave equation:

$$\frac{\partial^2 g}{\partial t^2} - \frac{\partial}{\partial z}\left(\sigma\left(\frac{\partial g}{\partial z}\right)\right) = 0$$

may be expressed as a p-system using the change of variables $u = \frac{\partial g}{\partial t}$, $v = \frac{\partial g}{\partial z}$ and $p(v) = -\sigma(v)$.

The last example is the *vibrating string.* Actually it is again a system of two conservation laws representing the canonical interdomain coupling between the kinetic energy and the elastic potential energy. However in this example, unlike the p-system, the *classical* choice of the state variables leads to express the total energy as a function of some of the *spatial derivatives* of the state variables. We shall analyze how the dynamic equations and the power balance are expressed in this case and we shall subsequently draw some conclusions on the choice of the state variables.

Example 4.3 (Vibrating string). Consider an elastic string subject to traction forces at its ends. The spatial variable z belongs to the interval $Z = [a, b] \subset \mathbb{R}$. Denote by $u(t, z)$ the displacement of the string and the velocity by $v(z, t) = \frac{\partial u}{\partial t}$. Using the vector of state variables $x(z, t) = (u, v)^T$, the dynamics of the vibrating string is described by the system of partial differential equations

$$\frac{\partial x}{\partial t} = \begin{pmatrix} v \\ \frac{1}{\mu}\frac{\partial}{\partial z}\left(T\frac{\partial u}{\partial z}\right) \end{pmatrix} \tag{4.14}$$

where the first equation is simply the definition of the velocity and the second one is Newton's second law.

The time variation of the state may be expressed as a function of the variational derivative of the total energy as in the preceeding examples. Indeed, define the total energy as $H(x) = U(u) + K(v)$, where U denotes the elastic potential energy and K the kinetic energy of the string. The elastic potential energy is given as a function of the *strain* $\epsilon(t, z) = \frac{\partial u}{\partial z}$

$$U(u) = \int_a^b \frac{1}{2} T \left(\frac{\partial u}{\partial z}\right)^2 dz \tag{4.15}$$

with T the elasticity modulus. The kinetic energy K is the following function of the velocity $v(z, t) = \frac{\partial u}{\partial t}$

$$K(v) = \int_a^b \frac{1}{2}\, \mu v(z, t)^2\, dz \tag{4.16}$$

Thus the total system (4.14) may be expressed as

$$\frac{\partial x}{\partial t} = \begin{pmatrix} 0 & \frac{1}{\mu} \\ -\frac{1}{\mu} & 0 \end{pmatrix} \begin{pmatrix} \frac{\delta H}{\delta u} \\ \frac{\delta H}{\delta v} \end{pmatrix} \tag{4.17}$$

where according to the definition of the variational derivative given in (4.4) one obtains

$$\frac{\delta H}{\delta u} = \frac{\delta U}{\delta u} = -\frac{\partial}{\partial z}\left(T\,\frac{\partial u}{\partial z}\right) \tag{4.18}$$

which is the elastic force and

$$\frac{\delta H}{\delta v} = \frac{\delta K}{\delta v} = \mu\, v \tag{4.19}$$

which is the momentum.

In the formulation of equation (4.17) there appears again an anti-diagonal skew-symmetric matrix which corresponds to the expression of a canonical interdomain coupling between the elastic energy domain and the kinetic energy domain. However the system is *not* expressed as a system of conservation laws since the rate of change of the state variables is a linear combination of the variational derivatives directly (and not of their spatial derivatives). Instead of being a simplification, this reveals a drawback for the case that there is energy flow through the boundary of the spatial domain. Indeed in this case, the *variational derivative has to be completed by a boundary term* since the Hamiltonian functional depends on the *spatial derivatives of the state.* For the elastic potential energy this becomes (integration by parts)

$$U(u+\epsilon\eta) = U(u) - \epsilon\int_a^b \frac{\partial}{\partial z}\left(T\,\frac{\partial u}{\partial z}\right)\eta\,dz + \epsilon\left[\eta\,\left(T\,\frac{\partial u}{\partial z}\right)\right]_a^b + O(\epsilon^2) \tag{4.20}$$

On the other hand, writing the system (4.14) as a second order equation yields the wave equation

$$\mu\frac{\partial^2 u}{\partial t^2} = \frac{\partial}{\partial z}\left(T\frac{\partial u}{\partial z}\right) \tag{4.21}$$

which according to Remark 4.1 may be alternatively expressed as a p-system.

In the sequel we shall formulate the vibrating string as a system of two conservation laws, which is however slightly different from the p-system formulated before. It differs from the p-system by the choice of the state variables in such a way that, first, the mass density may depend on the spatial variable z (which is not the case in the Hamiltonian density function defined in equation (4.10)), and secondly, that the variational derivatives of the total energy equal the co-energy variables.

Indeed, we take as vector of state variables

$$\alpha(z,t) = \begin{pmatrix} \epsilon \\ p \end{pmatrix} \tag{4.22}$$

where ϵ denotes the *strain* $\alpha_1 = \epsilon = \frac{\partial u}{\partial z}$ and p denotes the *momentum* $\alpha_2 = p = \mu v$. Recall that in these variables the total energy is written as

$$H_0 = \int_a^b \frac{1}{2} \left(T\, \alpha_1^2 + \frac{1}{\mu} \alpha_2^2 \right) dz \tag{4.23}$$

Notice that the energy functional now only depends on the state variables and *not* on their spatial derivatives. Furthermore, one may define the flux variables to be the *stress* $\beta_1 = \frac{\delta H_0}{\delta \alpha_1} = T\, \alpha_1$ and the *velocity* $\beta_2 = \frac{\delta H_0}{\delta \alpha_1} = \frac{\alpha_2}{\mu}$. In matrix notation, the fluxes are expressed as a function of the generating forces $\frac{\delta H_0}{\delta \alpha}$ by:

$$\beta = \begin{pmatrix} -\frac{\partial H_0}{\partial \epsilon} \\ -\frac{\partial H_0}{\partial p} \end{pmatrix} = \begin{pmatrix} 0 & -1 \\ -1 & 0 \end{pmatrix} \begin{pmatrix} \frac{\delta H_0}{\delta \alpha_1} \\ \frac{\delta H_0}{\delta \alpha_2} \end{pmatrix} = \begin{pmatrix} 0 & -1 \\ -1 & 0 \end{pmatrix} \frac{\delta H_0}{\delta \alpha} \tag{4.24}$$

Thus the model of the vibrating string may be expressed by the system of two conservation laws (as for the p-system):

$$\frac{\partial \alpha}{\partial t} = \begin{pmatrix} 0 & \frac{\partial}{\partial z} \\ \frac{\partial}{\partial z} & 0 \end{pmatrix} \frac{\delta H_0}{\delta \alpha} \tag{4.25}$$

which satisfies also the power balance equation (4.13).

4.2.2 Systems of Two Conservation Laws in Canonical Interaction

In this section we shall consider the general class of distributed-parameter systems consisting of two conservation laws with the canonical coupling presented as in the above examples of the p-system and the vibrating string. In the first part, for 1-dimensional spatial domains, we shall introduce the concept of *interconnection structure* and *port variables* which are fundamental to the definition of port-Hamiltonian systems. On this case we shall also introduce the notion of differential forms. In the second part we shall give the definition of systems of two conservation laws defined on $n-$*dimensional* spatial domains. We do not use the usual vector calculus formulation but express the systems in terms of differential forms [1] [16]. This leads to concise, coordinate independent formulations and unifies the notations for the various physical domains.

Interconnection Structure, Boundary Energy Flows and Port-Based Formulation for 1-D Spatial Domains

Interconnection Structure and Power Continuity

Let us consider the systems of two conservation laws arising from the modelling of two physical domains in canonical interaction as have been presented for the vibrating string and the p-system:

$$\frac{\partial \alpha}{\partial t} = \begin{pmatrix} 0 & \frac{\partial}{\partial z} \\ \frac{\partial}{\partial z} & 0 \end{pmatrix} \frac{\delta H_0}{\delta \alpha} \tag{4.26}$$

where $\alpha = (\alpha_1(z,t), \alpha_2(z,t))^T$. Let us now define an interconnection structure for this system in the sense of network [13] [4] or port-based modelling [23] [35]. Define the vector of *flow variables* to be the time variation of the state and denote it by:

$$f = \frac{\partial \alpha}{\partial t} \tag{4.27}$$

Define the vector of *effort variables* e to be the vector of the generating forces given as

$$e = \frac{\delta H_0}{\delta \alpha} \tag{4.28}$$

The flow and effort variables are *power-conjugated* since their product is the time-variation of the total energy:

$$\frac{d}{dt} H_0 = \int_a^b \left(\frac{\delta \mathcal{H}_0}{\delta \alpha_1} \frac{\partial \alpha_1}{\partial t} + \frac{\delta \mathcal{H}_0}{\delta \alpha_2} \frac{\partial \alpha_2}{\partial t} \right) dz = \int_a^b (e_1\, f_1 + e_2\, f_2)\, dz \tag{4.29}$$

where $\mathcal{H}_0$ denotes the density corresponding toH_0. Considering the right-hand side of the power balance equation (4.13) it is clear that the energy exchange of the system with its environment is determined by the flux variables restricted to the boundary of the domain. Therefore let us define two external boundary variables as follows:

$$\begin{pmatrix} f_\partial \\ e_\partial \end{pmatrix} = \begin{pmatrix} e_2 \\ e_1 \end{pmatrix} = \begin{pmatrix} \frac{\delta H_0}{\delta \alpha_2} \\ \frac{\delta H_0}{\delta \alpha_1} \end{pmatrix} = \begin{pmatrix} v \\ \sigma \end{pmatrix} \tag{4.30}$$

These boundary variables are also power-conjugated as their product $\beta_1\, \beta_2 = e_b\, f_b = \sigma\, v$ equals the right-hand side of the power balance equation (4.13). Considering the four power-conjugated variables $f_1, f_2, f_\partial, e_1, e_2, e_\partial$, the power balance equation (4.13) implies that their product is zero:

$$\int_a^b (e_1\, f_1 + e_2\, f_2)\, dz + e_\partial(b)\, f_\partial(b) - e_\partial(a)\, f_\partial(a) = 0 \tag{4.31}$$

This bilinear product between the power-conjugated variables is analogous to the product between the circuit variables expressing the *power continuity* relation in circuits and network models [13] [4]. Such products (or *pairings* are also central in the definition of implicit Hamiltonian systems [5] [7] and port-Hamiltonian systems in finite dimensions [35] [19]. In the forthcoming sections we shall show that this product will play the same role for infinite-dimensional port-Hamiltonian systems [20] [37].

The *interconnection structure* underlying the system (4.26) (analogous to Kirchhoff's laws for circuits) may now be summarized by (4.30) together with

$$f = \begin{pmatrix} 0 & \frac{\partial}{\partial z} \\ \frac{\partial}{\partial z} & 0 \end{pmatrix} e \tag{4.32}$$

Introduction to Differential Forms

Let us now introduce for the case of the 1−dimensional spatial domain the use of differential forms in the formulation of systems of conservation laws. Until now we have simply considered the state variables α and the flux variables β as functions on the space-time domain $Z \times I$. However considering the balance equation (4.3)

$$\frac{d}{dt}\int_a^b \alpha dz = \beta(a) - \beta(b)$$

associated with the conservation law (4.2) it becomes clear that they are of different nature. The state variables α correspond to conserved quantities through integration, while the flux variables β correspond to functions which can be evaluated at any point (for instance at the boundary points of the spatial domain). This distinction may be expressed by representing the variables as *differential forms*. For the case of one-dimensional spatial domains considered in this paragraph, the state variables are identified with differential forms of degree 1, which can be integrated along one-dimensional curves. The flux variables, on the other hand, are identified with differential forms of degree 0, that means functions evaluated at points of the spatial domain. The reader is referred to the following textbooks [1] [12] [16] for an exhaustive definition of differential forms that we shall use systematically in the rest of the paper.

Interconnection Structure, Boundary Energy Flows and Port-Based Formulation for N-Dimensional Spatial Domains

Systems of Two Conservation Laws with Canonical Interdomain Coupling

In this paragraph we shall give the general definition of the class of systems of conservation laws that we shall consider in the forthcoming sections. We first recall the expression of systems of conservation laws defined on n-dimensional spatial domains, and secondly generalize the systems of two conservation laws with canonical interdomain coupling as defined in the previous section 4.2.2 to the n-dimensional spatial domain.

Define the spatial domain of the considered distributed-parameter system as $Z \in \mathbb{R}^n$ being an n-dimensional smooth manifold with smooth $(n-1)$-dimensional boundary ∂Z. Denote by $\Omega^k(Z)$ the vector space of (differential) k-forms on Z (respectively by $\Omega^k(\partial Z)$the vector space of k-forms on ∂Z). Denote furthermore $\Omega = \bigoplus_{k \geq 0} \Omega^k(Z)$ the algebra of differential forms over Z and recall that it is endowed with an exterior product $\wedge$ and an exterior derivation d [1] [16].

Definition 4.1. *A* system of conservation laws *is defined by a set of* conserved quantities $\alpha_i \in \Omega^{k_i}(Z)$, $i \in \{1, \dots, N\}$ *where* $N \in \mathbb{N}, k_i \in \mathbb{N}$, *defining* *the state space* $\mathcal{X} = \bigotimes_{i=1,..,N} \Omega^{k_i}(Z)$. *They satisfy a set of* conservation laws

$$\frac{\partial \alpha_i}{\partial t} + d\beta_i = g_i \tag{4.33}$$

where $\beta_i \in \Omega^{k_i - 1}(Z)$ *denote the set of* fluxes *and* $g_i \in \Omega^{k_i}(Z)$ *denote the set of* distributed interaction forms. *Finally, the fluxes* β_i *are defined by the* closure equations

$$\beta_i = J(\alpha_i, z) \ , \ i = 1, .., N \tag{4.34}$$

The integral form of the conservation laws yield the following *balance equations*

$$\frac{d}{dt}\int_Z \alpha_i + \int_{\partial Z} \beta_i = \int_Z g_i \tag{4.35}$$

Remark 4.2. A common case is that the conserved quantities are 3-forms, that is, the balance equation is evaluated on volumes of the 3-dimensional space. Then, in vector calculus notation, the conserved quantities may be identified with vectors u_i on Z, the interaction terms g_i may also be considered as vectors, and the fluxes may be identified with vectors q_i. In this case the system of conservation laws takes the more familiar form:

$$\frac{\partial u_i}{\partial t}(z,t) + \mathrm{div}_z\, q_i = g_i \ , \ i = 1, .., n \tag{4.36}$$

However, systems of conservation laws may correspond to differential forms of any degree. Maxwell's equations provide a classical example where the conserved quantities are actually differential forms of degree 3 [12].

In the sequel, as in the case of the 1-dimensional spatial domain, we shall consider a particular class of systems of conservation laws where the fluxes, determined by the closure equations, are (linear) functions of the derivatives of some generating function. One may note again that this is in agreement with the general assumptions of irreversible thermodynamics [29] where the flux variables are (eventually nonlinear) functions of the generating forces, being the derivative of some generating functional. More precisely, we shall consider closure equations arising from the description of the canonical interaction of two physical domains (for instance the kinetic and elastic energy in the case of the vibrating string, or the electric and magnetic energy for electromagnetic fields) [20].

First recall the general definition of the *variational derivative* of a functional $H(\alpha)$ with respect to the differential form $\alpha \in \Omega^p(Z)$ (generalizing the definition given before).

Definition 4.2. *Consider a density function* $\mathcal{H} : \Omega^p(Z) \times Z \to \Omega^n(Z)$ *where* $p \in \{1, .., n\}$, *and denote by* $H := \int_Z \mathcal{H} \in \mathbb{R}$ *the associated functional. Then* *the uniquely defined differential form* $\frac{\delta H}{\delta \alpha} \in \Omega^{n-p}(Z)$ *which satisfies for all* $\Delta\alpha \in \Omega^p(Z)$ *and* $\varepsilon \in \mathbb{R}$ *:*

$$H(\alpha + \varepsilon \Delta\alpha) = \int_Z \mathcal{H}(\alpha + \varepsilon \Delta\alpha) = \int_Z \mathcal{H}(\alpha) + \varepsilon \int_Z \left[\frac{\delta H}{\delta \alpha} \wedge \Delta\alpha\right] + O\left(\varepsilon^2\right)$$

is called the variational derivative *of* H *with respect to* $\alpha \in \Omega^p(Z)$.

Now we define the generalization of the systems presented in the section 4.2.2 to spatial domains of arbitrary dimension.

Definition 4.3. Systems of two conservation laws with canonical interdomain coupling *are systems of two conservation laws involving a pair of conserved quantities* $\alpha_p \in \Omega^p(Z)$ *and* $\alpha_q \in \Omega^q(Z)$*, differential forms on the* n*-dimensional spatial domain* Z *of degree* p *and* q *respectively, where the integers* p *and* q *satisfy* $p + q = n + 1$*. The closure equations generated by a Hamiltonian density function* $\mathcal{H} : \Omega^p(Z) \times \Omega^q(Z) \times Z \to \Omega^n(Z)$ *resulting in the total Hamiltonian* $H := \int_Z \mathcal{H} \in \mathbb{R}$ *are given by:*

$$\begin{pmatrix} \beta_p \\ \beta_q \end{pmatrix} = \varepsilon \begin{pmatrix} 0 & (-1)^r \\ 1 & 0 \end{pmatrix} \begin{pmatrix} \frac{\delta H}{\delta \alpha_p} \\ \frac{\delta H}{\delta \alpha_q} \end{pmatrix} \tag{4.37}$$

where $r = p\,q + 1$, $\varepsilon \in \{-1, +1\}$ *depending on the sign convention of the considered physical domain.*

Remark 4.3. The total Hamiltonian $H(\alpha_q, \alpha_p)$ corresponds to the energy function of a physical system, the state variables α_i are called the *energy variables* and the variational derivatives $\frac{\delta H}{\delta \alpha_i}$ are called the *co-energy variables.*

Boundary Port Variables and the Power Continuity Relation

In the same way as for systems defined on 1-dimensional spatial domains, one may define for $n-$ spatial domains pairs of power conjugated variables. Define the *flow variables* to be the time-variation of the state denoted by

$$\begin{pmatrix} f_p \\ f_q \end{pmatrix} = \begin{pmatrix} \frac{\partial \alpha_p}{\partial t} \\ \frac{\partial \alpha_q}{\partial t} \end{pmatrix} \tag{4.38}$$

Furthermore, define the vector of *effort variables* to be the vector of the generating forces denoted by

$$\begin{pmatrix} e_p \\ e_q \end{pmatrix} = \begin{pmatrix} \frac{\delta H}{\delta \alpha_p} \\ \frac{\delta H}{\delta \alpha_q} \end{pmatrix} \tag{4.39}$$

The flow and effort variables are *power-conjugated* as their product is the time-variation of the Hamiltonian function:

$$\frac{dH}{dt} = \int_Z \left(\frac{\delta H}{\delta \alpha_p} \wedge \frac{\partial \alpha_p}{\partial t} + \frac{\delta H}{\delta \alpha_q} \wedge \frac{\partial \alpha_q}{\partial t} \right) = \int_Z (e_p \wedge f_p + e_q \wedge f_q) \tag{4.40}$$

Using the conservation laws (4.36), the closure relations (4.37) and the properties of the exterior derivative and Stokes' theorem, one may write the time-variation of the Hamiltonian as

$$\begin{aligned} \frac{dH}{dt} &= \int_Z (\varepsilon \beta_q \wedge (-d\beta_p) + (-1)^r \beta_p \wedge \varepsilon(-d\beta_q)) \\ &= -\varepsilon \int_Z \left(\beta_q \wedge d\beta_p + (-1)^{p\,q+1} (-1)^{(p-1)\,q} \beta_q \wedge d\beta_p \right) \\ &= -\varepsilon \int_Z (\beta_q \wedge d\beta_p + (-1)^q \beta_q \wedge d\beta_p) \\ &= -\varepsilon \int_{\partial Z} \beta_q \wedge \beta_p \end{aligned} \tag{4.41}$$

Finally we define *flow and effort variables on the boundary* of the system as the *restriction* of the flux variables to the boundary ∂Z of the domain Z:

$$\begin{pmatrix} f_\partial \\ e_\partial \end{pmatrix} = \begin{pmatrix} \beta_q|_{\partial Z} \\ \beta_p|_{\partial Z} \end{pmatrix} \tag{4.42}$$

They are also power conjugated variables as their product defined in (4.42) is the time variation of the Hamiltonian functional (the total energy of the physical system).

On the total space of power-conjugated variables, the differential forms (f_p, e_p) and (f_q, e_q) on the domain Z and the differential forms (f_∂, e_∂) defined on the boundary ∂Z, one may define an *interconnection structure*, underlying the system of two conservation laws with canonical interdomain coupling of Definition 4.3. This interconnection structure is defined by the equation (4.42) together with (combining the conservation laws (4.36) with the closure equation (4.37))

$$\begin{pmatrix} f_q \\ f_p \end{pmatrix} = \varepsilon \begin{pmatrix} 0 & (-1)^r \, d \\ d & 0 \end{pmatrix} \begin{pmatrix} e_q \\ e_p \end{pmatrix} \tag{4.43}$$

This interconnection is power-continuous in the sense that the power-conjugated variables related by (4.42) and (4.43) satisfy the *power continuity relation:*

$$\int_Z (e_p \wedge f_p + e_q \wedge f_q) + \varepsilon \int_{\partial Z} f_\partial \wedge e_\partial = 0 \tag{4.44}$$

This expression is the straightforward consequence of the two expressions of the variation of the Hamiltonian H in (4.40) (4.41).

In the next sections 4.3 and 4.4 we shall show how the above power-continuous interconnection structure can be formalized as a geometric structure, called Dirac structure, and how this leads to the definition of infinite-dimensional Hamiltonian systems with energy flows at the boundary of their spatial domain, called port-Hamiltonian systems.

4.3 Stokes-Dirac Structures

4.3.1 Dirac Structures

The notion of a Dirac structure was originally introduced in [5], [7] as a geometric structure generalizing both *symplectic* and *Poisson* structures. In e.g. [35], [19], [33], [2], [6], [34], [3], it was employed as the geometrical notion formalizing general *power-conserving interconnections*, thereby allowing the Hamiltonian formulation of interconnected and constrained mechanical and electrical systems.

A definition of Dirac structures (which is actually slightly more general than the one in [5], [7]) can be given as follows. Let $\mathcal{F}$ and $\mathcal{E}$ be linear spaces, equipped with a pairing, that is, a bilinear operation

$$\mathcal{F} \times \mathcal{E} \to L \tag{4.45}$$

with L a linear space. The pairing will be denoted by $< e|f > \in L,\ f \in \mathcal{F},\ e \in \mathcal{E}$.
By symmetrizing the pairing we obtain a symmetric bilinear form $\ll, \gg$ on $\mathcal{F} \times \mathcal{E}$, with values in L, defined as

$$\ll (f_1, e_1), (f_2, e_2) \gg := < e_1|f_2 > + < e_2|f_1 >, \quad (f_i, e_i) \in \mathcal{F} \times \mathcal{E} \tag{4.46}$$

Definition 4.4. *Let $\mathcal{F}$ and $\mathcal{E}$ be linear spaces with a pairing $< | >$. A Dirac structure is a linear subspace $D \subset \mathcal{F} \times \mathcal{E}$ such that $D = D^{\perp}$, with $\perp$ denoting the orthogonal complement with respect to the bilinear form $\ll, \gg$.*

Example 4.4. Let $\mathcal{F}$ be a linear space over $\mathbb{R}$. Let $\mathcal{E}$ be given as $\mathcal{F}^*$ (the space of linear functionals on $\mathcal{F}$), with pairing $< | >$ the duality product $< e|f > \in \mathbb{R}$.

(a) Let $J : \mathcal{E} \to \mathcal{F}$ be a skew-symmetric map. Then graph $J \subset \mathcal{F} \times \mathcal{E}$ is a Dirac structure.

(b) Let $\omega : \mathcal{F} \to \mathcal{E}$ be a skew-symmetric map. Then graph $\omega \subset \mathcal{F} \times \mathcal{E}$ is a Dirac structure.

(c) Let $V \subset \mathcal{F}$ be a finite-dimensional linear subspace. Then $V \times V^{orth} \subset \mathcal{F} \times \mathcal{E}$ is a Dirac structure, where $V^{orth} \subset \mathcal{E}$ is the annihilating subspace of V. The same holds if $\mathcal{F}$ is a topological vectorspace, $\mathcal{E}$ is the space of linear continuous functionals on $\mathcal{F}$, and V is a *closed* subspace of $\mathcal{F}$.

Example 4.5. Let M be a finite-dimensional manifold. Let $\mathcal{F} = V(M)$ denote the Lie algebra of smooth vector fields on M, and let $\mathcal{E} = \Omega^1(M)$ be the linear space of smooth 1-forms on M. Consider the usual pairing $< \alpha|X > = i_X \alpha$ between 1-forms α and vectorfields X; implying that L is the linear space of smooth functions on M.

(a) Let J be a Poisson structure on M, defining a skew-symmetric mapping $J : \Omega^1(M) \to V(M)$. Then graph $J \subset V(M) \times \Omega^1(M)$ is a Dirac structure.

(b) Let ω be a (pre-)symplectic structure on M, defining a skew-symmetric mapping $\omega : V(M) \to \Omega^1(M)$. Then graph $\omega \subset V(M) \times \Omega^1(M)$ is a Dirac structure.

(c) Let V be a constant-dimensional distribution on M, and let $annV$ be its annihilating co-distribution. Then $V \times annV$ is a Dirac structure.

Remark 4.4. Usually in Example 4.5 an additional integrability condition is imposed on the Dirac structure, cf. [5], [7]. In part (a) this condition is equivalent to the *Jacobi-identity* for the Poisson structure; in part (b) it is equivalent to the *closedness* of the presymplectic structure, while in part (c) it is equivalent to the *involutivity* of the distribution D. Integrability is equivalent to the existence of canonical coordinates, cf. [5], [7], [6]. Various formulations of integrability of Dirac structures and their implications have been worked out in [6]. For the developments of the current paper the notion of integrability is not crucial; see however the comment in the Conclusions.

From the defining property $D = D^\perp$ of a Dirac structure it directly follows that for any $(f, e) \in D$

$$0 = \ll (f, e), (f, e) \gg = 2 < e|f > \tag{4.47}$$

Thus if (f, e) is a pair of *power variables*(e.g., currents and voltages in an electric circuit context, or forces and velocities in a mechanical context), then the condition $(f, e) \in D$ implies *power-conservation* $< e|f >= 0$ (as do Kirchhoff's laws or Newton's third law). This is the starting point for the geometric formulation of general power-conserving interconnections in physical systems by Dirac structures as alluded to above.

4.3.2 Stokes-Dirac Structures

In this subsection we treat the underlying geometric framework for the Hamiltonian formulation of distributed-parameter systems on a bounded spatial domain, with non-zero energy flow through the boundary. The key concept is the introduction of a special type of Dirac structure on suitable spaces of differential forms on the spatial domain and its boundary, making use of Stokes' theorem. A preliminary treatment of this Dirac structure has been given in [20], [22].

Throughout, let Z be an n-dimensional smooth manifold with smooth $(n-1)$-dimensional boundary ∂Z, representing the space of *spatial variables.*

Denote by $\Omega^k(Z)$, $k = 0, 1, \cdots, n$, the space of exterior k-forms on Z, and by $\Omega^k(\partial Z)$, $k = 0, 1, \cdots, n-1$, the space of k-forms on ∂Z. (Note that $\Omega^0(Z)$, respectively $\Omega^0(\partial Z)$, is the space of smooth functions on Z, respectively ∂Z.) Clearly, $\Omega^k(Z)$ and $\Omega^k(\partial Z)$ are (infinite-dimensional) linear spaces (over $\mathbb{R}$). Furthermore, there is a natural pairing between $\Omega^k(Z)$ and $\Omega^{n-k}(Z)$ given by

$$< \beta|\alpha >:= \int_Z \beta \wedge \alpha \quad (\in \mathbb{R}) \tag{4.48}$$

with $\alpha \in \Omega^k(Z), \beta \in \Omega^{n-k}(Z)$, where $\wedge$ is the usual wedge product of differential forms yielding the n-form $\beta \wedge \alpha$. In fact, the pairing (4.48) is *non-degenerate* in the sense that if $< \beta|\alpha >= 0$ for all α, respectively for all β, then $\beta = 0$, respectively $\alpha = 0$.

Similarly, there is a pairing between $\Omega^k(\partial Z)$ and $\Omega^{n-1-k}(\partial Z)$ given by

$$< \beta|\alpha >:= \int_{\partial Z} \beta \wedge \alpha \tag{4.49}$$

with $\alpha \in \Omega^k(\partial Z), \beta \in \Omega^{n-1-k}(\partial Z)$. Now let us define the linear space

$$\mathcal{F}_{p,q} := \Omega^p(Z) \times \Omega^q(Z) \times \Omega^{n-p}(\partial Z), \tag{4.50}$$

for any pair p, q of positive integers satisfying

$$p + q = n + 1, \tag{4.51}$$

and correspondingly let us define

$$\mathcal{E}_{p,q} := \Omega^{n-p}(Z) \times \Omega^{n-q}(Z) \times \Omega^{n-q}(\partial Z). \tag{4.52}$$

Then the pairing (4.48) and (4.49) yields a (non-degenerate) pairing between $\mathcal{F}_{p,q}$ and $\mathcal{E}_{p,q}$ (note that by (4.51) $(n-p) + (n-q) = n - 1$). As before (see (4.46)), symmetrization of this pairing yields the following bilinear form on $\mathcal{F}_{p,q} \times \mathcal{E}_{p,q}$ with values in $\mathbb{R}$:

$$\begin{aligned} &\ll \left(f_p^1, f_q^1, f_b^1, e_p^1, e_q^1, e_b^1\right), \left(f_p^2, f_q^2, f_b^2, e_p^2, e_q^2, e_b^2\right) \gg := \\ &\int_Z \left[e_p^1 \wedge f_p^2 + e_q^1 \wedge f_q^2 + e_p^2 \wedge f_p^1 + e_q^2 \wedge f_q^1\right] + \int_{\partial Z} \left[e_b^1 \wedge f_b^2 + e_b^2 \wedge f_b^1\right] \end{aligned} \tag{4.53}$$

where for $i = 1, 2$

$$\begin{aligned} &f_p^i \in \Omega^p(Z), \; f_q^i \in \Omega^q(Z) \\ &e_p^i \in \Omega^{n-p}(Z), \; e_p^i \in \Omega^{n-q}(Z) \\ &f_b^i \in \Omega^{n-p}(\partial Z), \; e_b^i \in \Omega^{n-q}(\partial Z) \end{aligned} \tag{4.54}$$

The spaces of differential forms $\Omega^p(Z)$ and $\Omega^q(Z)$ will represent the energy variables of two different physical energy domains interacting with each other, while $\Omega^{n-p}(\partial Z)$ and $\Omega^{n-q}(\partial Z)$ will denote the boundary variables whose (wedge) product represents the boundary energy flow. For example, in

Maxwell's equations (Example 3.1) we will have $n = 3$ and $p = q = 2$; with $\Omega^p(Z) = \Omega^2(Z)$, respectively $\Omega^q(Z) = \Omega^2(Z)$, being the space of electric field inductions, respectively magnetic field inductions, and $\Omega^{n-p}(\partial Z) = \Omega^1(\partial Z)$ denoting the electric and magnetic field intensities at the boundary, with product the Poynting vector.

Theorem 4.1. *Consider $\mathcal{F}_{p,q}$ and $\mathcal{E}_{p,q}$ given in (4.50), (4.52) with p, q satisfying (4.51), and bilinear form $\ll, \gg$ given by (4.53). Define the following linear subspace D of $\mathcal{F}_{p,q} \times \mathcal{E}_{p,q}$*

$$D = \{(f_p, f_q, f_b, e_p, e_q, e_b) \in \mathcal{F}_{p,q} \times \mathcal{E}_{p,q} |$$

$$\begin{bmatrix} f_p \\ f_q \end{bmatrix} = \begin{bmatrix} 0 & (-1)^r d \\ d & 0 \end{bmatrix} \begin{bmatrix} e_p \\ e_q \end{bmatrix},$$

$$\begin{bmatrix} f_b \\ e_b \end{bmatrix} = \begin{bmatrix} 1 & 0 \\ 0 & -(-1)^{n-q} \end{bmatrix} \begin{bmatrix} e_{p|\partial Z} \\ e_{q|\partial Z} \end{bmatrix} \} \tag{4.55}$$

where $|_{\partial Z}$ denotes restriction to the boundary ∂Z, and $r := pq + 1$. Then $D = D^\perp$, that is, D is a Dirac structure.

For the proof of this theorem we refer to [37].

Remark 4.5. The spatial *compositionality* properties of the Stokes-Dirac structure immediately follow from its definition. Indeed, let Z_1, Z_2 be two n-dimensional manifolds with boundaries $\partial Z_1, \partial Z_2$, such that

$$\begin{aligned} \partial Z_1 &= \Gamma \cup \Gamma_1, \quad \Gamma \cap \Gamma_1 = \phi \\ \partial Z_2 &= \Gamma \cup \Gamma_2, \quad \Gamma \cap \Gamma_2 = \phi \end{aligned} \tag{4.56}$$

for certain $(n-1)$-dimensional manifolds $\Gamma, \Gamma_1, \Gamma_2$ (that is, Z_1 and Z_2 have boundary Γ in common). Then the Stokes-Dirac structures D_1, D_2 on Z_1, respectively Z_2, compose to the Stokes-Dirac structure on the manifold $Z_1 \cup Z_2$ with boundary $\Gamma_1 \cup \Gamma_2$ if we equate on Γ the boundary variables f_b^1 (corresponding to D_1) with $-f_b^2$ (corresponding to D_2). (Note that a minus sign is inserted in order to ensure that the power flowing *into* Z_1 via Γ is equal to the power flowing *out* of Z_2 via Γ.)

4.3.3 Poisson Brackets Associated to Stokes-Dirac Structures

Although Dirac structures strictly generalize Poisson structures we can associate a (pseudo-)Poisson structure to any Dirac structure, as defined in Section 2.1. Indeed, let $D \subset \mathcal{F} \times \mathcal{E}$ be a Dirac structure as given in Definition

4.4. Then we can define a skew-symmetric bilinear form on a subspace of $\mathcal{E}$; basically following [5], [7]. First, define the space of 'admissible efforts'

$$\mathcal{E}_{\text{adm}} = \{e \in \mathcal{E} | \exists f \in \mathcal{F} \text{ such that } (f,e) \in D\} \tag{4.57}$$

Then we define on $\mathcal{E}_{\text{adm}}$ the bilinear form

$$[e_1, e_2] := < e_1 | f_2 > \in L \tag{4.58}$$

where $f_2 \in \mathcal{F}$ is such that $(f_2, e_2) \in D$. This bilinear form is well-defined, since for any other $f_2' \in \mathcal{F}$ such that $(f_2', e_2) \in D$ we obtain by linearity $(f_2 - f_2', 0) \in D$, and hence

$$0 = \ll (f_1, e_1), (f_2 - f_2', 0) \gg = < e_1 | f_2 > - < e_1 | f_2' > \tag{4.59}$$

Furthermore, $[\ ,\ ]$ is *skew-symmetric* since for any $(f_1, e_1), (f_2, e_2) \in D$

$$0 = \ll (f_1, e_1), (f_2, e_2) \gg = < e_1 | f_2 > + < e_2 | f_1 > \tag{4.60}$$

Now, let us define on $\mathcal{F}$ the set of *admissible mappings*

$$\begin{aligned} K_{\text{adm}} = \{k : \mathcal{F} \to L | \forall a \in \mathcal{F}\ \exists e(k,a) \in \mathcal{E}_{\text{adm}} \\ \text{such that for all } \partial a \in \mathcal{F} \\ k(a + \partial a) = k(a) + < e(k,a) | \partial a > + O(\partial a)\} \end{aligned} \tag{4.61}$$

Note that $e(k,a)$ (if it exists) is uniquely defined modulo the following linear subspace of $\mathcal{E}$

$$\mathcal{E}_0 = \{e \in \mathcal{E} | < e | f > = 0 \quad \text{for all } f \in \mathcal{F}\} \tag{4.62}$$

We call $e(k,a)$ (in fact, its equivalence class) the *derivative* of k at a, and we denote it by $\delta k(a)$. We define on K_{adm} the following bracket

$$\{k_1, k_2\}_D(a) := [\delta k_1(a), \delta k_2(a)], \quad k_1, k_2 \in K_{\text{adm}} \tag{4.63}$$

which is clearly independent from the choice of the representants $\delta k_1(a), \delta k_2(a)$. By skew-symmetry of $[\ ,\ ]$ it immediately follows that also $\{,\}$ is *skew-symmetric*. The Jacobi-identity for $\{,\}_D$, however, is not automatically satisfied, and we call therefore $\{,\}_D$ a *pseudo-Poisson bracket.*

For the Stokes-Dirac structure D of Theorem 4.1, given in equation (4.55), the bracket takes the following form. The set of admissible functions K_{adm} consists of those functions

$$k : \Omega^p(z) \times \Omega^q(z) \times \Omega^{n-p}(\partial z) \to \mathbb{R} \tag{4.64}$$

whose derivatives

$$\delta k(a) = (\delta_p k(a),\ \delta_q k(a),\ \delta_b k(a)) \in \Omega^{n-p}(Z) \times \Omega^{n-q}(Z) \times \Omega^{n-q}(\partial Z) \quad (4.65)$$

satisfy (cf. the last line of (4.55))

$$\delta_b k(a) = -(-1)^{n-q} \delta_q k(a)|_{\partial Z} \quad (4.66)$$

Furthermore, the bracket on K_{adm} is given as (leaving out the arguments a)

$$\{k^1, k^2\}_D = \int_Z \left[\delta_p k^1 \wedge (-1)^r d(\delta_q k^2) + (\delta_q k^1) \wedge d(\delta_p k^2)\right] - \int_{\partial Z} (-1)^{n-q} (\delta_q k^1) \wedge (\delta_p k^2) \quad (4.67)$$

It follows from the general considerations above that this bracket is skew-symmetric. (This can be also directly checked using Stokes' theorem.) Furthermore, in this case it is straightforward to check that $\{,\}_D$ also satisfies the Jacobi-identity

$$\{\{k^1, k^2\}_D, k^3\}_D + \{\{k^2, k^3\}_D, k^1\}\} + \{k^3, k^1\}_D, k^2\}_D = 0 \quad (4.68)$$

for all $k^i \in K_{\text{adm}}$.

4.4 Hamiltonian Formulation of Distributed-Parameter Systems with Boundary Energy Flow

4.4.1 Boundary Port-Hamiltonian Systems

The definition of a distributed-parameter Hamiltonian system with respect to a Stokes-Dirac structure can now be stated as follows. Let Z be an n-dimensional manifold with boundary ∂Z, and let D be a Stokes-Dirac structure as in Subsection 2.2. Consider furthermore a *Hamiltonian density* (energy per volume element)

$$\mathcal{H} : \Omega^p(Z) \times \Omega^q(Z) \times Z \to \Omega^n(Z) \quad (4.69)$$

resulting in the total energy

$$H := \int_Z \mathcal{H} \in \mathbb{R} \quad (4.70)$$

Let $\alpha_p, \partial\alpha_p \in \Omega^p(Z)$, $\alpha_q, \partial\alpha_q \in \Omega^q(Z)$. Then (with $z \in Z$)

$$
\begin{aligned}
H(\alpha_p + \partial\alpha_p, \alpha_q + \partial\alpha_q) &= \\
\int_Z \mathcal{H}(\alpha_p + \partial\alpha_p, \alpha_q + \partial\alpha_q, z) &= \\
\int_Z \mathcal{H}(\alpha_p, \alpha_q, z) + \int_Z [\delta_p H \wedge \partial\alpha_p + \delta_q H \wedge \partial\alpha_q]& \\
+ \text{ higher order terms in } \partial\alpha_p, \partial\alpha_q& \qquad (4.71)
\end{aligned}
$$

for certain uniquely defined differential forms

$$
\begin{aligned}
\delta_p H &\in \Omega^{n-p}(Z) \\
\delta_q H &\in \Omega^{n-q}(Z)
\end{aligned}
\qquad (4.72)
$$

This means that $(\delta_p H, \delta_q H) \in \Omega^{n-p}(Z) \times \Omega^{n-q}(Z)$ can be regarded as the *variational derivative* of H at $(\alpha_p, \alpha_q) \in \Omega^p(Z) \times \Omega^q(Z)$.

Now consider a time-function

$$
(\alpha_p(t), \alpha_q(t)) \in \Omega^p(Z) \times \Omega^q(Z), \quad t \in \mathbb{R}, \qquad (4.73)
$$

and the Hamiltonian $H(\alpha_p(t), \alpha_q(t))$ evaluated along this trajectory. It follows that at any time t

$$
\frac{dH}{dt} = \int_Z \left[\delta_p H \wedge \frac{\partial \alpha_p}{\partial t} + \delta_q H \wedge \frac{\partial \alpha_q}{\partial t}\right] \qquad (4.74)
$$

The differential forms $\frac{\partial \alpha_p}{\partial t}, \frac{\partial \alpha_q}{\partial t}$ represent the generalized velocities of the energy variables α_p, α_q. They are connected to the Stokes-Dirac structure D by setting

$$
\begin{aligned}
f_p &= -\tfrac{\partial \alpha_p}{\partial t} \\
f_q &= -\tfrac{\partial \alpha_q}{\partial t}
\end{aligned}
\qquad (4.75)
$$

(again the minus sign is included to have a consistent energy flow description). Since the right-hand side of (4.74) is the rate of increase of the stored energy H, we set

$$
\begin{aligned}
e_p &= \delta_p H \\
e_q &= \delta_q H
\end{aligned}
\qquad (4.76)
$$

Now we come to the general Hamiltonian description of a distributed-parameter system with boundary energy flow. In order to emphasize that the boundary variables are regarded as *interconnection variables*, which can be interconnected to other systems and whose product represents *power*, we call these models *port*-Hamiltonian systems. (This terminology comes from network modelling, see e.g. [23], [35], [34].)

Definition 4.5. *The boundary port-Hamiltonian system with n-dimensional manifold of spatial variables Z, state space $\Omega^p(Z) \times \Omega^q(Z)$ (with $p+q = n+1$), Stokes-Dirac structure D given by (4.55), and Hamiltonian H, is given as (with $r = pq + 1$)*

$$\begin{bmatrix} -\frac{\partial \alpha_p}{\partial t} \\ -\frac{\partial \alpha_q}{\partial t} \end{bmatrix} = \begin{bmatrix} 0 & (-1)^r d \\ d & 0 \end{bmatrix} \begin{bmatrix} \delta_p H \\ \delta_q H \end{bmatrix}$$
$$\begin{bmatrix} f_b \\ e_b \end{bmatrix} = \begin{bmatrix} 1 & 0 \\ 0 & -(-1)^{n-q} \end{bmatrix} \begin{bmatrix} \delta_p H|_{\partial Z} \\ \delta_q H|_{\partial Z} \end{bmatrix} \tag{4.77}$$

By the power-conserving property (4.47) of any Dirac structure it immediately follows that for any $(f_p, f_q, f_b, e_p, e_q, e_b)$ in the Stokes-Dirac structure D

$$\int_Z [e_p \wedge f_p + e_q \wedge f_q] + \int_{\partial Z} e_b \wedge f_b = 0 \tag{4.78}$$

Hence by substitution of (4.75), (4.76) and using (4.74) we obtain

Proposition 4.1. *Consider the distributed-parameter port-Hamiltonian system (4.77). Then*

$$\frac{dH}{dt} = \int_{\partial Z} e_b \wedge f_b, \tag{4.79}$$

expressing that the increase in energy on the domain Z is equal to the power supplied to the system through the boundary ∂Z.

The system (4.77) can be called a (nonlinear) *boundary control* system in the sense of e.g. [9]. Indeed, we could interpret f_b as the boundary control inputs to the system, and e_b as the measured outputs (or the other way around). In Section 6 we shall further elaborate on this point of view.

4.4.2 Boundary Port-Hamiltonian Systems with Distributed Ports and Dissipation

Energy exchange through the boundary is not the only way a distributed-parameter system may interact with its environment. An example of this is provided by Maxwell's equations (Example 5.1), where interaction may also take place via the current density J, which directly affects the electric charge distribution in the domain Z. In order to cope with this situation we augment the spaces $\mathcal{F}_{p,q}, \mathcal{E}_{p,q}$ as defined in (4.50), (4.52) to

$$\mathcal{F}^a_{q,p} := \mathcal{F}_{p,q} \times \Omega^d(S)$$
$$\mathcal{E}^a_{q,p} := \mathcal{E}_{p,q} \times \Omega^{n-d}(S) \tag{4.80}$$

for some m-dimensional manifold S and some $d \in \{0, 1, \cdots, m\}$, with $f^d \in \Omega^d(S)$ denoting the externally supplied distributed control flow, and $e^d \in \Omega^{n-d}(S)$ the conjugate distributed quantity, corresponding to an energy exchange

$$\int_S e^d \wedge f^d \tag{4.81}$$

The Stokes-Dirac structure (4.55) is now extended to

$$\begin{bmatrix} f_p \\ f_q \end{bmatrix} = \begin{bmatrix} 0 & (-1)^r d \\ d & 0 \end{bmatrix} \begin{bmatrix} e_p \\ e_q \end{bmatrix} + G(f_d)$$
$$\begin{bmatrix} f_b \\ e_b \end{bmatrix} = \begin{bmatrix} 1 & 0 \\ 0 & -(-1)^{n-q} \end{bmatrix} \begin{bmatrix} e_{p|\partial Z} \\ e_{q|\partial Z} \end{bmatrix} \tag{4.82}$$
$$e_d = -G^* \begin{bmatrix} e_p \\ e_q \end{bmatrix}$$

with G denoting a linear map

$$G = \begin{pmatrix} G_p \\ G_q \end{pmatrix} : \Omega^d(S) \to \Omega^p(Z) \times \Omega^q(Z) \tag{4.83}$$

with dual map

$$G^* = (G_p^*, G_q^*) : \Omega^{n-p}(Z) \times \Omega^{n-q}(Z) \to \Omega^{n-d}(S) \tag{4.84}$$

satisfying

$$\int_Z [e_p \wedge G_p(f_d) + e_q \wedge G_q(f_d)] = \int_S [G_p^*(e_p) + G_q^*(e_q)] \wedge f_d \tag{4.85}$$

for all $e_p \in \Omega^{n-p}(Z), e_q \in \Omega^{n-q}(Z), f_d \in \Omega^d(S)$.
The following proposition can be easily checked.

Proposition 4.2. *Equations (4.82) determine a Dirac structure $D^a \subset \mathcal{F}^a_{p,q} \times \mathcal{E}^a_{p,q}$ with respect to the augmented bilinear form on $\mathcal{F}^a_{p,q} \times \mathcal{E}^a_{p,q}$ which is obtained by adding to the bilinear form (4.53) on $\mathcal{F}_{p,q} \times \mathcal{E}_{p,q}$ the term*

$$\int_S \left[e_d^1 \wedge f_d^2 + e_d^2 \wedge f_d^1\right] \tag{4.86}$$

By making now the substitutions (4.75), (4.76) into D^a given by (4.82) we obtain a port-Hamiltonian system with external variables (f_b, f_d, e_b, e_d), with

f_b, e_b the *boundary* external variables and f_d, e_d the *distributed* external variables. Furthermore, the energy balance (4.79) extends to

$$\frac{dH}{dt} = \int_{\partial Z} e_b \wedge f_b + \int_S e_d \wedge f_d, \tag{4.87}$$

with the first term on the right-hand side denoting the power flow through the boundary, and the second term denoting the distributed power flow.

Finally, *energy dissipation* can be incorporated in the framework of distributed-parameter port-Hamiltonian systems by *terminating* some of the ports (boundary or distributed) with a *resistive relation.* For example, for distributed dissipation, let $R : \Omega^{n-d}(S) \to \Omega^d(S)$ be a map satisfying

$$\int_S e_d \wedge R(e_d) \geq 0, \quad \forall e_d \in \Omega^{n-d}(S) \tag{4.88}$$

Then by adding the relation

$$f_d = -R(e_d) \tag{4.89}$$

to the port-Hamiltonian system defined with respect to the Dirac structure D^a, we obtain a port-Hamiltonian system *with dissipation*, satisfying the energy inequality

$$\frac{dH}{dt} = \int_{\partial Z} e_b \wedge f_b - \int_S e_d \wedge R(e_d) \leq \int_{\partial Z} e_b \wedge f_b \tag{4.90}$$

4.5 Examples

In this section we show how the framework of distributed-parameter port-Hamiltonian systems admits the representation of Maxwell's equations, the telegraph equations of an ideal transmission line, the vibrating string, and the Euler equations of an ideal isentropic fluid.

4.5.1 Maxwell's Equations

We closely follow the formulation of Maxwell's equations in terms of differential forms as presented in [12], and show how this directly leads to the formulation as a distributed-parameter port-Hamiltonian system.

Let $Z \subset \mathbb{R}^3$ be a 3-dimensional manifold with boundary ∂Z, defining the spatial domain, and consider the electromagnetic field in Z. The energy variables are the *electric field induction* 2-form $\alpha_p = \mathcal{D} \in \Omega^2(Z)$:

$$\mathcal{D} = \frac{1}{2} D_{ij}(t,z) dz^i \wedge dz^j \tag{4.91}$$

and the *magnetic field induction* 2-form $\alpha_q = \mathcal{B} \in \Omega^2(Z)$:

$$\mathcal{B} = \frac{1}{2} B_{ij}(t,z) dz^i \wedge dz^j \tag{4.92}$$

The corresponding Stokes-Dirac structure ($n = 3, p = 2, q = 2$) is given as (cf. (4.55))

$$\begin{bmatrix} f_p \\ f_q \end{bmatrix} = \begin{bmatrix} 0 & -d \\ d & 0 \end{bmatrix} \begin{bmatrix} e_p \\ e_q \end{bmatrix}, \begin{bmatrix} f_b \\ e_b \end{bmatrix} = \begin{bmatrix} 1 & 0 \\ 0 & 1 \end{bmatrix} \begin{bmatrix} e_{p|\partial Z} \\ e_{q|\partial Z} \end{bmatrix} \tag{4.93}$$

Usually in this case one does *not* start with the definition of the total energy (Hamiltonian) H, but instead with the co-energy variables $\delta_p H, \delta_q H$, given, respectively, as the electric field intensity $\mathcal{E} \in \Omega^1(Z)$:

$$\mathcal{E} = E_i(t,z) dz^i \tag{4.94}$$

and the magnetic field intensity $\mathcal{H} \in \Omega^1(Z)$:

$$\mathcal{H} = H_i(t,z) dz^i \tag{4.95}$$

They are related to the energy variables through the constitutive relations of the medium (or material equations)

$$\begin{aligned} *\mathcal{D} &= \epsilon \mathcal{E} \\ *\mathcal{B} &= \mu \mathcal{H} \end{aligned} \tag{4.96}$$

with the scalar functions $\epsilon(t,z)$ and $\mu(t,z)$ denoting the electric permittivity, respectively magnetic permeability, and $*$ denoting the Hodge star operator (corresponding to a Riemannian metric on Z), converting 2-forms into 1-forms. Then one *defines* the Hamiltonian H as

$$H = \int_Z \frac{1}{2} (\mathcal{E} \wedge \mathcal{D} + \mathcal{H} \wedge \mathcal{B}), \tag{4.97}$$

and one immediately verifies that $\delta_p H = \mathcal{E}, \delta_q H = \mathcal{H}$.
Nevertheless there are other cases (corresponding to a *nonlinear* theory of the electromagnetic field, such as the Born-Infeld theory, see e.g. [12]) where one starts with a more general Hamiltonian $H = \int_Z h$, with the energy density $h(\mathcal{D}, B)$ being a more general expression than $\frac{1}{2}(\epsilon^{-1} * \mathcal{D} \wedge \mathcal{D} + \mu^{-1} * \mathcal{B} \wedge \mathcal{B})$.

Assuming that there is no current in the medium Maxwell's equations can now be written as (see [12])

$$\begin{aligned} \frac{\partial \mathcal{D}}{\partial t} &= d\mathcal{H} \\ \frac{\partial \mathcal{B}}{\partial t} &= -d\mathcal{E} \end{aligned} \tag{4.98}$$

Explicitly taking into account the behavior at the boundary, Maxwell's equations on a domain $Z \subset \mathbb{R}^3$ are then represented as the port-Hamiltonian system with respect to the Stokes-Dirac structure given by (4.93), as

$$\begin{bmatrix} -\frac{\partial \mathcal{D}}{\partial t} \\ -\frac{\partial \mathcal{B}}{\partial t} \end{bmatrix} = \begin{bmatrix} 0 & -d \\ d & 0 \end{bmatrix} \begin{bmatrix} \delta_D H \\ \delta_B H \end{bmatrix}$$
$$\begin{bmatrix} f_b \\ e_b \end{bmatrix} = \begin{bmatrix} \delta_D H_{|\partial Z} \\ \delta_B H_{|\partial Z} \end{bmatrix} \tag{4.99}$$

Note that the first line of (4.98) is nothing else than (the differential version of) Ampère's law, while the second line of (4.98) is Faraday's law. Hence the Stokes-Dirac structure in (4.98), (4.99) expresses the basic physical laws connecting $\mathcal{D}, \mathcal{B}, \mathcal{H}$ and $\mathcal{E}$.
The energy-balance (4.79) in the case of Maxwell's equations takes the form

$$\frac{dH}{dt} = \int_{\partial Z} \delta_B H \wedge \delta_D H = \int_{\partial Z} \mathcal{H} \wedge \mathcal{E} = -\int_{\partial Z} \mathcal{E} \wedge \mathcal{H} \tag{4.100}$$

with $\mathcal{E} \wedge \mathcal{H}$ a 2-form corresponding to the *Poynting vector* (see [12]).

In the case of a non-zero *current density* we have to modify the first matrix equation of (4.99) to

$$\begin{bmatrix} -\frac{\partial D}{\partial t} \\ -\frac{\partial B}{\partial t} \end{bmatrix} = \begin{bmatrix} 0 & -d \\ d & 0 \end{bmatrix} \begin{bmatrix} \delta_D H \\ \delta_B H \end{bmatrix} + \begin{bmatrix} I \\ 0 \end{bmatrix} J \tag{4.101}$$

with I denoting the identity operator from $J \in \Omega^2(Z)$ to $\Omega^2(Z)$. (Thus, in the notation of (4.83), $f_d = J$, $S = Z$, and $\Omega^d(S) = \Omega^2(Z)$.) Furthermore, we add the equation

$$e_d = -[I\ 0] \begin{bmatrix} \delta_D H \\ \delta_B H \end{bmatrix} = -\mathcal{E}, \tag{4.102}$$

yielding the augmented energy balance

$$\frac{dH}{dt} = -\int_{\partial Z} \mathcal{E} \wedge \mathcal{H} - \int_Z \mathcal{E} \wedge J \tag{4.103}$$

which is known as *Poynting's theorem.*

Finally, in order to incorporate energy dissipation we write $J = J_d + \bar{J}$, and we impose Ohm's law

$$*J_d = \sigma \mathcal{E} \tag{4.104}$$

with $\sigma(t, z)$ the specific conductivity of the medium.

4.5.2 Telegraph Equations

Consider an ideal lossless transmission line with $Z = [0, 1] \subset \mathbb{R}$. The energy variables are the charge density 1-form $Q = Q(t, z)dz \in \Omega^1([0, 1])$, and the flux density 1-form $\varphi = \varphi(t,z)dz \in \Omega^1([0,1])$; thus $p = q = n = 1$. The total energy stored at time t in the transmission line is given as

$$H(Q,\varphi) = \int_0^1 \frac{1}{2}\left(\frac{Q^2(t,z)}{C(z)} + \frac{\varphi^2(t,z)}{L(z)}\right) dz \tag{4.105}$$

with co-energy variables

$$\begin{aligned} \delta_Q H &= \tfrac{Q(t,z)}{C(z)} = V(t,z) \quad \text{(voltage)} \\ \delta_Q H &= \tfrac{\varphi(t,z)}{L(z)} = I(t,z) \quad \text{(current)} \end{aligned} \tag{4.106}$$

where $C(z), L(z)$ are respectively the distributed capacitance and distributed inductance of the line.

The resulting port-Hamiltonian system is given by the telegraph equations

$$\begin{aligned} \tfrac{\partial Q}{\partial t} &= -\tfrac{\partial I}{\partial z} \\ \tfrac{\partial \varphi}{\partial t} &= -\tfrac{\partial V}{\partial z} \end{aligned} \tag{4.107}$$

together with the boundary variables

$$\begin{aligned} f_b^0(t) &= V(t,0), \quad f_b^1(t) = V(t,1) \\ e_b^0(t) &= -I(t,0), \; e_b^1(t) = -I(t,1) \end{aligned} \tag{4.108}$$

The resulting energy-balance is

$$\frac{dH}{dt} = \int_{\partial([0,1])} e_b f_b = -I(t,1)V(t,1) + I(t,0)V(t,0), \tag{4.109}$$

in accordance with (4.79).

4.5.3 Vibrating String

Consider an elastic string subject to traction forces at its ends. The spatial variable z belongs to the interval $Z = [0,1] \subset \mathbb{R}$. Let us denote by $u(t,z)$ the displacement of the string. The elastic potential energy is a function of the *strain* given by the 1-form

$$\alpha_q(t) = \epsilon(t,z)dz = \frac{\partial u}{\partial z}(t,z)dz \tag{4.110}$$

The associated co-energy variable is the *stress* given by the 0-form

$$\sigma = T * \alpha_q \tag{4.111}$$

with T the elasticity modulus and $*$ the Hodge star operator. Hence the potential energy is the quadratic function

$$U(\alpha_q) = \int_0^1 \sigma\alpha_q = \int_0^1 T * \alpha_q \wedge \alpha_q = \int_0^1 T\left(\frac{\partial u}{\partial z}\right)^2 dz \tag{4.112}$$

and $\sigma = \delta_q U$.

The kinetic energy K is a function of the kinetic *momentum* defined as the 1-form

$$\alpha_p(t) = p(t,z)dz \tag{4.113}$$

given by the quadratic function

$$K(\alpha_p) = \int_0^1 \frac{p^2}{\mu} dz \tag{4.114}$$

The associated co-energy variable is the *velocity* given by the 0-form

$$v = \frac{1}{\mu} * \alpha_p = \delta_p K \tag{4.115}$$

In this case the Dirac structure is the Stokes-Dirac structure for $n = p = q = 1$, with an *opposite sign* convention leading to the equations (with $H := U + K$)

$$\begin{aligned} \begin{bmatrix} -\frac{\partial \alpha_p}{\partial t} \\ -\frac{\partial \alpha_q}{\partial t} \end{bmatrix} &= \begin{bmatrix} 0 & -d \\ -d & 0 \end{bmatrix} \begin{bmatrix} \delta_p H \\ \delta_q H \end{bmatrix} \\ \begin{bmatrix} f_b \\ e_b \end{bmatrix} &= \begin{bmatrix} 1 & 0 \\ 0 & 1 \end{bmatrix} \begin{bmatrix} \delta_p H|_{\partial Z} \\ \delta_q H|_{\partial Z} \end{bmatrix} \end{aligned} \tag{4.116}$$

or, in more down-to-earth notation

$$\begin{aligned} \frac{\partial p}{\partial t} &= \frac{\partial \sigma}{\partial z} = \frac{\partial}{\partial z}(T\epsilon) \\ \frac{\partial \epsilon}{\partial t} &= \frac{\partial v}{\partial z} = \frac{\partial}{\partial z}\left(\frac{1}{\mu}p\right) \\ f_b &= v|_{\{0,1\}} \\ e_b &= \sigma|_{\{0,1\}} \end{aligned} \tag{4.117}$$

with boundary variables the velocity and stress at the ends of the string. Of course, by substituting $\epsilon = \frac{\partial u}{\partial z}$ into the 2nd equation of (4.117) one obtains $\frac{\partial}{\partial z}\left(\frac{\partial u}{\partial t} - \frac{p}{\mu}\right) = 0$, implying that

$$p = \mu \frac{\partial u}{\partial t} + \mu f(t) \tag{4.118}$$

for some function f, which may be set to zero. Substitution of (4.118) into the first equation of (4.117) then yields the wave equation

$$\mu\frac{\partial^2 u}{\partial t^2} = \frac{\partial}{\partial z}\left(T\frac{\partial u}{\partial z}\right) \tag{4.119}$$

4.6 Extension of Port-Hamiltonian Systems Defined on Stokes-Dirac Structures

4.6.1 Burger's Equations

Consider the inviscid Burger's equation as discussed in Section 2.1. Consider Z to be a bounded interval of $\mathbb{R}$, then *Burger's inviscid equations* are:

$$\frac{\partial u}{\partial t} + \frac{\partial}{\partial x}\left(\frac{u^2}{2}\right) = 0$$

which is a scalar convex conservation equation.

It may be formulated as a boundary control system as follows:

$$\begin{aligned} \frac{\partial u}{\partial t} &= -\frac{\partial}{\partial x}\left(\delta_u H\right) \\ w_b &= \delta_u H \mid_{\partial Z} \end{aligned}$$

where $\delta_u H$ denotes the variational derivative of the Hamiltonian functional $H(u) = \frac{1}{6}u^3$. Defining the power-conjugated variables to be $f = \frac{\partial u}{\partial t}$, $e = \delta_u H$ and on the boundary w_b, one may define an infinite-dimensional Dirac structure which is different from the the Stokes-Dirac structure. With regard to this Dirac structure the inviscid Burger's equation is represented as a distributed port-Hamiltonian system. For details we refer to [15].

4.6.2 Ideal Isentropic Fluid

Consider an ideal compressible isentropic fluid in three dimensions, described in Eulerian representation by the standard Euler equations

$$\begin{aligned} \tfrac{\partial \rho}{\partial t} &= -\nabla \cdot (\rho v) \\ \tfrac{\partial v}{\partial t} &= -v \cdot \nabla v - \tfrac{1}{\rho}\nabla p \end{aligned} \tag{4.120}$$

with $\rho(z,t) \in \mathbb{R}$ the mass density at the spatial position $z \in \mathbb{R}^3$ at time t, $v(z,t) \in \mathbb{R}^3$ the (Eulerian) velocity of the fluid at spatial position z and time t, and $p(z,t)$ the pressure function, derivable from an internal energy function $U(\rho)$ as

$$p(z,t) = \rho^2(z,t)\frac{\partial U}{\partial \rho}(\rho(z,t)) \tag{4.121}$$

Much innovative work has been done regarding the Hamiltonian formulation of (4.125) and more general cases; we refer in particular to [24, 17, 18, 25, 14]. However, in these treatments only *closed* fluid dynamical systems are being considered with no energy exchange through the boundary of the spatial domain. As a result, a formulation in terms of Poisson structures can be given, while as argued before, the general inclusion of boundary variables necessitates the use of Dirac structures.

The formulation of (4.120) as a port-Hamiltonian system is given as follows. Let $\mathcal{D} \subset \mathbb{R}^3$ be a given domain, filled with the fluid. We assume the existence of a Riemannian metric $<,>$ on $\mathcal{D}$; usually the standard Euclidean metric on $\mathbb{R}^3$. Let $Z \subset \mathcal{D}$ be any 3-dimensional manifold with boundary ∂Z.

We identify the mass-density ρ with a 3-form on Z (see e.g. [17, 18]), that is, with an element of $\Omega^3(Z)$. Furthermore, we identify the Eulerian vector field v with a 1-form on Z, that is, with an element of $\Omega^1(Z)$. (By the existence of the Riemannian metric on Z we can, by "index raising" or "index lowering", identify vector fields with 1-forms and vice versa.) The precise motivation for this choice of variables will become clear later on. As a result we consider as the carrier spaces for the port-Hamiltonian formulation of (4.120) the linear spaces $\mathcal{F}_{p,q}$ and $\mathcal{E}_{p,q}$ for $n = 3, p = 3, q = 1$; that is

$$\mathcal{F}_{p,q} = \Omega^3(Z) \times \Omega^1(Z) \times \Omega^0(\partial Z) \tag{4.122}$$

and

$$\mathcal{E}_{p,q} = \Omega^0(Z) \times \Omega^2(Z) \times \Omega^2(\partial Z) \tag{4.123}$$

Since $p+q = n+1$ we can define the corresponding Stokes-Dirac structure D given by (4.55) on $\mathcal{F}_{p,q} \times \mathcal{E}_{p,q}$. However, as will become clear later on, due to 3-dimensional convection we need to *modify* this Stokes-Dirac structure with an additional term into the following modified Stokes-Dirac structure

$$\begin{aligned}
D^m := \{(f_p, f_v, f_b, e_\rho, e_v, e_b) \in \\
\Omega^3(Z) \times \Omega^1(Z) \times \Omega^0(\partial Z) \times \Omega^0(Z) \times \Omega^2(Z) \times \Omega^2(\partial Z) \\
\begin{bmatrix} f_\rho \\ f_v \end{bmatrix} = \begin{bmatrix} de_v \\ de_\rho + \frac{1}{*\rho} * ((*dv) \wedge (*e_v)) \end{bmatrix} \\
\begin{bmatrix} f_b \\ e_b \end{bmatrix} = \begin{bmatrix} e_{\rho|\partial Z} \\ -e_{v|\partial Z} \end{bmatrix} \}
\end{aligned} \tag{4.124}$$

where $*$ denotes the Hodge star operator (corresponding to the Riemannian metric on Z), converting k-forms on Z to $(3-k)$-forms. A fundamental difference of the modified Stokes-Dirac structure D^m with respect to the standard

Stokes-Dirac structure D is that D^m explicitly depends on the energy variables ρ and v (via the terms $*\rho$ and dv in the additional term $\frac{1}{*\rho} * ((*dv) \wedge (*e_v))$. Completely similar to the proof of Theorem 5 it is shown that $(D^m(\rho, v))^{\perp} = D^m(\rho, v)$ for all ρ, v; the crucial additional observation is that the expression

$$e_v^2 \wedge *((*dv) \wedge (*e_v^1)) \tag{4.125}$$

is *skew-symmetric* in $e_v^1, e_v^2 \in \Omega^2(Z)$.

Remark 4.6. In the standard Euclidean metric, identifying via the Hodge star operator 2-forms β_i with 1-forms, and representing 1-forms as vectors, we have in vector calculus notation the equality

$$\beta_2 \wedge *(\alpha \wedge *\beta_1) = \alpha \cdot (\beta_1 \times \beta_2) \tag{4.126}$$

for all 2-forms β_1, β_2 and 1-forms α. This shows clearly the skew-symmetry of (4.125).

The Eulerian equations (4.120) for an ideal isentropic fluid are obtained in the port-Hamiltonian representation by considering the Hamiltonian

$$H(\rho, v) := \int_Z \left[\frac{1}{2} < v^\sharp, v^\sharp > \rho + U(*\rho)\rho \right] \tag{4.127}$$

with $v^\sharp$ the vector field corresponding to the 1-form v ("index lowering"), and $U(*\rho)$ the potential energy. Indeed, by making the substitutions (4.75), (4.76) in D^m, and noting that

$$\text{grad } H = (\delta_\rho H, \delta_v H) = \left(\frac{1}{2} < v^\sharp, v^\sharp > + \frac{\partial}{\partial \tilde{\rho}} (\tilde{\rho} U(\tilde{\rho})) \, , \; i_{v^\sharp} \rho \right) \tag{4.128}$$

with $\tilde{\rho} := *\rho$, the port-Hamiltonian system takes the form

$$\begin{aligned} -\tfrac{\partial \rho}{\partial t} &= d(i_{v^\sharp} \rho) \\ -\tfrac{\partial v}{\partial t} &= d\left(\tfrac{1}{2} < v^\sharp, v^\sharp > + w(*\rho)\right) + \tfrac{1}{*\rho}\left((*dv) \wedge (*i_{v^\sharp} \rho)\right) \\ f_b &= \left[\tfrac{1}{2} < v^\sharp, v^\sharp > + w(*\rho)\right]_{|\partial Z} \\ e_b &= -i_{v^\sharp} \rho_{|\partial Z} \end{aligned} \tag{4.129}$$

with

$$w(\tilde{\rho}) := \frac{\partial}{\partial \tilde{\rho}} (\tilde{\rho} U(\tilde{\rho})) \tag{4.130}$$

the *enthalpy*. The expression $\delta_\rho H = \frac{1}{2} < v^\sharp, v^\sharp > + w(\tilde{\rho})$ is known as the *Bernoulli function.*

The first two equations of (4.129) can be seen to represent the Eulerian equations (4.120). The first equation corresponds to the basic law of *mass-balance*

$$\frac{d}{dt}\int_{\varphi_t(V)} \rho = 0 \tag{4.131}$$

where V denotes an arbitrary volume in Z, and φ_t is the flow of the fluid (transforming the material volume V at $t = 0$ to the volume $\varphi_t(V)$ at time t). Indeed, (4.131) for any V is equivalent to

$$\frac{\partial \rho}{\partial t} + L_{v^\sharp}\rho = 0 \tag{4.132}$$

Since by Cartan's magical formula $L_{v^\sharp}\rho = d(i_{v^\sharp}\rho) + i_{v^\sharp}d\rho = d(i_{v^\sharp}\rho)$ (since $d\rho = 0$) this yields the first line of (4.129). It also makes clear the interpretation of ρ as a 3-form on Z.

For the identification of the second equation of (4.129) with the second equation of (4.125) we note the following (see [36] for further details). Interpret $\nabla\cdot$ in (4.120) as the covariant derivative corresponding to the assumed Riemannian metric $<,>$ on Z. For a vector field u on Z, let $u^\flat$ denote the corresponding 1-form $u^\flat := i_u <,>$ ("index raising"). The covariant derivative ∇ is related to the Lie derivative by the following formula (see for a proof [14], p. 202)

$$L_u u^\flat = (\nabla_u u)^\flat + \frac{1}{2} d < u, u > \tag{4.133}$$

Since by Cartan's magical formula $L_u u^\flat = i_u du^\flat + d(i_u u^\flat) = i_u du^\flat + d < u, u >$, (4.133) can be also written as

$$(\nabla_u u)^\flat = i_u du^\flat + \frac{1}{2} d < u, u > \tag{4.134}$$

(This is the coordinate-free analog of the well-known vector calculus formula $u \cdot \nabla u = \text{ curl } u \times u + \frac{1}{2}\nabla |u|^2$.) Furthermore we have the identity

$$i_{v^\sharp} dv = \frac{1}{*\rho} * ((*dv) \wedge (*i_{v^\sharp}\rho)) \tag{4.135}$$

Finally, we have the following well-known relation between enthalpy and pressure (obtained from (4.126) and (4.130))

$$\frac{1}{\tilde{\rho}} dp = d(w(\tilde{\rho})). \tag{4.136}$$

Hence by (4.134) (with $u = v^\sharp$), (4.110) and (4.136), we may rewrite the 2nd equation of (4.129) as

$$-\frac{\partial v}{\partial t} = \left(\nabla_{v^\sharp} v^\sharp\right)^\flat + \frac{1}{*\rho} dp \tag{4.137}$$

which is the coordinate-free formulation of the 2nd equation of (4.120).

The boundary variables f_b and e_b given in (4.129) are respectively the *stagnation pressure* at the boundary divided by ρ, and the (incoming) *mass flow* through the boundary. The energy-balance (4.79) can be written out as

$$\begin{aligned}\frac{dH}{dt} &= \int_{\partial Z} e_b \wedge f_b = -\int_{\partial Z} i_{v^\sharp}\rho \wedge \left[\tfrac{1}{2} < v^\sharp, v^\sharp > + w(*\rho)\right] \\ &= -\int_{\partial Z} i_{v^\sharp} \left[\tfrac{1}{2} < v^\sharp, v^\sharp > \rho + w(*\rho)\rho\right] \\ &= -\int_{\partial Z} i_{v^\sharp} \left[\tfrac{1}{2} < v^\sharp, v^\sharp > \rho + U(*\rho)\rho\right] - \int_{\partial Z} i_{v^\sharp}(*p)\end{aligned} \tag{4.138}$$

where for the last equality we have used the relation (following from (4.121), (4.130))

$$w(*\rho)\rho = U(*\rho)\rho + *p \tag{4.139}$$

The first term in the last line of (4.138) corresponds to the convected energy through the boundary ∂Z, while the second term is (minus) the external work (static pressure times velocity).

Usually, the second line of the Euler equations (4.120) (or equivalently equation (4.137)) is obtained from the basic conservation law of momentum-balance together with the first line of (4.120). Alternatively, emphasizing the interpretation of v as a 1-form, we may obtain it from *Kelvin's circulation theorem*

$$\frac{d}{dt}\int_{\varphi_t(C)} v = 0 \tag{4.140}$$

where C denotes any *closed* contour. Indeed, (4.140) for any closed C is equivalent to the 1-form $\frac{\partial v}{\partial t} + L_{v^\sharp} v$ being *closed*. By (4.133) this is equivalent to requiring

$$\frac{\partial v}{\partial t} + \left(\nabla_{v^\sharp} v^\sharp\right)^\flat \tag{4.141}$$

to be closed, that is

$$\frac{\partial v}{\partial t} + \left(\nabla_{v^\sharp} v^\sharp\right)^\flat = -dk \tag{4.142}$$

for some (possibly locally defined) $k : Z \to \mathbb{R}$. Now additionally requiring that this function k depends on z through ρ, that is

$$k(z) = w(\rho(z)) \tag{4.143}$$

for some function w, we recover (4.137) with $\frac{1}{*\rho}dp$ replaced by dw (the differential of the enthalpy).

Remark 4.7. In the case of a one- or two-dimensional fluid flow the extra term in the Dirac structure D^m as compared with the standard Stokes-Dirac structure D vanishes, and so in these cases we are back to the standard definition

of a distributed-parameter port-Hamiltonian system (with ρ being a 1-form, respectively, a 2-form).

Furthermore, if in the 3-dimensional case the 2-form $dv(t)$ happens to be zero at a certain time-instant $t = t_0$ (*irrotational flow*), then it *continues* to be zero for all time $t \geq t_0$. Hence also in this case the extra term (4.125) in the modified Stokes-Dirac structure D^m vanishes, and the port-Hamiltonian system describing the Euler equations reduces to the standard distributed-parameter port-Hamiltonian system given in Definition 4.5.

Remark 4.8. For the modified Stokes-Dirac structure D^m given in (4.124) the space of admissible mappings K_{adm} given in equation (4.61) is the same as for the Stokes-Dirac structure, but the resulting skew-symmetric bracket has an additional term:

$$\{k^1, k^2\}D^m = \int_Z [(\delta_p k^1) \wedge (-1)^r d(\delta_q k^2) + (\delta_q k^1) \wedge d(\delta_p k^2)$$

$$+\frac{1}{*\rho}\delta_v k^1 \wedge *((*dv) \wedge (*\delta_v k^2))] - \int_{\partial Z} (-1)^{n-q} (\delta_q k^1) \wedge (\delta_p k^2) \quad (4.144)$$

(For the skew-symmetry of the additional term see (4.125) and Remark 4.6.)

4.7 Conserved Quantities

Let us consider the distributed-parameter port-Hamiltonian system Σ, as defined in Definition 4.5, on an n-dimensional spatial domain Z having state space $\Omega^p(Z) \times \Omega^q(Z)$ (with $p+q = n+1$) and Stokes-Dirac structure D given by (4.55).
Conservation laws for Σ, which are *independent* from the Hamiltonian H, are obtained as follows. Let

$$C : \Omega^p(Z) \times \Omega^q(Z) \times Z \to \mathbb{R} \quad (4.145)$$

be a function satisfying

$$d(\delta_p C) = 0, \quad d(\delta_q C) = 0, \quad (4.146)$$

where $d(\delta_p C)$, $d(\delta_q C)$ are defined similarly to (4.72). Then the time-derivative of C along the trajectories of Σ is given as (in view of (4.146), and using similar calculations as in the proof of Theorem 4.1

$$\begin{aligned}\frac{d}{dt}C &= \int_Z \delta_p C \wedge \dot{\alpha}_p + \int_Z \delta_q C \wedge \dot{\alpha}_q \\ &= -\int_Z \delta_p C \wedge (-1)^r d(\delta_q H) - \int_Z \delta_q C \wedge d(\delta_p H) \\ &= -(-1)^{n-q}\int_Z d(\delta_q H \wedge \delta_p C) - (-1)^{n-q}\int_Z d(\delta_q C \wedge \delta_p H) \\ &= \int_{\partial Z} e_b \wedge f_b^C + \int_{\partial Z} e_b^c \wedge f_b \end{aligned} \tag{4.147}$$

where we have denoted, in analogy with (4.55),

$$f_b^C := \delta_p C|_{\partial Z}, \quad e_b^C := -(-1)^{n-q}\delta_q C|_{\partial Z} \tag{4.148}$$

In particular, if additionally to (4.146) the function C satisfies

$$\delta_p C|_{\partial Z} = 0, \quad \delta_q C|_{\partial Z} = 0 \tag{4.149}$$

then $\frac{dC}{dt} = 0$ along the system trajectories of Σ for any Hamiltonian H. Therefore a function C satisfying (4.146), (4.149) is called a *Casimir* function. If C satisfies (4.146) but not (4.149) then C is called a *conservation law* for Σ: its time-derivative is determined by the boundary conditions of Σ.

Example 4.6. In the case of the telegraph equations (Example 5.2) the total charge

$$C_Q = \int_0^1 Q(t,z)dz$$

as well as the total magnetic flux

$$C_\varphi = \int_0^1 \varphi(t,z)dz$$

are both conservation laws. Indeed

$$\frac{d}{dt}C_Q = -\int_0^1 \frac{\partial I}{\partial z} = I(0) - I(1)$$

$$\frac{d}{dt}C_\varphi = -\int_0^1 \frac{\partial V}{\partial z}dz = V(0) - V(1)$$

Similarly, in the case of the vibrating string (Example 5.3) conservation laws are $= \int_0^1 \epsilon(t,z)dz = u(t,1) - u(t,0)$,

$$\frac{d}{dt}\int_0^1 \epsilon(t,z)dz = \frac{d}{dt}(u(t,1) - u(t,0)) = v(t,1) - v(t,0)$$

$$\frac{d}{dt}\int_0^1 p(t,z)dz = \sigma(t,1) - \sigma(t,0)$$

□

Conservation laws C for Σ which are *dependent* on the Hamiltonian H are obtained by replacing (4.146) by the weaker condition

$$\delta_q H \wedge d(\delta_p C) + (-1)^r \delta_p H \wedge d(\delta_q C) = 0 \tag{4.150}$$

Indeed, it immediately follows from the computation in (4.147) that under this condition (4.147) continues to hold.

In the case of the modifies Stokes-Dirac structure D^m defined in (4.124), for any function $C : \Omega^3(Z) \times \Omega^1(Z) \times Z \to \mathbb{R}$ satisfying

$$\delta_v H \wedge d(\delta_p C) + \delta_\rho H \wedge d(\delta_v C) = 0, \quad \rho \in \Omega^3(Z),\ v \in \Omega^1(Z) \tag{4.151}$$

equation (4.147) takes the form

$$\begin{aligned}\frac{d}{dt}C &= \int_Z \delta_\rho C \wedge d(\delta_v H) + \int_Z \delta_v C \wedge \left[d(\delta_\rho H) + \frac{1}{*\rho} * ((*dv) \wedge (*\delta_v H))\right] \\ &= \int_{\partial Z} \delta_\rho C \wedge \delta_v H + \int_{\partial Z} \delta_v C \wedge \delta_\rho H \\ &\quad + \int_Z \frac{1}{*\rho} \delta_v C * ((*dv) \wedge (*\delta_v H))\end{aligned} \tag{4.152}$$

Hence we conclude that in order to obtain a conservation law we need to impose an extra condition eliminating the last $\int_Z$ integral. A specific example of a conservation law in this case is the *helicity*

$$C = \int_Z v \wedge dv \tag{4.153}$$

with time-derivative

$$\frac{d}{dt}C = -\int_{\partial Z} f_b \wedge dv \tag{4.154}$$

A *second* class of conserved quantities corresponding to the Stokes-Dirac structure D (4.55) is identified by noting that by (4.77)

$$\begin{aligned}-d\left(\frac{\partial \alpha_p}{\partial t}\right) &= (-1)^r d(d\delta_q H) = 0 \\ -d\left(\frac{\partial \alpha_q}{\partial t}\right) &= d(d\delta_p H) = 0\end{aligned} \tag{4.155}$$

and thus the differential forms $d\alpha_p$ and $d\alpha_q$ do not depend on time. Therefore, the component functions of $d\alpha_p$ and $d\alpha_q$ are conserved quantities of any port-Hamiltonian system corresponding to D.

Example 4.7. In the case of Maxwell's equations (Example 5.1) this yields that $d\mathcal{D}$ and $d\mathcal{B}$ are constant 3-forms. The 3-form $d\mathcal{D}$ is the *charge density* (Gauss'

electric law), while by Gauss' magnetic law $d\mathcal{B}$ is actually zero.
In the case of an ideal isentropic fluid (Section 6.2) for which the vorticity $dv(t_0, z)$ is *zero* at a certain time t_0 we obtain by the same reasoning (since the additional term in the Stokes-Dirac structure D^m is zero for t_0) that $dv(t, z)$ is *zero* for all $t \geq t_0$ (irrotational flow); cf. Remark 4.7.

4.8 Conclusions and Final Remarks

In this paper we have exposed a framework for the compositional modelling of distributed-parameter systems, based on our papers [37, 20, 22]. This allows the Hamiltonian formulation of a large class of distributed-parameter systems with boundary energy- ow, including the examples of the telegraph equations, Maxwell's equations, vibrating strings and ideal isentropic fluids. It has been argued that in order to incorporate boundary variables into this formulation the notion of a Dirac structure provides the appropriate generalization of the more commonly used notion of a Poisson structure for evolution equations. The employed Dirac structure is based on Stokes' theorem, and emphasizes the geometrical content of the variables as being differential k-forms. From a physical point of view the Stokes-Dirac structure captures the balance laws inherent to the system, like Faraday's and Ampère's law (in Maxwell's equations), or mass-balance (in the case of an ideal fluid). This situation is quite similar to the lumped-parameter case where the Dirac structure incorporates the topological interconnection laws (Kirchhoff's laws, Newton's third law) and other interconnection constraints (see e.g. [19] [19] [35]). Hence the starting point for the Hamiltonian description is different from the more common approach of deriving Hamiltonian equations from a variational principle and its resulting Lagrangian equations, or (very much related) a Hamiltonian formulation starting from a state space being a co-tangent bundle endowed with its natural symplectic structure. In the case of Maxwell's equations this results in the use of the basic physical variables D and B (the electric and magnetic field inductions), instead of the use of the variable D (or E) together with the vector potential A (with $dA = B$) in the symplectic formulation of Maxwell's equations. It should be of interest to compare both approaches more closely, also in the context of the natural multi-symplectic structures which have been formulated for the Hamiltonian formulation of Lagrangian field equations.

A prominent and favorable property of Dirac structures is that they are closed under power-conserving interconnection. This has been formally proven in the finite-dimensional case, but the result carries through to the infinite-dimensional case as well. It is a property of fundamental importance since it enables to link port-Hamiltonian systems (lumped- or distributed-parameter) to each other to obtain an interconnected port-Hamiltonian system with total energy being the sum of the Hamiltonians of its constituent parts. Clearly, this is equally important in modelling (coupling e.g. solid components with fluid components, or finite-dimensional electric components with transmission lines), as in control. First of all, it enables to formulate directly distributed-parameter systems with constraints as (implicit) Hamiltonian systems, like this has been done in the finite-dimensional case for mechanical systems with kinematic constraints, multi-body systems, and general electrical networks. Secondly, from the control perspective the notion of feedback control can be understood on its most basic level as the coupling of given physical components with additional control components (being themselves physical systems, or software components linked to sensors and actuators). A preliminary study from this point of view of a control scheme involving transmission lines has been provided in [30]. Among others, this opens up the way for the application of passivity-based control techniques, which have been proven to be very effective for the control of lumped-parameter physical systems modelled as port-Hamiltonian systems.

References

1. R. Abraham and J. E. Marsden (March 1994) *Foundations of Mechanics.* Addison, ii edition. ISBN 0-8053-0102-X.
2. A.M.Bloch and P.E.Crouch (1999) Representation of Dirac structures on vector spaces and nonlinear lcv-circuits. In H.Hermes G. Ferraya, R.Gardner and H.Sussman, editors, *Proc. of Symposia in Pure mathematics, Differential Geometry and Control Theory,* volume 64, pages 103–117.
3. A.J. van der Schaft G. Blankenstein (2001) Symmetry and reduction in implicit generalized hamiltonian systems. *Rep. Math. Phys.,* 47:57–100.
4. P.C. Breedveld (Feb 1984) *Physical Systems Theory in Terms of Bond Graphs.* PhD thesis, Technische Hogeschool Twente, Enschede, The Netherlands, ISBN 90-90005999-4.
5. T.J. Courant (1990) Dirac manifolds. *Trans. American Math. Soc. 319,* pages 631–661.
6. M. Dalsmo and A.J. van der Schaft (1999) On representations and integrability of mathematical structures in energy-conserving physical systems. *SIAM Journal of Control and Optimization,* 37(1):54–91.
7. I. Dorfman (1993) *Dirac structures and integrability of nonlinear evolution equations.* John Wiley.
8. D. Eberard and B.M. Maschke (2004) An extension of port Hamiltonian systems to irreversible systems. In *Proc. Int. Conf. on Nonlinear Systems' Theory and Control, NOLCOS'04,* Stuttgart, Germany. Submitted.
9. H.O. Fattorini (1968) Boundary control systems. *SIAM J. Control and Opt.,* (6):349–385.
10. E. Godlewsky and P. Raviart (1996) *Numerical Approximation of Hyperbolic Systems of Conservation Laws,* volume 118 of *Applied Mathematical Sciences.* Springer Verlag, New-York, USA.
11. T.E. Ratiu A. Weinstein D.D. Holm, J.E. Marsden (1985) Nonlinear stability of fluid and plasma equilibria. *Phys.Rep.,* 123:1–116.

12. A. Jamiolkowski R.S. Ingarden (1985) *Classical Electrodynamics.* PWN-Polish Sc. Publ. Elsevier, Warszawa, Poland.
13. Dean C. Karnopp, Donald L. Margolis, and Ronald C. Rosenberg (1990) *System Dynamics, A Unified Approach.* Wiley.
14. B.A. Khesin V. I. Arnold (1998) *Topological Methods in Hydrodynamics*, volume 125 of *Applied Mathematical Sciences.* Springer Verlag, New York, USA.
15. Y. Le Gorrec, H. Zwart, B.M. Maschke (in preparation).
16. (1987) Paulette Libermann and Charles-Michel Marle. *Symplectic Geometry and Analytical Mechanics.* D. Reidel Publishing Company, Dordrecht, Holland ISBN 90-277-2438-5.
17. A. Weinstein J.E. Marsden, T. Ratiu (1984) Reduction and Hamiltonian structures on duals of semidirect product Lie algebras. *AMS Contemporary Mathematics*, 28:55–100.
18. A. Weinstein J.E. Marsden, T. Ratiu (1984) Semidirect products and reduction in Mechanics. *Trans. American Math. Society*, (281):147–177.
19. B.M. Maschke and A.J. van der Schaft (1997) *Modelling and Control of Mechanical Systems*, chapter Interconnected Mechanical systems. Part 1 and 2, pages 1–30. Imperial College Press, London. ISBN 1-86094-058-7.
20. B.M. Maschke and A.J. van der Schaft (2000) Port controlled Hamiltonian representation of distributed parameter systems. In R. Ortega N.E. Leonard, editor, *Proc. IFAC Workshop on modeling and Control of Lagrangian and Hamiltonian Systems*, Princeton, USA.
21. A.J. van der Schaft, G. Escobar, B.M. Maschke and R. Ortega (July 1999) An energy-based derivation of lyapunov functions for forced systems with application to stabilizing control. In *Proc. 14th IFAC World Congress*, volume E, pages 409–414, Beijing, China.
22. A.J. van der Schaft B.M. Maschke (2000) *Nonlinear Control in the Year 2000*, chapter Hamiltonian representation of distributed parameter systems with boundary energy flow, pages 137–142. Springer-Verlag, eds. A. Isidori, F. lamnabhi-lagarrigue, W. Respondek.
23. B.M. Maschke, A.J. van der Schaft, and P.C. Breedveld (1992) An intrinsic Hamiltonian formulation of network dynamics: Non-standard poisson structures and gyrators. *Journal of the Franklin institute*, 329(5):923–966. Printed in Great Britain.
24. P.J. Morrison & J.M. Greene (1980) Noncanonical Hamiltonian density formulation of hydrodynamics and ideal magnetohydrodynamics. *Phys. Rev. Letters*, (45):790–794.
25. P.J. Morrison (1998) Hamiltonian description of the ideal fluid. *Rev. Mod. Phys.*, (70):467–521.
26. Peter J. Olver (1993) *Applications of Lie Groups to Differential Equations*, volume 107 of *Graduate texts in mathematics.* Springer, New-York, ii edition. ISBN 0-387-94007-3.
27. R. Ortega, A.J. van der Schaft, I. Mareels and B. Maschke (2001) Putting energy back in control. *IEEE Control Systems Magazine*, 21(2):18– 32.
28. B. Maschke, R. Ortega, A.J. van der Schaft and G. Escobar (2002) Interconnection and damping assignment: passivity-based control of port-controlled Hamiltonian systems. *Automatica*, 38:585–596.
29. I. Prigogine (1962) *Introduction to Thermodynamics of Irreversible Processes.* John Wiley and Sons.

30. H. Rodriguez, A.J. van der Schaft and R. Ortega (2001) On stabilization of nonlinear distributed parameter port-controlled Hamiltonian systems via energy shaping. Proc. 40th IEEE Conf. Decision and Control (CDC), Orlando FL, pp.131–136.
31. D. Serre (1999) *Systems of Conservation Laws.* Cambridge University Press, Cambridge, U.K.
32. S. Stramigioli, B.M. Maschke, and A.J. van der Schaft (1998) Passive output feedback and port interconnection. In *Proc NOLCOS 1998*, Enschede, The Netherlands.
33. A.J. van der Schaft (1994) Implicit Hamiltonian systems with symmetry. *Reports on Mathematical Physics*, 41:203–221.
34. A.J. van der Schaft (1996) *L_2-Gain and Passivity Techniques in Nonlinear Control.* Springer Communications and Control Engineering series. Springer-Verlag, London, 2nd revised and enlarged edition, 2000. first edition Lect. Notes in Control and Inf. Sciences, vol. 218, Springer-Verlag, Berlin.
35. A.J. van der Schaft and B.M. Maschke (1995) The Hamiltonian formulation of energy conserving physical systems with external ports. *Archiv für Elektronik und Übertragungstechnik*, 49(5/6):362–371.
36. A.J. van der Schaft and B.M. Maschke (Dec. 2001) Fluid dynamical systems as hamiltonian boundary control systems. In *Proc. 40th IEEE Conf. on Decision and Control*, pages 4497–4502, Orlando, USA.
37. A.J. van der Schaft and B.M. Maschke (2002) Hamiltonian formulation of distributed parameter systems with boundary energy flow. *J. of Geometry and Physics*, 42:166–174.

5

Algebraic Analysis of Control Systems Defined by Partial Differential Equations

Jean-François Pommaret

CERMICS/Ecole Nationale des Ponts et Chaussées, 6/8 ave Blaise Pascal, Cité Descartes, 77455 Marne-la-Vallée CEDEX 2, FRANCE. E-mail: pommaret@cermics.enpc.fr

The present chapter contains the material taught within the module P2 of FAP 2004. The purpose of this intensive course is first to provide an introduction to "*algebraic analysis*". This fashionable though quite difficult domain of pure mathematics today has been pioneered by V.P. Palamodov, M. Kashiwara and B. Malgrange around 1970, after the work of D.C. Spencer on the formal theory of systems of partial differential equations. We shall then focus on its application to control theory in order to study linear control systems defined by partial differential equations with constant or variable coefficients, also called multidimensional control systems, by means of new methods from module theory and homological algebra. We shall revisit a few basic concepts and prove, in particular, that controllability, contrary to a well established engineering tradition or intuition, is an intrinsic structural property of a control system, not depending on the choice of inputs and outputs among the control variables or even on the presentation of the control system. Our exposition will be rather elementary as we shall insist on the main ideas and methods while illustrating them through explicit examples. Meanwhile, we want to stress out the fact that these new techniques bring striking results even on classical control systems of Kalman type!

5.1 Introduction

We start recalling and revisiting the important controllability concept in classical control theory. For this, we shall adopt standard notations for the *input* $u = (u^1, ..., u^p)$, the *state* $x = (x^1, ..., x^n)$ and the *output* $y = (y^1, ..., y^m)$, the dot indicating time derivative.
A control system with coefficients in a constant field k (say $\mathbb{Q}, \mathbb{R}, \mathbb{C}$ in general) is said to be in "*Kalman form*" if it can be written as $\dot{x} = Ax + Bu$, where A is a constant $n \times n$ matrix and B is a constant $n \times p$ matrix with maximum

F. Lamnabhi-Lagarrigue et al. (Eds.): Adv. Top. in Cntrl. Sys. Theory, LNCIS 311, pp. 155–223, 2005

rank p, where we understand that A and B have coefficients in k. If we should like to characterize such a system of ordinary differential (OD) equations, we could say that it is linear, first order, with no equation of order zero and with no derivative of u appearing in the OD equations.

Definition 5.1. *The above control system is said to be controllable if, starting from any initial point x_0 at time $t = 0$, it is possible to find an input such that the corresponding trajectory is passing through an arbitrary final point x_T at a final time T.*

The following celebrated test has been given by E.R. Kalman in 1963 [6]:

Theorem 5.1. *The above control system is controllable if and only if the controllability matrix $(B, AB, A^2B, ..., A^{n-1}B)$ has maximum rank n.*

Of course, in general, starting with $rk(B) = l_0 = p$, the rank will increase successively and we may set $rk(B, AB) = l_0 + l_1, ..., rk(B, AB, ..., A^{n-1}B) = l_0 + ... + l_{n-1}$ as the rank will not increase anymore because of the Cayley-Hamilton theorem saying that A^n is a linear combination of the previous lower powers of A.
In order to provide a preview of the type of techniques that will be used in the sequel, we provide a new short homological proof of the following technical result [25]:

Proposition 5.1. $p = l_0 \geq l_1 \geq ... \geq l_{n-1} \geq 0$.

Proof. Denoting also by B the vector space over k generated by the column vectors $(b_1, ..., b_p)$ of B, we may introduce the vector spaces $S_i = B + AB + ... + A^iB$ over k for $i = 0, 1, ..., n-1$ and obtain $dim(S_i/S_{i-1}) = l_i$. Denoting also by A the multiplication by A, we get $AS_i = AB + A^2B + ... + A^{i+1}B \subseteq S_{i+1}$ and $S_{i+1} = AS_i + B$. In this very elementary situation that will be generalized in the sequel, we shall say that a square is "commutative" if the composition of the maps (matrices) along two sides of the square is equal to the composition of the maps along the two other opposite sides, in a standard way. Similarly, a chain of maps will be called a sequence if the composition of two successive maps (matrices) is zero. Accordingly, a sequence (line or column) will be said to be "exact" if the kernel of any map is equal to the image of the preceding map, a "0" on the left meaning injectivity (monomorphism) and a "0" on the right meaning surjectivity (epimorphism). We have the following commutative and exact diagram, namely a diagram where all squares are commutative while all lines and columns are exact:

$$\begin{array}{ccccccc}
 & & 0 & & 0 & & \\
 & & \downarrow & & \downarrow & & \\
0 & \longrightarrow & AS_{i-1} \cap B & \longrightarrow & AS_i \cap B & & \\
 & & \downarrow & & \downarrow & & \\
0 & \longrightarrow & B & = & B & \longrightarrow & 0 \\
 & & \downarrow & & \downarrow & & \\
 & & B/(AS_{i-1} \cap B) & \longrightarrow & B/(AS_i \cap B) & \longrightarrow & 0 \\
 & & \downarrow & & \downarrow & & \\
 & & 0 & & 0 & &
\end{array}$$

Using the snake theorem (see subsection on homological algebra) or counting the dimensions, we notice that the cokernel $(AS_i \cap B)/(AS_{i-1} \cap B)$ of the upper map is isomorphic to the kernel of the lower map. We use the lower line of this diagram as the right column of the following commutative and exact diagram:

$$\begin{array}{ccccccccccc}
 & & 0 & & 0 & & 0 & & & & \\
 & & \downarrow & & \downarrow & & \downarrow & & & & \\
0 & \longrightarrow & S_{i-1} \cap ker(A) & \longrightarrow & S_{i-1} & \xrightarrow{A} & S_i & \longrightarrow & B/(AS_{i-1} \cap B) & \longrightarrow & 0 \\
 & & \downarrow & & \downarrow & & \downarrow & & \downarrow & & \\
0 & \longrightarrow & S_i \cap ker(A) & \longrightarrow & S_i & \xrightarrow{A} & S_{i+1} & \longrightarrow & B/(AS_i \cap B) & \longrightarrow & 0 \\
 & & \downarrow & & \downarrow & & \downarrow & & \downarrow & & \\
0 & \longrightarrow & K_i & \longrightarrow & S_i/S_{i-1} & \longrightarrow & S_{i+1}/S_i & \longrightarrow & 0 & & \\
 & & & & \downarrow & & \downarrow & & & & \\
 & & & & 0 & & 0 & & & &
\end{array}$$

where K_i is the kernel of the induced lower central map and we have used the relation $S_{i+1}/AS_i \simeq B/(AS_i \cap B)$. Counting the dimensions, we get $l_i = l_{i+1} + dim(K_i) \geq l_{i+1}$. Finally, again using the snake theorem, we get the long exact sequence:

$$0 \to S_{i-1} \cap ker(A) \to S_i \cap ker(A) \to K_i \to B/(AS_{i-1} \cap B) \to B/(AS_i \cap B) \to 0$$

and thus the short exact sequence:

$$0 \to (S_i \cap ker(A))/(S_{i-1} \cap ker(A)) \to K_i \to (AS_i \cap B)/(AS_{i-1} \cap B) \to 0$$

where we have used the previous isomorphism, as a way to compute $dim(K_i)$. ∎

The reader will have noticed how tricky is such a proof that could be quite tedious otherwise, though we advise the reader to draw pictures of the various spaces involved and their inclusions in order to understand the meaning of the respective quotients.

Surprisingly, through this engineering setting, if we understand that the l_i, called "*controllability indices*", can be determined by means of elementary computer algebra (rank of matrices), it seems that we are very far from being able to extend these indices to a more general framework. Indeed, not a single of the previous results can be extended to systems of partial differential (PD) equations, or even to systems of OD equations containing the derivatives of the inputs or having variable coefficients. Also, it seems strange that controllability, defined in a purely *functional way*, could be tested in a purely *formal way*. Finally, it seems that controllability is highly depending on the choice of inputs and outputs among the control variables, according to a well established engineering tradition.

In order to provide a first feeling that the proper concept of controllability must be revisited, we provide a short but illuminating example:

Example 5.1. Let us consider the system of two OD equations:

$$\dot{y}^1 - ay^2 - \dot{y}^3 = 0 \quad , \quad y^1 - \dot{y}^2 + \dot{y}^3 = 0$$

depending on a constant parameter a.
First choosing the control variables to be $y^1 = x^1, y^2 = x^2, y^3 = u$ and setting $\bar{x}^1 = x^1 - u, \bar{x}^2 = x^2 - u$, we get the Kalman form:

$$\dot{\bar{x}}^1 = a\bar{x}^2 + au \quad , \quad \dot{\bar{x}}^2 = \bar{x}^1 + u$$

and the system is controllable, with controllability indices $(1, 1)$, if and only if $a \neq 0, a \neq 1$.
Now choosing the control variables to be $y^1 = x^1, y^2 = u, y^3 = x^2$, and setting anew $\bar{x}^1 = x^1 - u, \bar{x}^2 = x^2 - u$, *though with a totally different meaning*, we get the totally different Kalman form:

$$\dot{\bar{x}}^1 = -\bar{x}^1 + (a-1)u \quad , \quad \dot{\bar{x}}^2 = -\bar{x}^1 - u$$

and this new system is controllable, with the same controllability indices, if and only if $a \neq 0, a \neq 1$ too.

It follows from this example that *controllability must be a structural property* of a control system, neither depending on the choice of inputs and outputs among the control variables, nor even on the presentation of the control system (change of the control variables eventually leading to change the order of the system). The next definition is crucial for revisiting controllability and extending it to systems of PD equations [16, 18, 19]. It stems from the fact that, in engineering sciences, a measurement apparatus (thermometer, manometer,...) is always measuring a scalar quantity (temperature, pressure,...).

Definition 5.2. *An autonomous (or torsion) element is an observable, that is to say a linear combination of the control variables and their derivatives, which satisfies at least one (and thus only one of minimum order when there is only one independent variable) OD or PD equation by itself. An observable satisfying no OD or PD equation for itself will be said to be "free".*

This leads to the following formal definition:

Definition 5.3. *A control system is controllable if it has no nontrivial autonomous element, that is any observable is free.*

Example 5.2. In the preceding example, if $a = 0$, setting $z = y^1 - y^3$, we get $\dot{z} = 0$. Also, if $a = 1$, setting now $z = y^1 - y^2$ and adding the first OD equation to the second, we get $\dot{z} + z = 0$.

Though it does not seem evident at first sight, we have:

Proposition 5.2. *The preceding Definition is coherent with the Kalman test.*

Proof. Using the given OD equations and their derivatives in order to compute $\dot{x}, \ddot{x}, ...$ from the arbitrary/parametric $x, u, \dot{u}, \ddot{u}, ...$ we could for simplicity imagine that $z = \lambda x + \mu_0 u + \mu_1 \dot{u}$ does satisfy the single OD equation $\dot{z} + \nu z = 0$. Differentiating z and substituting, we obtain at once the necessary conditions $\mu_1 = 0 \Rightarrow \mu_0 = 0$. However, we notice that, if z is an autonomous element, then of course $\dot{z}, \ddot{z}, ...$ are again autonomous elements. Hence, from $z = \lambda x$ we successively get $\dot{z} = \lambda A x + \lambda B u \Rightarrow \lambda B = 0$, $\ddot{z} = \lambda A^2 x + \lambda A B u \Rightarrow \lambda A B u = 0$, ... It follows that there are as many autonomous elements linearly independent over k as the corank of the controllability matrix. Conversely, we notice that, if there exists at least one autonomous element, the control system is surely not controllable in any sense as there is no way to control the OD equation satisfied by this autonomous element. ∎

However, if a control system is given by high order input/output OD equations, the search for autonomous elements is no longer simple as one cannot use the above computation which is essentially based on the fact that no derivative of the input does appear in the OD equations of the Kalman form. We now raise another problem.

Ordinary differential (OD) control theory studies input/output relations defined by systems of ordinary differential (OD) equations. In this case, with standard notations, if a control system is defined by input/state/output relations:

$$\dot{x} = Ax + Bu \quad , \quad y = Cx + Du$$

with $dim(x) = n$, this system is "*controllable*" if $rk(B, AB, ..., A^{n-1}B) = n$, as we already said, and "*observable*" if $rk(\tilde{C}, \tilde{A}\tilde{C}, ..., \tilde{A}^{n-1}\tilde{C}) = n$ where the tilde sign indicates the transpose of a matrix [25]. Accordingly, the so-called "*dual system*":

$$\dot{x}_a = -\tilde{A}x_a - \tilde{C}u_a \quad , \quad y_a = \tilde{B}x_a + \tilde{D}u_a$$

is controllable (observable) if and only if the given system is observable (controllable). However, and despite many attempts, such a dual definition still seems purely artificial as one cannot avoid introducing the state. The same method could be applied to delay systems with constant coefficients. One must nevertheless notice that, if we do not want to define observability as a property "dual" to controllability, the standard meaning, namely the possibility to recover the state from the input and the output is clear. Indeed, by differentiation, we successively get $y = Cx + ...$, $\dot{y} = CAX + ...$, $\ddot{y} = CA^2x + ...$, and so on, where the missing terms only involve the input and its derivatives.

Hence, if the derivatives of the inputs do appear in the control system, for example in the SISO system $\dot{x} - \dot{u} = 0$, not a word is left from the original functional definition of controllability which is only valid for systems in "*Kalman form*" as any input satisfying $u(T) - u(0) = x(T) - x(0)$ is convenient. The same comment can be made for the corresponding duality.

More generally, "*partial differential (PD) control theory*" will study input/output relations defined by systems of partial differential (PD) equations. At first sight, we have no longer a way to generalize the Kalman form and not a word of the preceding approach is left as, in most cases, the number of arbitrary parametric derivatives playing the rôle of state could be infinite. However, even if the definition of autonomous elements is still meaningful though we have no longer any way to test it, we also understand that a good definition of controllability and duality should also be valid for control systems with variable coefficients. A similar comment can be made for the definition of the transfer matrix.

Example 5.3. Denoting by $y_i^k = d_i y^k$ for $i = 1, 2$ and $k = 1, 2, 3$ the formal derivatives of the three differential indeterminates y^1, y^2, y^3, we consider the system of three PD equations for three unknowns and two independent variables [5, 18]:

$$\left\{ \begin{array}{llll} y_2^2 & +y_2^3 & -y_1^3 & -y_1^2 = 0 \\ y_2^1 & -y_2^3 & -y_1^3 & -y_1^2 = 0 \\ y_1^1 & & -2y_1^3 & -y_1^2 = 0 \end{array} \right.$$

One can check that one among (y^1, y^2, y^3) can be given arbitrarily like in the preceding example. Also, setting $z = y^1 - y^2 - 2y^3$, we get both $z_1 = 0, z_2 = 0$ and z is an autonomous element. Then one can prove that any other autonomous element can be expressible by means of a differential operator acting on z which is therefore a generator (exercise). Accordingly, in the present situation, any autonomous element is a constant multiple of z but no other analogy can be exhibited.

Keeping aside these problems for the moment, let us now turn for a few pages to the formal theory of systems of OD or PD equations.

In 1920, M. Janet provided an effective algorithm for looking at the formal (power series) solutions of systems of ordinary differential (OD) or partial differential (PD) equations [5]. The interesting point is that this algorithm also allows to determine the *compatibility conditions* $\mathcal{D}_1\eta = 0$ for solving (*formally* again but this word will not be repeated) inhomogeneous systems of the form $\mathcal{D}\xi = \eta$ when $\mathcal{D}$ is an OD or PD operator and ξ, η certain functions. Similarly, one can also determine the compatibility conditions $\mathcal{D}_2\zeta = 0$ for solving $\mathcal{D}_1\eta = \zeta$, and so on. With no loss of generality, this construction of a "*differential sequence*" can be done in such a canonical way that we successively obtain $\mathcal{D}_1, \mathcal{D}_2, ..., \mathcal{D}_n$ from $\mathcal{D}$ and $\mathcal{D}_n$ is surjective when n is the number of independent variables.

With no reference to the above work, D.C. Spencer developed, from 1965 to 1975, the *formal theory* of systems of PD equations by relating the preceding results to *homological algebra* and *jet theory* [24]. However, this tool has been *largely ignored* by mathematicians and, "*a fortiori*", by engineers or even physicists. Therefore, the module theoretic counterpart, today known as "*algebraic analysis*", which has been pioneered around 1970 by V.P. Palamodov for the constant coefficient case [14], then by M. Kashiwara [7] and B. Malgrange [11] for the variable coefficient case, as it heavily depends on the previous difficult work and looks like even more abstract, has been *totally ignored* within the range of any application before 1990, when U. Oberst revealed its importance for control theory [13].

The purpose of this lecture will essentially be to fill in this gap by explaining, in a self-contained way on a few explicit examples, what is the powerfulness of this new approach for understanding both the structural and input/output properties of linear PD control systems, also called multidimensional or simply n-dimensional. Meanwhile, the reader will evaluate the price to pay for such a better understanding. Needless to say that many results obtained could not even be imagined without this new approach, dating back to 1986 when we gave for the first time the formal definition of controllability of a control system [16] but now largely acknowledged by the control community [26, 27].

5.2 Motivating Examples

As we always use to say, the difficulty in studying systems of PD equations is not only of a functional nature (unicity and existence of solutions,...) but also of a formal nature (integrability and compatibility conditions,...). This is the reason for which the study of algebraic analysis is at once touching delicate points of differential geometry, the main one being *formal integrability*. Hence, forgetting about control theory for a few pages, we now explain this concept and other related ones on a few tricky motivating examples. It will therefore be a first striking challenge for the reader to wonder what certain of these academic/engineering examples have to do with controllability!

Example 5.4. With two independent variables (x^1, x^2), one unknown y and standard notations for PD equations and computer algebra (MACSYMA, MAPLE, MATHEMATICA,...), we consider the following third order system of PD equations with second member (u, v):

$$\begin{cases} Py \equiv d_{222}y + x^2 y = u \\ Qy \equiv d_2 y + d_1 y \quad = v \end{cases}$$

where P and Q are PD operators with coefficients in the (differential) field $K = \mathbb{Q}(x^1, x^2)$ of rational functions in x^1 and x^2. We check the identity $QP - PQ \equiv 1$ and obtain easily:

$$y = Qu - Pv = d_2 u + d_1 u - d_{222} v - x^2 v$$

Substituting in the previous PD equations, we therefore obtain the two generating 6th-order compatibility conditions for (u, v) in the form:

$$\begin{cases} A \equiv PQu - P^2 v - u = 0 \\ B \equiv Q^2 u - QPv - v = 0 \end{cases}$$

These two compatibility conditions are not differentially independent as we check at once:

$$\begin{aligned} QA &\equiv QPQu - QP^2 v - Qu \\ &\equiv (1 + PQ)Qu - (1 + PQ)Pv - Qu \\ &\equiv PQ^2 u - PQPv - Pv \\ &\equiv PB \end{aligned}$$

Finally, setting $u = 0, v = 0$, we notice that the preceding homogeneous system can be written in the form $\mathcal{D}y = 0$ and admits the only solution $y = 0$.

Example 5.5. Again with two independent variables (x^1, x^2) and one unknown y, let us consider the following second order system with constant coefficients:

$$\begin{cases} Py \equiv d_{22}y \qquad = u \\ Qy \equiv d_{12}y - y = v \end{cases}$$

where now P and Q are PD operators with coefficients in the subfield $k = \mathbb{Q}$ of constants of K. We obtain at once:

$$y = d_{11}u - d_{12}v - v$$

and could hope to obtain the 4th-order generating compatibility conditions by substitution, that is to say:

$$\left\{\begin{array}{ll} A \equiv d_{1122}u - d_{1222}v - d_{22}v - u = 0 \\ B \equiv d_{1112}u - d_{11}u - d_{1122}v & = 0 \end{array}\right.$$

However, *in this particular case*, we notice that there is an unexpected *unique second order* generating compatibility condition of the form:

$$C \equiv d_{12}u - u - d_{22}v = 0$$

as we now have indeed $PQ - QP = 0$ both with $A \equiv d_{12}C + C$ and $B \equiv d_{11}C$, a result leading to $C \equiv d_{22}B - d_{12}A + A$. Accordingly, the systems $A = 0, B = 0$ on one side and $C = 0$ on the other side are completely different though they have the same solutions in u, v.
Finally, setting $u = 0, v = 0$, we notice that the preceding homogeneous system can be written in the form $\mathcal{D}y = 0$ and admits the only solution $y = 0$, like the preceding example.

Remark 5.1. It is only in the last section of this chapter that we shall understand why such a specific result could only happen for systems of PD equations with constant coefficients. Indeed, this result will be shown to be a straightforward consequence of the famous Quillen-Suslin theorem which can be considered as one of the most difficult theorems of mathematics [23].

Example 5.6. (See [18] for more details on this interesting example first provided by M. Janet in [5]). With $n = 3$, let us consider the second order system:

$$\left\{\begin{array}{ll} Py \equiv d_{33}y - x^2 d_{11}y = u \\ Qy \equiv d_{22}y & = v \end{array}\right.$$

where now P and Q are PD operators with coefficients in the (differential) field $K = \mathbb{Q}(x^1, x^2, x^3)$. Introducing as before:

$$d_{112}y = \frac{1}{2}(d_{33}v - x^2 d_{11}v - d_{22}u) = w$$

we finally get the two following compatibility conditions:

$$\begin{array}{ll} A \equiv d_{233}v - x^2 d_{112}v - 3d_{11}v - d_{222}u & = 0 \\ B \equiv d_{3333}w - 2x^2 d_{1133}w + (x^2)^2 d_{1111}w - d_{11233}u + x^2 d_{11112}u - d_{1111}u & = 0 \end{array}$$

These two compatibility conditions of respective orders 3 and 6 are differentially dependent as one checks at once through computer algebra:

$$d_{3333}A - 2x^2 d_{1133}A + (x^2)^2 d_{1111}A - 2d_2 B = 0$$

However, and contrary to the two preceding motivating examples, we now have no way to know whether A and B are the only two generating compatibility conditions.

Finally, in the present situation, the space of solutions of the system $\mathcal{D}y = 0$ can be seen to be expressible by polynomials in (x^1, x^2, x^3) and has dimension 12 over the constants.

Having in mind the preceding examples and a computer algebra framework, we are in a position to explain in a simple manner how systems of PD equations can be divided into two classes, namely the "*good*" ones and the "*bad*" ones. This terminology will lead, in the next section, to an intrinsic/coordinate-free formulation where the good systems will be called "*formally integrable*" systems.
For understanding the difference, let us plug a given system of order q and use computer algebra to have all possible information about the various derivatives of the unknowns at the orders $q+1, q+2, ...$, by differentiating the given PD equations successively once, twice, ... and so on. In the case of the first example with $q = 3$, differentiating once we get the two new second order PD equation $d_{12}y + d_{11}y = 0, d_{22}y + d_{12}y = 0$. Similarly, differentiating twice we get four linearly independent PD equations of strict order three, namely $d_{222}y + x^2 y = 0, d_{122}y - x^2 y = 0, d_{112}y + x^2 y = 0, d_{111}y - x^2 y = 0$ though we had only one initially. Finally, differentiating three times, we even get the zero order equation $y = 0$, that we need again to differentiate once in order to get separately $d_1 y = 0, d_2 y = 0$, and so on. In the case of the second example, the things are similar and we understand that, in general, we have to differentiate $r + s(r)$ times in order to know all the possible information on all the derivatives up to order $q + r$. The situation is even more tricky in the case of the third example as, differentiating once with respect to x^1, x^2, x^3 respectively, we get 6 linearly independent PD equations of strict order 3, with no information "backwards" on the order 2 as in the preceding situations. However, we are not finished with the order 3 as indeed, differentiating twice, that is once more, we get the new equation $d_{112}y = 0$ but we are not sure that,differentiating 100 times will not produce new equations. In fact, in order to be convinced about the difficulty of this problem and how highly it depends on the coefficients, we ask the reader, as an exercise, to work out this example with $(x^2)^r$ in place of x^2 for $r = 1, 2, 3, ...$ successively.
It is essential to notice that, contrary to the use of Gröbner bases, the preceding results do not depend on any change of both the independent or the dependent variables among them as the order is unchanged.
Contrary to the previous (very) bad examples, in the case of good examples, differentiating just r times will provide all information on the derivatives up to order $q + r$.

Example 5.7. With all indices ranging from 1 to n, let us consider the so-called Killing system:

$$\frac{1}{2}(\mathcal{L}(\xi)\omega)_{ij} \equiv \frac{1}{2}(\omega_{rj}\partial_i\xi^r + \omega_{ir}\partial_j\xi^r + \xi^r\partial_r\omega_{ij}) = \epsilon_{ij}$$

where $\omega_{ij} = 1$ if $i = j$ and 0 if $i \neq j$ and we have used the standard Einstein implicit summation on the up and down indices. This system is used in continuum mechanics to define the (small) strain tensor ϵ from the displacement vector ξ by using the Lie derivative of the euclidean metric ω along the vector field ξ. If we look for a displacement providing no strain, we have just to set $\epsilon = 0$ and to notice (tricky exercise!) that differentiating once provides the $n^2(n+1)/2$ second order PD equations $\partial_{ij}\xi^k = 0$ that can be easily integrated. Hence we get $\xi^k = A^k_r x^r + B^k$ with A an arbitrary skew-symmetric constant matrix and B an arbitrary vector, that is to say we recognize an arbitrary infinitesimal rigid motion. This is a good system indeed as the $n+1$ derivatives up to order one are arbitrary, the $n(n+1)/2$ given equations provide all information on the n^2 derivatives of strict order one and all the other derivatives of order ≥ 2 vanish.
More generally, any linear system with constant coefficients having only derivatives of the same order appearing in it is automatically good.
In order to look for the compatibility conditions for ϵ, we notice that the compatibility conditions of order r are obtained by Cramer's rule through the elimination of the $n(n+r)!/(n-1)!(r+1)!$ derivatives of strict order $r+1$ of the ξ from the $n(n+1)(n+r-1)!/2(n-1)!r!$ derivatives of strict order r of the ϵ. Accordingly, we get no compatibility condition of order one and $n^2(n+1)^2/4 - n^2(n+1)(n+2)/6 = n^2(n^2-1)/12$ compatibility conditions of order two. For the case $n = 2$ of 2-dimensional plane elasticity, we get the only second order compatibility condition :

$$\partial_{11}\epsilon_{22} + \partial_{22}\epsilon_{11} - 2\partial_{12}\epsilon_{12} = 0$$

that will be used in the sequel.

We now explain the close relationship existing between the search for compatibility conditions and the study of autonomous elements. Indeed, in order to look for the self-governing PD equations of a given autonomous element, we just need to write out the system of PD equations with zero second member, add the expression of the autonomous element as a left member/operator, keeping the autonomous element as a right member and look for the set of compatibility conditions of the total system thus obtained.

Example 5.8. With respect to the system:

$$d_{22}y = 0 \qquad , \qquad d_{12}y = 0$$

if we add $d_1 y = z$ we get $d_2 z = 0$ while if we add $d_2 y = z$, we get $d_2 z = 0, d_1 z = 0$.

This result proves that, contrary to the OD case where a given autonomous element always satisfies a single generating OD equation, in the PD case the situation can be far more complicate. In fact, there is a way to classify autonomous elements according to the type of system they satisfy. This result, leading to the concept of "*purity*", involves a lot of delicate homological algebra and is out of the scope of this chapter. For more details, we invite the interested reader to look at [19, Section V.1.3].

In view of the importance of autonomous elements in differential algebra [16, 18], we first notice that the corresponding concept can be easily extended to nonlinear systems and we have the following proposition:

Proposition 5.3. *The sum, product, quotient and derivatives of two autonomous elements are again autonomous elements.*

Proof. We only prove this result for the sum as the other proofs are similar. As an autonomous element satisfies at least one OD or PD equation of order q for itself, the number of its arbitrary/parametric derivatives up to order $q+r$ is at most equal to:

$$(n+q+r)!/n!(q+r)! - (n+r)!/n!r! = \frac{q}{(n-1)!}r^{n-1} + \ldots$$

and this number is multiplied by 2 in the sum while the number of derivatives of the sum up to order r is $(n+r)!/n!r! = \frac{1}{n!}r^n + \ldots$. Accordingly, when r is large enough, this last number is greater than the preceding number and the sum cannot be free. ∎

We end this section presenting two examples where any observable is autonomous.

Example 5.9. (More details concerning the Bénard problem can be found in [18, p. 366]). When a viscous liquid is in between two horizontal parallel plates with distance L between them, the lower heated one being at temperature T_0 while the upper one is at temperature T_1 with $T_0 - T_1 = AgL > 0$ where $\mathbf{g}$ is the vertical gravity, its stationary evolution is governed by the following linear Boussinesq system of five PD equations:

$$\boldsymbol{\nabla} \cdot \mathbf{v} = 0$$

$$\eta \Delta \mathbf{v} - \boldsymbol{\nabla} \pi - \alpha a \theta \mathbf{g} = 0$$

$$\lambda \Delta \theta - \frac{A}{g} \mathbf{g} \cdot \mathbf{v} = 0$$

describing successively the continuity equation, the three Navier-Stokes equations and the heat equation. In this system, the mass per unit volume is $\rho = a(1 - \alpha(T - T_0))$ with $a = \rho(T_0)$, η is the viscosity coefficient, π and θ are the respective perturbations of the pressure and temperature around an equilibrium state and $\lambda = \kappa a c$ where κ is the thermal conductivity and c is the heating coefficient at constant pressure and temperature T_0.

We prove that θ is an autonomous element.
Taking into account the continuity equation while remembering a well known vector identity, we obtain at once:

$$\nabla \wedge \nabla \wedge \mathbf{v} = \nabla(\nabla \cdot \mathbf{v}) - \Delta \mathbf{v} = -\Delta \mathbf{v}$$

Now, applying twice the curl operator to the Navier-Stokes equations for eliminating π and projecting onto the vertical axis x^3, we get:

$$\eta \Delta \Delta v^3 + \alpha a g(d_{11} + d_{22})\theta = 0$$

Using the heat equation in the form $\lambda \Delta \theta = A v^3$ and introducing the dimensionless Rayleigh number $R = g\alpha A L^4 a^2 c/\eta\kappa$, we finally obtain the following sixth order PD equation:

$$\Delta\Delta\Delta\theta - \frac{R}{L^4}(d_{11} + d_{22})\theta = 0$$

The same equation is satisfied by v^3 but we notice that the vertical component $\zeta = d_1 v^2 - d_2 v^1$ only satisfies $\Delta\zeta = 0$.

Example 5.10. The Euler equations for an incompressible fluid with speed $\mathbf{v}$, pressure p and mass per unit volume set to 1, are made by the nonlinear system:

$$\nabla \cdot \mathbf{v} = 0 \quad , \quad d_t \mathbf{v} + (\mathbf{v} \cdot \nabla)\mathbf{v} + \nabla p = 0$$

For a 1-dimensional flow, we get $d_x v = 0, d_t v + v d_x v + d_x p = 0$ and thus both v and p are autonomous because $d_x v = 0, d_{xx} p = 0$. For a 2-dimensional flow, v^1 is autonomous but its highly nonlinear fifth order decoupling PD equation is covering one full page of book [19, p. 40].

We end this section with a nonlinear example, showing out that the study of linear systems can be of some help for studying the structural properties of nonlinear systems by means of linearization, only if we are able to deal with linear systems with variable coefficients.

Example 5.11. Let us consider the single input/single output (SISO) system $u\dot{y} - \dot{u} = a = cst$ and ask about its controllability (see [18] for more details). Of course, if $a = 0$, setting $z = y - log u$, we get $\dot{z} = 0$ and the system cannot be controlled as there is one autonomous element. Introducing the variations $U = \delta u, Y = \delta y$, the generic linearization (not to be confused with the linearization at a specific solution) becomes:

$$u\dot{Y} - \dot{U} + \dot{y}U = 0$$

as the constant parameter a is untouched. It seems that this system is no longer depending on a but the reader must not forget that $u = u(t)$ and

$y = y(t)$ are solutions of the given nonlinear system which definitively depends on a. According to the previous comments, it just remains to study under what condition on a the above linear system with variable coefficients is controllable. The use of such a study is as follows. If the linearized system is controllable, the nonlinear system is surely controllable because, otherwise, it should have at least one autonomous element, the linearization of which should satisfy the corresponding linearized decoupling equation. The converse is not evident as an autonomous element for the linearized system may not necessarily come from the linearization of an autonomous element for the nonlinear system. In fact, one can prove that that this converse is only true for OD systems but false in general for PD systems as counterexamples may exist [19].
In the present example, when $a = 0$, we easily check that $\delta z = Z = Y - \frac{1}{u}U$ satisfies $\dot{Z} = 0$ and the problem is thus to obtain this critical value $a = 0$ directly from the linear system, a result highly not evident at first sight, even on this elementary example.

5.3 Algebraic Analysis

It becomes clear from the examples of the second section that there is a need for classifying the properties of systems of PD equations in a way that does not depend on their presentations and *this is the purpose of algebraic analysis.*

5.3.1 Module Theory

Before entering the heart of the paper, we need a few technical definitions and results from commutative algebra [4, 8, 12, 23].

First of all, we start defining rings and modules.

Definition 5.4. *A ring A is a non-empty set with two associative binary operations called addition and multiplication, respectively sending $a, b \in A$ to $a + b \in A$ and $ab \in A$ in such a way that A becomes an abelian group for the multiplication, so that A has a zero element denoted by 0, every $a \in A$ has an additive inverse denoted by $-a$ and the multiplication is distributive over the addition, that is to say $a(b + c) = ab + ac, (a + b)c = ac + bc, \forall a, b, c \in A$.*
A ring A is said to be unitary if it has a (unique) element $1 \in A$ such that $1a = a1 = a, \forall a \in A$ and commutative if $ab = ba, \forall a, b \in A$.
A non-zero element $a \in A$ is called a zero-divisor if one can find a non-zero $b \in A$ such that $ab = 0$ and a ring is called an integral domain if it has no zero-divisor.

Definition 5.5. *A ring K is called a field if every non-zero element $a \in K$ is a unit, that is one can find an element $b \in K$, called inverse of a, such that $ab = 1 \in K$.*

Definition 5.6. *A left module M over a ring A or simply an A-module is a set of elements $x, y, z, ...$ which is an abelian group for an addition $(x, y) \rightarrow x + y$ with a left action $A \times M \rightarrow M : (a, x) \rightarrow ax$ satisfying:*
- $a(x + y) = ax + ay, \forall a \in A, \forall x, y \in M$
- $a(bx) = (ab)x, \forall a, b \in A, \forall x \in M$
- $(a + b)x = ax + bx, \forall a, b \in A, \forall x \in M$
- $1x = x, \forall x \in M$

The set of modules over a ring A will be denoted by $mod(A)$. A module over a field is called a vector space. Right modules could be similarly defined from right actions (exercise).

Definition 5.7. *A map $f : M \rightarrow N$ between two A-modules is called a homomorphism over A if $f(x + y) = f(x) + f(y), \forall x, y \in M$ and $f(ax) = af(x), \forall a \in A, \forall x \in M$. We successively define:*
- $ker(f) = \{x \in M | f(x) = 0\}$
- $im(f) = \{y \in N | \exists x \in M, f(x) = y\}$
- $coker(f) = N/im(f)$.

Definition 5.8. *We say that a chain of modules and homomorphisms is a sequence if the composition of two successive such homomorphisms is zero. A sequence is said to be exact if the kernel of each map is equal to the image of the map preceding it. An injective homomorphism is called a monomorphism, a surjective homomorphism is called an epimorphism and a bijective homomorphism is called an isomorphism. A short exact sequence is an exact sequence made by a monomorphism followed by an epimorphism.*

The proof of the following proposition is left to the reader as an exercise:

Proposition 5.4. *If one has a short exact sequence:*

$$0 \longrightarrow M' \xrightarrow{f} M \xrightarrow{g} M'' \longrightarrow 0$$

then the following conditions are equivalent:
- *There exists a monomorphism $v : M'' \rightarrow M$ such that $g \circ v = id_{M''}$.*
- *There exists an epimorphism $u : M \rightarrow M'$ such that $u \circ f = id_{M'}$.*
- *There exist isomorphisms $\varphi = (u, g) : M \rightarrow M' \oplus M''$ and $\psi = f + v : M' \oplus M'' \rightarrow M$ that are inverse to each other and provide an isomorphism $M \simeq M' \oplus M''$.*

Definition 5.9. *In the above situation, we say that the short exact sequence splits and $u(v)$ is called a lift for $f(g)$. In particular we have the relation: $f \circ u + v \circ g = id_M$.*

Definition 5.10. *A left (right) ideal $\mathfrak{a}$ in a ring A is a submodule of A considered as a left (right) module over itself. When the inclusion $\mathfrak{a} \subset A$ is strict, we say that $\mathfrak{a}$ is a proper ideal of A.*

Lemma 5.1. *If $\mathfrak{a}$ is an ideal in a ring A, the set of elements $rad(\mathfrak{a}) = \{a \in A | \exists n \in \mathbb{N}, a^n \in \mathfrak{a}\}$ is an ideal of A containing $\mathfrak{a}$ and called the radical of $\mathfrak{a}$. An ideal is called perfect or radical if it is equal to its radical.*

Definition 5.11. *For any subset $S \subset A$, the smallest ideal containing S is called the ideal generated by S. An ideal generated by a single element is called a principal ideal and a ring is called a principal ideal ring if any ideal is principal. The simplest example is that of polynomial rings in one indeterminate over a field. When $\mathfrak{a}$ and $\mathfrak{b}$ are two ideals of A, we shall denote by $\mathfrak{a} + \mathfrak{b}$ ($\mathfrak{ab}$) the ideal generated by all the sums $a + b$ (products ab) with $a \in \mathfrak{a}, b \in \mathfrak{b}$.*

Definition 5.12. *An ideal $\mathfrak{p}$ of a ring A is called a prime ideal if, whenever $ab \in \mathfrak{p}$ ($aAb \in \mathfrak{p}$ in the non-commutative case) then either $a \in \mathfrak{p}$ or $b \in \mathfrak{p}$. The set of proper prime ideals of A is denoted by $spec(A)$ and called the spectrum of A.*

Definition 5.13. *The annihilator of a module M in A is the ideal $ann_A(M)$ of A made by all the elements $a \in A$ such that $ax = 0, \forall x \in M$.*

From now on, all rings considered will be unitary integral domains, that is rings containing 1 and having no zero-divisor. For the sake of clarity, as a few results will also be valid for modules over non-commutative rings, we shall denote by ${}_AM_B$ a module M which is a left module for A with operation $(a, x) \rightarrow ax$ and a right module for B with operation $(x, b) \rightarrow xb$. In the commutative case, lower indices are not needed. If $M = {}_AM$ and $N = {}_AN$ are two left A-modules, the set of A-linear maps $f : M \rightarrow N$ will be denoted by $hom_A(M, N)$ or simply $hom(M, N)$ when there will be no confusion and there is a canonical isomorphism $hom(A, M) \simeq M : f \rightarrow f(1)$ with inverse $x \rightarrow (a \rightarrow ax)$. When A is commutative, $hom(M, N)$ is again an A-module for the law $(bf)(x) = f(bx)$ as we have indeed:

$$(bf)(ax) = f(bax) = f(abx) = af(bx) = a(bf)(x).$$

In the non-commutative case things are much more complicate and we have:

Lemma 5.2. *Given ${}_AM_B$ and ${}_AN$, then $hom_A(M, N)$ becomes a left module over B for the law $(bf)(x) = f(xb)$.*

Proof. We just need to check the two relations:

$$(bf)(ax) = f(axb) = af(xb) = a(bf)(x),$$

$$(b'(b''f))(x) = (b''f)(xb') = f(xb'b'') = ((b'b'')f)(x).$$

∎

A similar result can be obtained (exercise) with ${}_AM$ and ${}_AN_B$, where $hom_A(M,N)$ now becomes a right B-module for the law $(fb)(x) = f(x)b$.

Now we recall that a sequence of modules and maps is exact if the kernel of any map is equal to the image of the map preceding it and we have:

Theorem 5.2. *If M, M', M'' are A-modules, the sequence:*

$$M' \xrightarrow{f} M \xrightarrow{g} M'' \to 0$$

is exact if and only if the sequence:

$$0 \to hom(M'', N) \to hom(M, N) \to hom(M', N)$$

is exact for any A-module N.

Proof. Let us consider homomorphisms $h : M \to N$, $h' : M' \to N$, $h'' : M'' \to N$ such that $h'' \circ g = h$, $h \circ f = h'$. If $h = 0$, then $h'' \circ g = 0$ implies $h''(x'') = 0, \forall x'' \in M''$ because g is surjective and we can find $x \in M$ such that $x'' = g(x)$. Then $h''(x'') = h''(g(x)) = h'' \circ g(x) = 0$. Now, if $h' = 0$, we have $h \circ f = 0$ and h factors through g because the initial sequence is exact. Hence there exists $h'' : M'' \to N$ such that $h = h'' \circ g$ and the second sequence is exact.
We let the reader prove the converse as an exercise. ∎

Similarly, one can prove (exercise):

Corollary 5.1. *The short exact sequence:*

$$0 \to M' \to M \to M'' \to 0$$

splits if and only if the short exact sequence:

$$0 \to hom(M'', N) \to hom(M, N) \to hom(M', N) \to 0$$

is exact for any module N.

Definition 5.14. *If M is a module over a ring A, a system of generators of M over A is a family $\{x_i\}_{i \in I}$ of elements of M such that any element of M can be written $x = \sum_{i \in I} a_i x_i$ with only a finite number of nonzero a_i.*

Definition 5.15. *An A-module is called noetherian if every submodule of M (and thus M itself) is finitely generated.*

One has the following technical lemma (exercise):

Lemma 5.3. *In a short exact sequence of modules, the central module is noetherian if and only if the two other modules are noetherian.*

We obtain in particular:

Proposition 5.5. *If A is a noetherian ring and M is a finitely generated module over A, then M is noetherian.*

Proof. Applying the lemma to the short exact sequence $0 \rightarrow A^{r-1} \rightarrow A^r \rightarrow A \rightarrow 0$ where the epimorphism on the right is the projection onto the first factor, we deduce by induction that A^r is noetherian. Now, if M is generated by $\{x_1, ..., x_r\}$, there is an epimorphism $A^r \rightarrow M : (1, 0, ..., 0) \rightarrow x_1, ..., \{0, ..., 0, 1\} \rightarrow x_r$ and M is noetherian because of the lemma. ∎

In the preceding situation, the kernel of the epimorphism $A^r \rightarrow M$ is also finitely generated, say by $\{y_1, ..., y_s\}$ and we therefore obtain the exact sequence $A^s \rightarrow A^r \rightarrow M \rightarrow 0$ that can be extended inductively to the left.

Definition 5.16. *In this case, we say that M is finitely presented.*

We now present the technique of *localization* that will play a central rôle and is used to introduce rings and modules of fractions. We shall define the procedure in the non-commutative case but the reader will discover that, in the commutative case, localization is just the formal counterpart superseding Laplace transform as there is no longer any technical assumption on the initial data. Indeed, it is well known that, if the Laplace transform is $f(t) \rightarrow \hat{f}(s) = \int_0^\infty e^{-st} f(t)dt$, then $df(t)/dt \rightarrow s\hat{f}(s)$ only if we suppose that $f(0) = 0$. Of course, the achievement of introducing rational functions through the transfer matrix is the heart of the procedure which is also valid in the multidimensional (PD) case. However, it is essential to notice that only the localization technique can be applied to systems with variable coefficients. We start with a basic definition:

Definition 5.17. *A subset S of a ring A is said to be multiplicatively closed if $\forall s, t \in S \Rightarrow st \in S$ and $1 \in S$.*

Example 5.12. We provide a few useful cases:
- $\forall a \in A$, we may consider $S_a = \{1, a, a^2, ...\}$.
- $\forall \mathfrak{p} \in spec(A)$, we may consider $S = A - \mathfrak{p}$.
- For any ideal $\mathfrak{a} \in A$, we may consider $S = \{1 + a \mid a \in \mathfrak{a}\}$.
- We may consider the set S of non-zerodivisors of A. In particular, if A is an integral domain, we may consider $S = A - \{0\}$.

In a general way, whenever A is a non-commutative ring, that is $ab \neq ba$ when $a, b \in A$, we shall set the following definition:

Definition 5.18. *By a left ring of fractions or left localization of a non-commutative ring A with respect to a multiplicatively closed subset S of A, we mean a ring denoted by $S^{-1}A$ and a homomorphism $\theta = \theta_S : A \rightarrow S^{-1}A$ such that:*
1) $\theta(s)$ is invertible in $S^{-1}A, \forall s \in S$.
2) Each element of $S^{-1}A$ or fraction has the form $\theta(s)^{-1}\theta(a)$ for some $s \in S, a \in A$.
3) $ker(\theta) = \{a \in A \mid \exists s \in S, sa = 0\}$.
A right ring of fractions or right localization can be similarly defined.

In actual practice, a fraction will be simply written $s^{-1}a$ and we have to distinguish carefully $s^{-1}a$ from as^{-1}. We shall recover the standard notation a/s of the commutative case when two fractions a/s and b/t can be reduced to the same denominator $st = ts$. The following proposition is essential and will be completed by two technical lemmas that will be used for constructing localizations.

Proposition 5.6. *If there exists a left localization of A with respect to S, then we must have:*
1) $Sa \cap As \neq 0, \forall a \in A, \forall s \in S$.
2) If $s \in S$ and $a \in A$ are such that $as = 0$, then there exists $t \in S$ such that $ta = 0$.

Proof . The element $\theta(a)\theta(s)^{-1}$ in $S^{-1}A$ must be of the form $\theta(t)^{-1}\theta(b)$ for some $t \in S, b \in A$. Accordingly, $\theta(a)\theta(s)^{-1} = \theta(t)^{-1}\theta(b) \Rightarrow \theta(t)\theta(a) = \theta(b)\theta(s)$ and thus $\theta(ta - bs) = 0 \Rightarrow \exists u \in S, uta = ubs$ with $ut \in S, ub \in A$. Finally, $as = 0 \Rightarrow \theta(a)\theta(s) = 0 \Rightarrow \theta(a) = 0$ because $\theta(s)$ is invertible in $S^{-1}A$. Hence $\exists t \in S$ such that $ta = 0$. ∎

Definition 5.19. *A set S satisfying the condition 1) is called a left Ore set.*

Lemma 5.4. *If S is a left Ore set in a noetherian ring, then S also satisfies the condition 2) of the preceding lemma.*

Proof . Let $s \in S$ and $a \in A$ be such that $as = 0$. Denoting the left annihilator by $lann$, we have $lann(s^n) \subseteq lann(s^{n+1})$ for each integer n. As A is noetherian, $lann(s^n) = lann(s^{n+1})$ for $n \gg 0$. Using the left Ore condition, we can find $t \in S, b \in A$ such that $ta = bs^n$ and thus $bs^{n+1} = tas = 0 \Rightarrow b \in lann(s^{n+1}) = lann(s^n) \Rightarrow ta = 0$. We notice that $S^{-1}A$ is the zero ring if and only if $0 \in S$ and the second condition is not needed when A is an integral domain. ∎

Lemma 5.5. *If S is a left Ore set in a ring A, then $As \cap At \cap S \neq 0, \forall s,t \in S$ and two fractions can be brought to the same denominator.*

Proof. From the left Ore condition, we can find $u \in S$ and $a \in A$ such that $us = at \in S$. More generally, we can find $u, v \in A$ such that $us = vt \in S$ and we successively get:

$$\theta(us)^{-1}\theta(ua) = \theta(s)^{-1}\theta(u)^{-1}\theta(u)\theta(a) = \theta(s)^{-1}\theta(a), \forall a \in A$$

$$\theta(vt)^{-1}\theta(vb) = \theta(t)^{-1}\theta(v)^{-1}\theta(v)\theta(b) = \theta(t)^{-1}\theta(b)$$

so that the two fractions $\theta(s)^{-1}\theta(a)$ and $\theta(t)^{-1}\theta(b)$ can be brought to the same denominator $\theta(us) = \theta(vt)$. ∎

We are now in position to construct the ring of fractions $S^{-1}A$ whenever S satisfies the two conditions of the last proposition. For this, using the preceding lemmas, let us define an equivalence relation on $S \times A$ by saying that $(s,a) \sim (t,b)$ if one can find $u, v \in S$ such that $us = vt \in S$ and $ua = vb$. Such a relation is clearly reflexive and symmetric, thus we only need to prove that it is transitive. So let $(s_1, a_1) \sim (s_2, a_2)$ and $(s_2, a_2) \sim (s_3, a_3)$. Then we can find $u_1, u_2 \in A$ such that $u_1 s_1 = u_2 s_2 \in S$ and $u_1 a_1 = u_2 a_2$. Also we can find $v_2, v_3 \in A$ such that $v_2 s_2 = v_3 s_3 \in S$ and $v_2 a = v_3 a_3$. Now, from the Ore condition, one can find $w_1, w_3 \in A$ such that $w_1 u_1 s_1 = w_3 v_3 s_3 \in S$ and thus $w_1 u_2 s_2 = w_3 v_2 s_2 \in S$, that is to say $(w_1 u_2 - w_3 v_2) s_2 = 0$. Hence, unless A is an integral domain, using the second condition of the last proposition, we can find $t \in S$ such that $t(w_1 u_2 - w_3 v_2) = 0 \Rightarrow t w_1 u_2 = t w_3 v_2$. Hence, changing w_1 and w_3 if necessary, we may finally assume that $w_1 u_2 = w_3 v_2 \Rightarrow w_1 u_1 a_1 = w_1 u_2 a_2 = w_3 v_2 a_2 = w_3 v_3 a_3$ as wished. We finally define $S^{-1}A$ to be the quotient of $S \times A$ by the above equivalence relation with $\theta : A \rightarrow S^{-1}A : a \rightarrow 1^{-1}a$.
The sum $(s,a) + (t,b)$ will be defined to be $(us = vt, ua + vb)$ and the product $(s,a) \times (t,b)$ will be defined to be (st, ab).
A similar approach can be used in order to define and construct modules of fractions whenever S satisfies the two conditions of the last proposition. For this we need a preliminary lemma:

Lemma 5.6. *If S is a left Ore set in a ring A and M is a left module over A, the set:*

$$t_S(M) = \{x \in M \mid \exists s \in S, sx = 0\}$$

is a submodule of M called the S-torsion submodule of M.

Proof. If $x, y \in t_S(M)$, we may find $s, t \in S$ such that $sx = 0, ty = 0$. Now, we can find $u, v \in A$ such that $us = vt \in S$ and we successively get $us(x + y) = usx + vty = 0 \Rightarrow x + y \in t_S(M)$. Also, $\forall a \in A$, using the Ore condition for S, we can find $b \in A, t \in S$ such that $ta = bs$ and we get $tax = bsx = 0 \Rightarrow ax \in t_S(M)$. ∎

Definition 5.20. *By a left module of fractions or left localization of M with respect to S, we mean a left module $S^{-1}M$ over $S^{-1}A$ both with a homomorphism $\theta = \theta_S : M \to S^{-1}M : x \to 1^{-1}x$ such that:*
1) Each element of $S^{-1}M$ has the form $s^{-1}\theta(x)$ for $s \in S, x \in M$.
2) $ker(\theta_S) = t_S(M)$.

In order to construct $S^{-1}M$, we shall define an equivalence relation on $S \times M$ by saying that $(s, x) \sim (t, y)$ if there exists $u, v \in A$ such that $us = vt \in S$ and $ux = vy$. Checking that this relation is reflexive, symmetric and transitive can be done as before (exercise) and we define $S^{-1}M$ to be the quotient of $S \times M$ by this equivalence relation.
The main property of localization is expressed by the following theorem.

Theorem 5.3. *If one has an exact sequence:*

$$M' \xrightarrow{f} M \xrightarrow{g} M''$$

then one also has the exact sequence:

$$S^{-1}M' \xrightarrow{S^{-1}f} S^{-1}M \xrightarrow{S^{-1}g} S^{-1}M''$$

where $S^{-1}f(s^{-1}x) = s^{-1}f(x)$.

Proof. As $g \circ f = 0$, we also have $S^{-1}g \circ S^{-1}f = 0$ and thus $im(S^{-1}f) \subseteq ker(S^{-1}g)$.
In order to prove the reverse inclusion, let $s^{-1}x \in ker(S^{-1}g)$. We have therefore $s^{-1}g(x) = 0$ in $S^{-1}M''$ and there exists $t \in S$ such that $tg(x) = g(tx) = 0$ in M''. As the initial sequence is exact, we can find $x' \in M'$ such that $tx = f(x')$. Accordingly, in $S^{-1}M$ we have $s^{-1}x = s^{-1}t^{-1}tx = (ts)^{-1}tx = (ts)^{-1}f(x') = S^{-1}f((ts)^{-1}x')$ and thus $ker(S^{-1}g) \subseteq im(S^{-1}f)$.

∎

The proof of the following corollary is left to the reader as an exercise.

Corollary 5.2. *If M' and M'' are submodules of an A-module M and S is a multiplicatively closed subset of A, we have the relations:*
- $S^{-1}(M' \cap M'') = (S^{-1}M') \cap (S^{-1}M'')$.
- $S^{-1}(M' + M'') = (S^{-1}M') + (S^{-1}M'')$.
- $S^{-1}(M' \oplus M'') = (S^{-1}M') \oplus (S^{-1}M'')$.
- $S^{-1}(M/M') = (S^{-1}M)/(S^{-1}M')$.

We now turn to the definition and brief study of tensor products of modules over rings that will not be necessarily commutative unless stated explicitly.
Let $M = M_A$ be a right A-module and $N = {}_AN$ be a left A-module. We may introduce the free $\mathbb{Z}$-module made by finite formal linear combinations of elements of $M \times N$ with coefficients in $\mathbb{Z}$.

Definition 5.21. *The tensor product of M and N over A is the $\mathbb{Z}$-module $M\otimes_A N$ obtained by quotienting the above $\mathbb{Z}$-module by the submodule generated by the elements of the form:*

$$(x+x',y)-(x,y)-(x',y),(x,y+y')-(x,y)-(x,y'),(xa,y)-(x,ay)$$

and the image of (x,y) will be denoted by $x\otimes y$.

It follows from the definition that we have the relations:

$$(x+x')\otimes y = x\otimes y + x'\otimes y, x\otimes(y+y') = x\otimes y + x\otimes y', xa\otimes y = x\otimes ay$$

and there is a canonical isomorphism $M\otimes_A A \simeq M, A\otimes_A N \simeq N$. When A is commutative, we may use left modules only and $M\otimes_A N$ becomes a left A-module.

Example 5.13. If $A=\mathbb{Z}, M=\mathbb{Z}/2\mathbb{Z}$ and $N=\mathbb{Z}/3\mathbb{Z}$, we have $(\mathbb{Z}/2\mathbb{Z})\otimes_{\mathbb{Z}}(\mathbb{Z}/3\mathbb{Z}) = 0$ because $x\otimes y = 3(x\otimes y)-2(x\otimes y) = x\otimes 3y - 2x\otimes y = 0-0=0$.

As a link with localization, we let the reader prove that the multiplication map $S^{-1}A\times M \to S^{-1}M$ given by $(s^{-1}a,x)\to s^{-1}ax$ induces an isomorphism $S^{-1}A\otimes_A M \to S^{-1}M$ of modules over $S^{-1}A$ when$S^{-1}A$ is considered as a right module over A and M as a left module over A.
When A is a commutative integral domain and $S=A-\{0\}$, the field $K=Q(A)=S^{-1}A$ is called the field of fractions of A and we have the short exact sequence:

$$0\longrightarrow A\longrightarrow K\longrightarrow K/A\longrightarrow 0$$

If now M is a left A-module, we may tensor this sequence by M on the right with $A\otimes M = M$ but we do not get in general an exact sequence. The defect of exactness *on the left* is nothing else but the *torsion submodule* $t(M)=\{m\in M \mid \exists 0\neq a\in A, am=0\}\subseteq M$ and we have the long exact sequence:

$$0\longrightarrow t(M)\longrightarrow M\longrightarrow K\otimes_A M\longrightarrow K/A\otimes_A M\longrightarrow 0$$

as we may describe the central map as follows:

$$m\longrightarrow 1\otimes m = \frac{a}{a}\otimes m = \frac{1}{a}\otimes am \quad , \quad \forall a\neq 0$$

Such a result, based on the localization technique, allows to understand why controllability has to do with the so-called "simplification" of the *transfer matrix*. In particular, a module M is said to be a *torsion module* if $t(M)=M$ and a *torsion-free module* if $t(M)=0$.

Definition 5.22. *A module in $mod(A)$ is called a free module if it has a basis, that is a system of generators linearly independent over A. When a module F is free, the number of generators in a basis, and thus in any basis (exercise), is called the rank of F over A and is denoted by $rank_A(F)$. In particular, if F is free of finite rank r, then $F \simeq A^r$.*

More generally, if M is any module over a ring A and F is a maximum free submodule of M, then $M/F = T$ is a torsion module. Indeed, if $x \in M, x \notin F$, then one can find $a \in A$ such that $ax \in F$ because, otherwise, $F \subset \{F, x\}$ should be free submodules of M with a strict inclusion. In that case, the *rank* of M is by definition the rank of F over A and one has equivalently :

Lemma 5.7. $rk_A(M) = dim_K(K \otimes_A M)$.

Proof. Taking the tensor product by K over A of the short exact sequence $0 \rightarrow F \rightarrow M \rightarrow T \rightarrow 0$, we get an isomorphism $K \otimes_A F \simeq K \otimes_A M$ because $K \otimes_A T = 0$ (exercise) and the lemma follows from the definition of the rank. ∎

We now provide two proofs of the *additivity property of the rank*, the second one being also valid for non-commutative rings.

Proposition 5.7. *If $0 \rightarrow M' \xrightarrow{f} M \xrightarrow{g} M'' \rightarrow 0$ is a short exact sequence of modules over a ring A, then we have $rk_A(M) = rk_A(M') + rk_A(M'')$.*

Proof 1: Using localization with respect to the multiplicatively closed subset $S = A - \{0\}$, this proposition is just a straight consequence of the definition of rank and the fact that localization preserves exactness.
Proof 2: Let us consider the following diagram with exact left/right columns and central row:

$$
\begin{array}{ccccccc}
 & & 0 & & 0 & & 0 & \\
 & & \downarrow & & \downarrow & & \downarrow & \\
0 \rightarrow & & F' & \rightarrow & F' \oplus F'' & \rightarrow & F'' & \rightarrow 0 \\
 & & \downarrow i' & & \downarrow i & & \downarrow i'' & \\
0 \rightarrow & & M' & \xrightarrow{f} & M & \xrightarrow{g} & M'' & \rightarrow 0 \\
 & & \downarrow p' & & \downarrow p & & \downarrow p'' & \\
0 \rightarrow & & T' & \rightarrow & T & \rightarrow & T'' & \rightarrow 0 \\
 & & \downarrow & & \downarrow & & \downarrow & \\
 & & 0 & & 0 & & 0 &
\end{array}
$$

where $F'(F'')$ is a maximum free submodule of $M'(M'')$ and $T' = M'/F'(T'' = M''/F'')$ is a torsion module. Pulling back by g the image under i'' of a basis of F'', we may obtain by linearity a map $\sigma : F'' \rightarrow M$ and we define $i = f \circ i' \circ \pi' + \sigma \circ \pi''$ where $\pi' : F' \oplus F'' \rightarrow F'$ and $\pi'' : F' \oplus F'' \rightarrow F''$ are the canonical projections on each factor of the direct sum. We have $i|_{F'} = f \circ i'$ and $g \circ i = g \circ \sigma \circ \pi'' = i'' \circ \pi''$. Hence, the diagram is commutative and thus exact with $rk_A(F' \oplus F'') = rk_A(F') + rk_A(F'')$ trivially. Finally, if T' and T'' are torsion modules, it is easy to check that T is a torsion module too and $F' \oplus F''$ is thus a maximum free submodule of M. ∎

Definition 5.23. *If $f : M \rightarrow N$ is any morphism, the rank of f will be defined to be $rk_A(f) = rk_A(im(f))$.*

We provide a few additional properties of the rank that will be used in the sequel. For this we shall set $M^* = hom_A(M, A)$ and, for any morphism $f : M \rightarrow N$ we shall denote by $f^* : N^* \rightarrow M^*$ the corresponding morphism which is such that $f^*(h) = h \circ f, \forall h \in hom_A(N, A)$.

Proposition 5.8. *When A is a commutative integral domain and M is a finitely presented module over A, then $rk_A(M) = rk_A(M^*)$.*

Proof. Applying $hom_A(\bullet, A)$ to the short exact sequence in the proof of the preceding lemma while taking into account $T^* = 0$, we get a monomorphism $0 \rightarrow M^* \rightarrow F^*$ and obtain therefore $rk_A(M^*) \leq rk_A(F^*)$. However, as $F \simeq A^r$ with $r < \infty$ because M is finitely generated, we get $F^* \simeq A^r$ too because $A^* \simeq A$. It follows that $rk_A(M^*) \leq rk_A(F^*) = rk_A(F) = rk_A(M)$ and thus $rk_A(M^*) \leq rk_A(M)$.
Now, if $F_1 \xrightarrow{d} F_0 \rightarrow M \rightarrow 0$ is a finite presentation of M, applying $hom_A(\bullet, A)$ to this presentation, we get the ker/coker exact sequence:

$$0 \leftarrow N \leftarrow F_1^* \xleftarrow{d^*} F_0^* \leftarrow M^* \leftarrow 0$$

Applying $hom_A(\bullet, A)$ to this sequence while taking into account the isomorphisms $F_0^{**} \simeq F_0, F_1^{**} \simeq F_1$, we get the ker/coker exact sequence:

$$0 \rightarrow N^* \rightarrow F_1 \xrightarrow{d} F_0 \rightarrow M \rightarrow 0$$

Counting the ranks, we obtain:

$$\begin{aligned} rk_A(N) - rk_A(M^*) = rk_A(F_1^*) - rk_A(F_0^*) &= rk_A(F_1) - rk_A(F_0) \\ &= rk_A(N^*) - rk_A(M) \end{aligned}$$

and thus:

$$(rk_A(M) - rk_A(M^*)) + (rk_A(N) - rk_A(N^*)) = 0$$

As both two numbers in this sum are non-negative, they must be zero and we finally get $rk_A(M) = rk_A(M^*), rk_A(N) = rk_A(N^*)$. ∎

Corollary 5.3. *Under the condition of the proposition, we have* $rk_A(f) = rk_A(f^*)$.

Proof. Introducing the *ker/coker* exact sequence:

$$0 \to K \to M \xrightarrow{f} N \to Q \to 0$$

we have: $rk_A(f) + rk_A(Q) = rk_A(N)$. Applying $hom_A(\bullet, A)$ and taking into account Theorem 5.2, we have the exact sequence:

$$0 \to Q^* \to N^* \xrightarrow{f^*} M^*$$

and thus : $rk_A(f^*) + rk_A(Q^*) = rk_A(N^*)$. Using the preceding proposition, we get $rk_A(Q) = rk_A(Q^*)$ and $rk_A(N) = rk_A(N^*)$, that is to say $rk_A(f) = rk_A(f^*)$. ∎

5.3.2 Homological Algebra

Having in mind the introductory section, we now need a few definitions and results from homological algebra [4, 12, 23]. In all that follows, $A, B, C, ...$ are modules over a ring A or vector spaces over a field k and the linear maps are making the diagrams commutative.
We start recalling the well known Cramer's rule for linear systems through the exactness of the ker/coker sequence for modules. We introduce the notations $rk = rank$, $nb = number$, $dim = dimension$, $ker = kernel$, $im = image$, $coker = cokernel$. When $\Phi : A \to B$ is a linear map (homomorphism), we introduce the so-called ker/coker exact sequence:

$$0 \longrightarrow ker(\Phi) \longrightarrow A \xrightarrow{\Phi} B \longrightarrow coker(\Phi) \longrightarrow 0$$

where $coker(\Phi) = B/im(\Phi)$.
In the case of vector spaces over a field k, we successively have $rk(\Phi) = dim(im(\Phi))$, $dim(ker(\Phi)) = dim(A) - rk(\Phi)$, $dim(coker(\Phi)) = dim(B) - rk(\Phi) = nb$ of compatibility conditions, and obtain by subtraction:

$$dim(ker(\Phi)) - dim(A) + dim(B) - dim(coker(\Phi)) = 0$$

In the case of modules, using localization, we may replace the dimension by the rank and obtain the same relations because of the additive property of the rank. The following theorem is essential:

Theorem 5.4 (Snake). *When one has the following commutative diagram resulting from the the two central vertical short exact sequences by exhibiting the three corresponding horizontal ker/coker exact sequences:*

$$\begin{array}{ccccccccccc}
 & & 0 & & 0 & & 0 & & & & \\
 & & \downarrow & & \downarrow & & \downarrow & & & & \\
0 & \longrightarrow & K & \longrightarrow & A & \longrightarrow & A' & \longrightarrow & Q & \longrightarrow & 0 \\
 & & \downarrow & & \downarrow \Phi & & \downarrow \Phi' & & \downarrow & & \\
0 & \longrightarrow & L & \longrightarrow & B & \longrightarrow & B' & \longrightarrow & R & \longrightarrow & 0 \\
 & & \downarrow & & \downarrow \Psi & & \downarrow \Psi' & & \downarrow & & \\
0 & \longrightarrow & M & \longrightarrow & C & \longrightarrow & C' & \longrightarrow & S & \longrightarrow & 0 \\
 & & & & \downarrow & & \downarrow & & \downarrow & & \\
 & & & & 0 & & 0 & & 0 & &
\end{array}$$

then there exists a connecting map $M \longrightarrow Q$ both with a long exact sequence:

$$0 \longrightarrow K \longrightarrow L \longrightarrow M \longrightarrow Q \longrightarrow R \longrightarrow S \longrightarrow 0.$$

Proof. We start constructing the connecting map by using the following succession of elements:

$$\begin{array}{ccccc}
 & & a & \cdots & a' & \longrightarrow & q \\
 & & \vdots & & \downarrow & & \\
 & & b & \longrightarrow & b' & & \\
 & & \downarrow & & \vdots & & \\
m & \longrightarrow & c & \cdots & 0 & &
\end{array}$$

Indeed, starting with $m \in M$, we may identify it with $c \in C$ in the kernel of the next horizontal map. As Ψ is an epimorphism, we may find $b \in B$ such that $c = \Psi(b)$ and apply the next horizontal map to get $b' \in B'$ in the kernel of Ψ' by the commutativity of the lower square. Accordingly, there is a unique $a' \in A'$ such that $b' = \Phi'(a')$ and we may finally project a' to $q \in Q$. The map is well defined because, if we take another lift for c in B, it will differ from b by the image under Φ of a certain $a \in A$ having zero image in Q by composition. The remaining of the proof is similar and left to the reader as an exercise. The above explicit procedure will not be repeated. ∎

We may now introduce *cohomology theory* through the following definition.

Definition 5.24. *If one has a sequence $A \xrightarrow{\Phi} B \xrightarrow{\Psi} C$, then one may introduce coboundary $= im(\Phi) \subseteq ker(\Psi) =$ cocycle $\subseteq B$ and define the cohomology at B to be the quotient cocycle/coboundary.*

Theorem 5.5. *The following commutative diagram where the two central vertical sequences are long exact sequences and the horizontal lines are ker/coker exact sequences:*

$$
\begin{array}{ccccccccccc}
 & & 0 & & 0 & & 0 & & & & \\
 & & \downarrow & & \downarrow & & \downarrow & & & & \\
0 & \longrightarrow & K & \longrightarrow & A & \longrightarrow & A' & \longrightarrow & Q & \longrightarrow & 0 \\
 & & \downarrow & & \downarrow \Phi & & \downarrow \Phi' & & \downarrow & & \\
0 & \longrightarrow & L & \longrightarrow & B & \longrightarrow & B' & \longrightarrow & R & \longrightarrow & 0 \\
\cdots & \cdots & \downarrow & \cdots & \downarrow \Psi & \cdots & \downarrow \Psi' & \cdots & \downarrow & \cdots & \cdots\cdots\cdots\ cut \\
0 & \longrightarrow & M & \longrightarrow & C & \longrightarrow & C' & \longrightarrow & S & \longrightarrow & 0 \\
 & & \downarrow & & \downarrow \Omega & & \downarrow \Omega' & & \downarrow & & \\
0 & \longrightarrow & N & \longrightarrow & D & \longrightarrow & D' & \longrightarrow & T & \longrightarrow & 0 \\
 & & & & \downarrow & & \downarrow & & \downarrow & & \\
 & & & & 0 & & 0 & & 0 & &
\end{array}
$$

induces an isomorphism between the cohomology at M in the left vertical column and the kernel of the morphism $Q \to R$ in the right vertical column.

Proof. Let us "cut" the preceding diagram into the following two commutative and exact diagrams by taking into account the relations $im(\Psi) = ker(\Omega), im(\Psi') = ker(\Omega')$:

$$
\begin{array}{ccccccccccc}
 & & 0 & & 0 & & 0 & & & & \\
 & & \downarrow & & \downarrow & & \downarrow & & & & \\
0 & \longrightarrow & K & \longrightarrow & A & \longrightarrow & A' & \longrightarrow & Q & \longrightarrow & 0 \\
 & & \downarrow & & \downarrow \Phi & & \downarrow \Phi' & & \downarrow & & \\
0 & \longrightarrow & L & \longrightarrow & B & \longrightarrow & B' & \longrightarrow & R & \longrightarrow & 0 \\
 & & \downarrow & & \downarrow \Psi & & \downarrow \Psi' & & & & \\
0 & \longrightarrow & cocycle & \longrightarrow & im\,\Psi & \longrightarrow & im\,\Psi' & & & & \\
 & & & & \downarrow & & \downarrow & & & & \\
 & & & & 0 & & 0 & & & &
\end{array}
$$

$$
\begin{array}{ccccccc}
 & & 0 & & 0 & & 0 \\
 & & \downarrow & & \downarrow & & \downarrow \\
0 & \longrightarrow & cocycle & \longrightarrow & ker\,\Omega & \longrightarrow & ker\,\Omega' \\
 & & \downarrow & & \downarrow & & \downarrow \\
0 & \longrightarrow & M & \longrightarrow & C & \longrightarrow & C' \\
 & & \downarrow & & \downarrow \Omega & & \downarrow \Omega' \\
0 & \longrightarrow & N & \longrightarrow & D & \longrightarrow & D' \\
 & & & & \downarrow & & \downarrow \\
 & & & & 0 & & 0
\end{array}
$$

Finally, using the snake theorem, we successively obtain:

$$
\begin{array}{lcr}
\Longrightarrow \exists & 0 \longrightarrow K \longrightarrow L \stackrel{\Psi}{\longrightarrow} cocycle \longrightarrow Q \stackrel{\Phi'}{\longrightarrow} R & exact \\
\Longrightarrow \exists\, 0 & \longrightarrow coboundary \longrightarrow cocycle \longrightarrow ker\,(Q \longrightarrow R) \longrightarrow 0 & exact \\
\Longrightarrow & cohomology\ at\ M \simeq ker\,(Q \longrightarrow R) &
\end{array}
$$

■

We now introduce the *extension functor* in an elementary manner, using the standard notation $hom_A(M, A) = M^*$. First of all, by a *free resolution* of an A-module M, we understand a long exact sequence:

$$... \xrightarrow{d_2} F_1 \xrightarrow{d_1} F_0 \longrightarrow M \longrightarrow 0$$

where $F_0, F_1, ...$are free modules, that is to say modules isomorphic to powers of A and $M = coker(d_1) = F_0/im(d_1)$. We may *take out* M and obtain the *deleted sequence*:

$$... \xrightarrow{d_2} F_1 \xrightarrow{d_1} F_0 \longrightarrow 0$$

which is of course no longer exact. If N is any other A-module, we may apply the functor $hom_A(\bullet, N)$ and obtain the sequence:

$$... \xleftarrow{d_2^*} hom_A(F_1, N) \xleftarrow{d_1^*} hom_A(F_0, N) \longleftarrow 0$$

in order to state:

Definition 5.25. $ext_A^0(M, N) = ker(d_1^*) = hom_A(M, N)$,
$ext_A^i(M, N) = ker(d_{i+1}^*)/im(d_i^*), \forall i \geq 1$

One can prove that the extension modules do not depend on the resolution of M chosen and have the following two main properties, the first of which only is classical [19, 22, 23].

Proposition 5.9. *If* $0 \to M' \to M \to M'' \to 0$ *is a short exact sequence of A-modules, then we have the following connecting long exact sequence:*

$$\begin{aligned} 0 \to hom_A(M'', N) \to hom_A(M, N) \\ \to hom_A(M', N) \to ext_A^1(M'', N) \to ... \end{aligned}$$

of extension modules.

Proposition 5.10. $ext_A^i(M, A)$ *is a torsion module,* $\forall i \geq 1$.

Proof 1. Let F be a maximal free submodule of M. From the short exact sequence:

$$0 \longrightarrow F \longrightarrow M \longrightarrow M/F \longrightarrow 0$$

where M/F is a torsion module, we obtain the long exact sequence:

$$\begin{aligned} ... \to ext_A^{i-1}(F, A) \to ext_A^i(M/F, A) \\ \to ext_A^i(M, A) \to ext_A^i(F, A) \to ... \end{aligned}$$

From the definitions, we obtain $ext_A^i(F, A) = 0, \forall i \geq 1$ and thus $ext_A^i(M, A) \simeq ext_A^i(M/F, A), \forall i \geq 2$. Now it is known that the tensor by the field K of any exact sequence is again an exact sequence. Accordingly, we have from the definition:

$$K\otimes_A ext^i_A(M/F,A) \simeq ext^i_A(M/F,K) \\ \simeq ext^i_K(K\otimes_A M/F,K) = 0, \forall i \geq 1$$

and we finally obtain from the above sequence $K\otimes_A ext^i_A(M,A) = 0 \Rightarrow ext^i_A(M,A)$ torsion, $\forall i \geq 1$. ■

Proof 2. Having in mind that $B_i = im(d^*_i)$ and $Z_i = ker(d^*_{i+1})$, we obtain $rk(B_i) = rk(d^*_i) = rk(d_i)$ and $rk(Z_i) = rk(F^*_i) - rk(d^*_{i+1}) = rk(F_i) - rk(d_{i+1})$. However, we started from a resolution, that is an exact sequence in which $rk(d_i) + rk(d_{i+1}) = rk(F_i)$. It follows that $rk(B_i) = rk(Z_i)$ and thus $rk(H_i) = rk(Z_i) - rk(B_i) = 0$, that is to say $ext^i_A(M,A)$ is a torsion module for $i \geq 1$, $\forall M \in mod(A)$. ■

As we have seen in the motivating Examples 5.4, 5.5, 5.6, the same module may have many very different presentations. In particular, we have [8, 19, 23]:

Lemma 5.8 (Schanuel). *If $F'_1 \xrightarrow{d'_1} F'_0 \to M \to 0$ and $F''_1 \xrightarrow{d''_1} F''_0 \to M \to 0$ are two presentations of M, there exists a presentation $F_1 \xrightarrow{d_1} F_0 \to M \to 0$ of M projecting onto the preceding ones.*

Definition 5.26. *An A-module P is projective if there exists a free module F and another (thus projective) module Q such that $P \oplus Q \simeq F$. Any free module is projective.*

Proposition 5.11. *The short exact sequence*

$$0 \longrightarrow M' \xrightarrow{f} M \xrightarrow{g} M'' \longrightarrow 0$$

splits whenever M'' is projective.

Proposition 5.12. *When P is a projective module and N is any module, we have $ext^i_A(P,N) = 0, \forall i \geq 1$.*

Proposition 5.13. *When P is a projective module, applying $hom_A(P,\bullet)$ to any short exact sequence gives a short exact sequence.*

5.3.3 System Theory

We recall a few basic facts from jet theory and system theory [11, 18, 24].

Let X be a manifold of dimension n with local coordinates $x = (x^1, ..., x^n)$ and E be a vector bundle over X with local coordinates (x^i, y^k), where

$i = 1, ..., n$ for the independent variables, $k = 1, ..., m$ for the dependent variables, and projection $(x, y) \rightarrow x$. A (local) section $\xi : X \rightarrow E : x \rightarrow (x, \xi(x))$ is defined locally by $y^k = \xi^k(x)$. Under any change of local coordinates $(x, y) \rightarrow (\bar{x} = \varphi(x), \bar{y} = A(x)y)$ the section changes according to $\bar{y}^l = \bar{\xi}^l(\bar{x})$ in such a way that $\bar{\xi}^l(\varphi(x)) \equiv A^l_k(x)\xi^k(x)$ and we may differentiate successively each member in order to obtain, though in a more and more tedious way, the transition rules for the derivatives $\xi^k(x), \partial_i\xi^k(x), \partial_{ij}\xi^k(x), ...$ up to order q. As usual, we shall denote by $J_q(E)$ and call q-*jet bundle* the vector bundle over X with the same transition rules and local *jet coordinates* (x, y_q) with $y_q = (y^k, y^k_i, y^k_{ij}, ...)$ or, more generally y^k_μ with $1 \leq \mid \mu \mid \leq q$ where $\mu = (\mu_1, ..., \mu_n)$ is a multi-index of length $\mid \mu \mid = \mu_1 + ... + \mu_n$ and $\mu + 1_i = (\mu_1, ..., \mu_{i-1}, \mu_i + 1, \mu_{i+1}, ..., \mu_n)$. The reader must not forget that the above definitions are standard ones in physics or mechanics because of the use of tensors in electromagnetism or elasticity.

Definition 5.27. *A system of PD equations on E is a vector sub-bundle $R_q \subset J_q(E)$ locally defined by a constant rank system of linear equations $A^{\tau\mu}_k(x)y^k_\mu = 0$.*

Substituting the derivatives of a section in place of the corresponding jet coordinates, then differentiating once with respect to x^i and substituting the jet coordinates, we get the *first prolongation* $R_{q+1} \subset J_{q+1}(E)$, defined by the previous equations and by the new equations $A^{\tau\mu}_k(x)y^k_{\mu+1_i} + \partial_i A^{\tau\mu}_k(x)y^k_\mu = 0$, and, more generally, the r-prolongations $R_{q+r} \subset J_{q+r}(E)$ which need not be vector bundles ($xy_x - y = 0 \Longrightarrow xy_{xx} = 0$).

Definition 5.28. *R_q is said to be formally integrable if the R_{q+r} are vector bundles and all the generating PD equations of order $q + r$ are obtained by prolonging R_q exactly r-times only, $\forall r \geq 0$.*

The modern way to deal with a linear system of PDE is to use jet coordinates instead of derivatives in order to define a vector sub-bundle $R_q \subset J_q(E)$ by a system of local (*nondifferential*) equations $A^{\tau\mu}_k(x)y^k_\mu = 0$ where we have used Einstein summation. The *r-prolongation* $\rho_r(R_q) = J_r(R_q) \cap J_{q+r}(E) \subset J_r(J_q(E))$ or simply R_{q+r} will be obtained by substituting derivatives instead of jet coordinates, differentiating r times in the usual way and substituting again jet coordinates. The projections $\pi^{q+r+s}_{q+r} : J_{q+r+s}(E) \longrightarrow J_{q+r}(E)$ induce maps $\pi^{q+r+s}_{q+r} : R_{q+r+s} \longrightarrow R_{q+r}$ which are not in general surjective and we may introduce the families of vector spaces $R^{(s)}_{q+r} = \pi^{q+r+s}_{q+r}(R_{q+r+s}) \subseteq R_{q+r}$ which may not be vector bundles for any $r, s \geq 0$.

The *symbol* $g_q = R_q \cap S_qT^* \otimes E \subset J_q(E)$ of R_q is defined by the linear equations:

$$A_k^{\tau\mu}(x)v_\mu^k = 0 \qquad\qquad |\mu| = q$$

We let the reader check that the symbol g_{q+r} of R_{q+r} is similarly defined by the linear equations:

$$A_k^{\tau\mu}(x)v_{\mu+\nu}^k = 0 \qquad\qquad |\mu| = q, |\nu| = r$$

Now, to any set (P^τ) of polynomials of degree q in $k[\chi]$ with $P^\tau = A^{\tau\mu}\chi_\mu$ and $\chi_\mu = (\chi_1)^{\mu_1}...(\chi_n)^{\mu_n}$ we may associate a trivial vector bundle with fiber dimension one and the linear system R_q defined "locally" by the equations:

$$A^{\tau\mu}y_\mu = 0 \qquad\qquad 0 \leq |\mu| \leq q$$

on the condition to consider it with locally constant coefficicnts. We notice that such a property insures all the regularity conditions needed for the applications of the formal theory and allows to associate a linear system of PD equations in one unknown with constant coefficients to any system of polynomials generating an ideal in $k[\chi]$.

In any case, from now on we shall suppose that the various symbols and projections already defined are vector bundles over k, with a slight abuse of language. Apart from the situation coming from pure algebra as we saw, such a case is quite rare in practice and only happens usually in the study of transitive Lie pseudogroups of transformations [18].

If we introduce the cotangent bundle $T^* = T^*(X)$ with corresponding tensor, exterior and symmetric products respectively denoted by $\otimes, \Lambda, S$, we may define the Spencer map:

$$\delta : \Lambda^s T^* \otimes g_{q+r+1} \longrightarrow \Lambda^{s+1}T^* \otimes g_{q+r}$$

by the following local formulas on families of forms:

$$(\delta_i v)_\mu^k = v_{\mu+1_i}^k, \qquad\qquad (\delta v)_\mu^k = dx^i \wedge \delta_i v$$

One has:

$$((\delta \circ \delta)v)_\mu^k = dx^i \wedge dx^j \wedge v_{\mu+1_i+1_j}^k \equiv 0$$

and thus $\delta \circ \delta = 0$. The cohomology at $\Lambda^s T^* \otimes g_{q+r}$ of the corresponding sequence is denoted by $H_{q+r}^s(g_q)$ as it only depends on g_q which is said to be *s-acyclic* if $H_{q+r}^1 = ... = H_{q+r}^s = 0, \forall r \geq 0$, *involutive* if it is n-acyclic and *finite type* if $g_{q+r} = 0$ for r big enough. Contrary to 2-acyclicity, involutivity can be checked by means of a finite algorithm and g_{q+r} becomes involutive for r big enough whenever g_q is not involutive.

Indeed, let us define locally:

$$(g_q)^i = \{v \in g_q \mid \delta_1 v = 0, ..., \delta_i v = 0\}$$

with $(g_q)^0 = g_q, (g_q)^n = 0$ and introduce the local *characters*:

$$\alpha_q^i = dim(g_q)^{i-1} - dim(g_q)^i$$

Then g_q is involutive if and only if there exists a system of local coordinates, called *δ-regular*, such that:

$$dim(g_{q+1}) = \alpha_q^1 + 2\alpha_q^2 + ... + n\alpha_q^n$$

or, equivalently, if we have the following short exact sequences:

$$0 \longrightarrow (g_{q+1})^i \longrightarrow (g_{q+1})^{i-1} \xrightarrow{\delta_i} (g_q)^{i-1} \longrightarrow 0$$

The latter definition is the modern version of the *multiplicative* and *nonmultiplicative* variables used by Janet that we now recall and illustrate [5].

For this, let us order the v_μ^k according to the lexicographic order on μ, k and let us solve in cascade the various linear equations defining g_q with respect to the highest v_μ^k each time. Then, let us associate with each such solved equation the multiplicative variables $(x^1, ..., x^i)$ if the highest v_μ^k is of *class* i, that is to say is such that $\mu_1 = ... = \mu_{i-1} = 0$ with $\mu_i \neq 0$. Of course, this choice will highly depend on the local coordinates that we can always change linearly. One can prove that the following definition is equivalent to the previous one.

Definition 5.29. *g_q is said to be involutive if its first prolongation with respect to the only multiplicative variables is producing g_{q+1}. In that case, the system of coordinates is said to be δ-regular.*

Remark 5.2. The case of a finite type symbol is the only situation where one can test 2-acyclicity because one can prove easily that the prolongation of a finite type symbol becomes involutive only when the symbol becomes zero [18, 19]. Also, when $n = 2$, we notice that the symbol of the system $y_{11} = 0, y_{12} = 0$ is involutive though (x^1, x^2) is not δ-regular as we may exchange x^1 and x^2 to check the preceding definition.

Example 5.14. With $n = 3, q = 2, m = 1$, we provide an example of a symbol which is finite type and 2-acyclic but not involutive.
Let us consider the symbol g_2 defined by the 3 equations:

$$v_{33} - v_{11} = 0, \qquad v_{23} = 0, \qquad v_{22} - v_{11} = 0$$

We easily obtain $dim(g_2) = 3$ with only parametric jets v_{11}, v_{12}, v_{13}, then $dim(g_3) = 1$ with only parametric jet v_{111} and, surprisingly, $g_4 = 0$. It follows that g_2 is finite type but cannot be involutive because, if it were so, all the Spencer δ-sequences should be exact, in particular the one finishing at $\Lambda^3 T^* \otimes g_3$. Counting the dimensions in this sequence, we should obtain $g_3 = 0$ and a contradiction. A similar comment should prove that g_3 is of course finite type but not involutive too but we now prove that g_3 is 2-acyclic. For this, using the sequence:

$$0 \to g_5 \to T^* \otimes g_4 \to \Lambda^2 T^* \otimes g_3 \to \Lambda^3 T^* \otimes g_2 \to 0$$

and the fact that $g_4 = 0 \Rightarrow g_5 = 0$, we just need to prove that the last map on the right is an isomorphism. However, the kernel of this map is defined by the 3 equations:

$$v_{11,123} = v_{111,23} + v_{112,31} + v_{113,12} = 0$$

$$v_{12,123} = v_{121,23} + v_{122,31} + v_{123,12} = 0$$

$$v_{13,123} = v_{131,23} + v_{132,31} + v_{133,12} = 0$$

that is to say:

$$v_{111,23} = 0, \qquad v_{111,31} = 0, \qquad v_{111,12} = 0$$

Hence, the last map on the right, being a monomorphism between two spaces of the same dimension 3 is an isomorphism and $H_3^2(g_2) = 0$.

The key (absolutely nontrivial) theorem from which all results can be obtained is the following one that can also be extended to nonlinear systems (cf. [18, 24])

Theorem 5.6. *If g_q is 2-acyclic, then $(R_q^{(1)})_{+r} = R_{q+r}^{(1)}, \forall r \geq 0$.*

Definition 5.30. *A system is said to be formally integrable if $R_{q+r}^{(1)} = R_{q+r}$, $\forall r \geq 0$ and involutive if it is formally integrable with an involutive symbol.*

Accordingly, we have the following *criterion for formal integrability* which is crucial for applications (cf. [18, 19, 24]):

Corollary 5.4. *If g_q is 2-acyclic and $R_q^{(1)} = R_q$, then R_q is formally integrable.*

A delicate inductive use of this criterion provides [18, 19]:

Corollary 5.5. *There is a finite algorithm providing two integers $r, s \geq 0$ such that the system $R_{q+r}^{(s)}$ is formally integrable (involutive) with the same solutions as R_q .*

Example 5.15. In order to help the reader maturing these new concepts, we illustrate the preceding effective results by showing out, in the case of the Janet motivating example 5.6, that $r = 3, s = 2$ with, totally unexpectedly, $g_5^{(2)} = 0$.

For this, if $\Phi : E \rightarrow F$ is a morphism between vector bundles, we define the r-prolongation $\rho_r(\Phi) : J_{q+r}(E) \rightarrow J_r(F)$ by means of the local induction $\Phi : A_k^{\tau\mu}(x)y_\mu^k = u^\tau \Rightarrow \rho_r(\Phi) : A_k^{\tau\mu}(x)y_{\mu+\nu}^k + ... + \partial_\nu A_k^{\tau\mu}(x)y_\mu^k$ for $0 \leq\mid \nu \mid\leq r$, according to the well known Leibnitz rules of derivations.

Also, if $\Phi : E \rightarrow F$ is a morphism of vector bundles, we may introduce, as in the Subsection 5.3.2 $K = ker(\Phi), Q = coker(\Phi) = F/im(\Phi)$ in the following ker-coker long exact sequence of vector bundles:

$$0 \rightarrow K \rightarrow E \rightarrow F \rightarrow Q \rightarrow 0$$

where one checks the equality $dim(K) - dim(E) + dim(F) - dim(Q) = 0$ because $dim(E) - dim(K) = dim(im(\Phi)) = dim(F) - dim(Q)$.

Finally, introducing $J_{q+r}^q(E) = ker(\pi_q^{q+r})$, it is easy to check that $J_q^{q-1}(E) = S_qT^* \otimes E$ and we recall the various above results and definitions in the following commutative diagram $diag(q, r, s)$ where we have set $F = J_q(E)/R_q$:

$$\begin{array}{ccccccccccc}
 & & 0 & & 0 & & & 0 & & & \\
 & & \downarrow & & \downarrow & & & \downarrow & & & \\
0 & \longrightarrow & R_{q+r+s}^{q+r} & \longrightarrow & J_{q+r+s}^{q+r}(E) & \longrightarrow & & J_{r+s}^r(F) & & & \\
 & & \downarrow & & \downarrow & & & \downarrow & & & \\
 & \longrightarrow & R_{q+r+s} & \longrightarrow & J_{q+r+s}(E) & \overset{\rho_{r+s}(\Phi)}{\longrightarrow} & & J_{r+s}(F) & \longrightarrow & Q_{r+s} & \longrightarrow 0 \\
 & & \downarrow & & \downarrow \pi_{q+r}^{q+r+s} & & & \downarrow \pi_r^{r+s} & & \downarrow & \\
0 & \longrightarrow & R_{q+r} & \longrightarrow & J_{q+r}(E) & \overset{\rho_r(\Phi)}{\longrightarrow} & & J_r(F) & \longrightarrow & Q_r & \longrightarrow 0 \\
 & & & & \downarrow & & & \downarrow & & \downarrow & \\
 & & & & 0 & & & 0 & & &
\end{array}$$

We have $g_{q+r} = R_{q+r}^{q+r-1}$ and the short exact sequences $seq(q, r, s)$:

$$0 \longrightarrow R_{q+r+s}^{q+r} \longrightarrow R_{q+r+s} \longrightarrow R_{q+r}^{(s)} \longrightarrow 0$$

In the Janet example, $n = 3, dim(E) = m = 1, dim(F) = 2, q = 2$. We remember that $dim(J_q(E)) = (n + q)!/n!q!$ and we can easily use computer algebra in order to obtain the numbers $dim(im(\rho_r(\Phi))) = dim(J_{q+r}(E)) - dim(R_{q+r})$ for $r = 0, 1, ..., 6$, that is successively 2, 8, 20, 39, 66, 102, 147. We are now ready for applying inductively the preceding theorem and criterion

of formal integrability to R_2 until we reach $R_5^{(2)}$ and prove that it is formally integrable with involutive (in fact zero!) symbol or simply involutive.
Let us consider R_2 with $dim(R_2) = 10 - 2 = 8$. Then $seq(2,0,1)$ gives: $dim(R_2^{(1)}) = dim(R_3) - dim(g_3) = 12 - (10-6) = 8 = dim(R_2)$ and thus $R_2^{(1)} = R_2$. However, we have for g_2:

$$\left\{\begin{array}{ll} v_{33} - x^2 v_{11} = 0 & \\ v_{22} \qquad\qquad = 0 & \end{array}\right. \boxed{\begin{array}{ccc} x^1 & x^2 & x^3 \\ x^1 & x^2 & \cdot \end{array}}$$

and g_2 is not involutive because we have for g_3:

$$\left\{\begin{array}{rl} v_{333} - x^2 v_{113} &= 0 \\ v_{233} - x^2 v_{112} &= 0 \\ v_{223} &= 0 \\ v_{222} &= 0 \\ v_{133} - x^2 v_{111} &= 0 \\ v_{122} &= 0 \end{array}\right. \boxed{\begin{array}{ccc} x^1 & x^2 & x^3 \\ x^1 & x^2 & \cdot \\ x^1 & x^2 & \cdot \\ x^1 & x^2 & \cdot \\ x^1 & \cdot & \cdot \\ x^1 & \cdot & \cdot \end{array}}$$

and $dim(g_3) = 10 - 6 = 4$ instead of $10 - 5 = 5$. Hence R_2 is not formally integrable. Indeed, $R_3^{(1)} \subset R_3$ because $seq(2,1,1)$ gives: $dim(R_3^{(1)}) = dim(R_4) - dim(g_4) = (35-20) - (15-11) = 15 - 4 = 11 < 12$ and we may start afresh with $R_3^{(1)}$.
Now we notice that g_3 is involutive because $dim(g_4) = 15 - 11 = 4 = 15 - nb$ of multiplicative variables for g_3. We may thus apply the prolongation theorem to R_3 and get $(R_3^{(1)})_{+r} = R_{3+r}^{(1)}$. In particular, if we want to apply the formal integrability criterion to $R_3^{(1)}$, we must study $g_3^{(1)}$:

$$\left\{\begin{array}{rl} v_{333} - x^2 v_{113} &= 0 \\ v_{233} &= 0 \\ v_{223} &= 0 \\ v_{222} &= 0 \\ v_{133} - x^2 v_{111} &= 0 \\ v_{122} &= 0 \\ v_{112} &= 0 \end{array}\right. \boxed{\begin{array}{ccc} x^1 & x^2 & x^3 \\ x^1 & x^2 & \cdot \\ x^1 & x^2 & \cdot \\ x^1 & x^2 & \cdot \\ x^1 & \cdot & \cdot \\ x^1 & \cdot & \cdot \\ x^1 & \cdot & \cdot \end{array}}$$

$g_3^{(1)}$ is not involutive because of the non-multiplicative variable x^3 for $v_{112} = 0$. However, its first prolongation $g_4^{(1)}$ is involutive (exercise). Hence, if we want to check the criterion, we have:

$$\pi_4^5((R_4^{(1)})_{+1}) = \pi_4^5((R_3^{(1)})_{+2}) = \pi_4^5(R_5^{(1)}) = R_4^{(2)} \subseteq R_4^{(1)}$$

$$dim(R_4^{(1)}) = dim(R_5) - dim(g_5) = (56-39) - 4 = 17 - 4 = 13$$

Now the reader may check by himself that $R_4^{(2)}$ is obtained from $R_4^{(1)}$ by adding the equation $y_{1111} = 0$ and thus $dim(R_4^{(2)}) = dim(R_4^{(1)}) - 1 = 12$. We may start afresh with $R_4^{(2)}$ with symbol $g_4^{(2)}$ given by:

$$\left\{\begin{array}{ll|lll} v_{3333} & = 0 & x^1 & x^2 & x^3 \\ v_{2333} & = 0 & x^1 & x^2 & . \\ v_{2233} & = 0 & x^1 & x^2 & . \\ v_{2223} & = 0 & x^1 & x^2 & . \\ v_{2222} & = 0 & x^1 & x^2 & . \\ v_{1333} - x^2 v_{1113} & = 0 & x^1 & . & . \\ v_{1233} & = 0 & x^1 & . & . \\ v_{1223} & = 0 & x^1 & . & . \\ v_{1222} & = 0 & x^1 & . & . \\ v_{1133} & = 0 & x^1 & . & . \\ v_{1123} & = 0 & x^1 & . & . \\ v_{1122} & = 0 & x^1 & . & . \\ v_{1112} & = 0 & x^1 & . & . \\ v_{1111} & = 0 & x^1 & . & . \end{array}\right.$$

We notice that $g_4^{(2)}$ is not involutive because of the non-multiplicative variable x^3 for $v_{1111} = 0$. Its first prolongation is zero and thus trivially involutive. However, we have:

$$(R_4^{(2)})_{+1} = ((R_4^{(1)})^{(1)})_{+1} = ((R_4^{(1)})_{+1})^{(1)} = (R_5^{(1)})^{(1)} = R_5^{(2)}$$

because $g_4^{(1)}$ and g_3 are involutive. But we have $\pi_4^5(R_5^{(2)}) = R_4^{(3)}$ and we deduce from $diag(2,3,2)$:

$$dim(R_5^{(2)}) = dim(R_7) - dim(R_7^5) = (120 - 102) - (64 - 58) = 18 - 6 = 12$$

while we deduce from $diag(2,2,3)$:

$$dim(R_4^{(3)}) = dim(R_7) - dim(R_7^4) = (120 - 102) - (85 - 79) = 18 - 6 = 12.$$

It follows that $dim(g_5^{(2)}) = dim(R_5^{(2)}) - dim(R_4^{(3)}) = 12 - 12 = 0$ and $g_5^{(2)} = 0$ is trivially involutive. However, we have:

$$(R_5^{(2)})_{+1} = (R_4^{(2)})_{+2} = ((R_4^{(1)})^{(1)})_{+2} = ((R_4^{(1)})_{+2})^{(1)} = (R_6^{(1)})^{(1)} = R_6^{(2)}$$

and $\pi_5^6(R_6^{(2)}) = R_5^{(3)}$. Using similarly $diag(2,3,3)$, we get:

$$dim(R_5^{(3)}) = dim(R_8) - dim(R_8^5) = (165 - 147) - (109 - 103) = 18 - 6 = 12$$

and it follows that $R_5^{(2)}$ is involutive with zero symbol. As a matter of fact, using $diag(2,4,2)$, we get:

$$dim(R_6^{(2)}) = dim(R_8) - dim(R_8^6) = (165 - 147) - (81 - 75) = 18 - 6 = 12$$

and we check that $R_6^{(2)} \simeq R_5^{(2)}$ while $R_5^{(3)} = R_5^{(2)}$
We have thus proved that the space of solutions of R_2 is a 12-dimensional vector space over the constants as it coincides with the space of solutions of the involutive system $R_5^{(2)}$ which has zero symbol.
We finally use this result in order to look for the generating compatibility conditions. Indeed, introducing u and v as second members of R_2, we may therefore add to $R_5^{(2)}$ second members involving the derivatives of u and v up to order $7 - 2 = 5$. Now, as $R_5^{(2)}$ is involutive, the compatibility conditions for the second members only become first order because the criterion only involves one prolongation. Accordingly, the compatibility conditions for R_2 only involves the derivatives of u and v up to order $1 + 5 = 6$ and we have successively:

$$Q_1 = Q_2 = 0, dim(Q_3) = 1, ..., dim(Q_6) = 21 = dim(J_3(Q_3) + 1.$$

We now specify the correspondence:

$$SYSTEM \Leftrightarrow OPERATOR \Leftrightarrow MODULE$$

in order to show later on that certain concepts, which are clear in one framework, may become quite obscure in the others and conversely (check this for the formal integrability and torsion concepts for example!).

Having a system of order q, say $R_q \subset J_q(E)$, we can introduce the canonical projection $\Phi : J_q(E) \longrightarrow J_q(E)/R_q = F$ and define a linear differential operator $\mathcal{D} : E \longrightarrow F : \xi(x) \longrightarrow \eta^\tau(x) = A_k^{\tau\mu}(x)\partial_\mu \xi^k(x)$. When $\mathcal{D}$ is given, the compatibility conditions for solving $\mathcal{D}\xi = \eta$ can be described in operator form by $\mathcal{D}_1\eta = 0$ and so on. In general (see the preceding examples), if a system is not formally integrable, it is possible to obtain a formally integrable system, having the same solutions, by "*saturating*" conveniently the given PD equations through the adjunction of new PD equations obtained by various prolongations and *such a procedure must absolutely be done before looking for the compatibility conditions.*

Starting with the work of M. Janet in 1920, effective tests have been provided for checking formal integrability and computer algebra packages dealing with Gröbner bases can be used for such a purpose [4]. However, for reasons that will become clear later on, formal integrability is not sufficient for having certain canonical forms of systems and tests. We have already introduced the Spencer cohomology of the symbol as an intrinsic/coordinate free homological tool for the previous test and now, *passing from the symbol to the system*, we shall provide an equivalent local description, more useful in practice. For this, *changing linearly the independent variables if necessary*, we shall solve the maximum number β of equations, called *equations of class n*, with respect to the jets of order q and class n. Then, we shall solve the maximum number of remaining equations, called *equations of class* $n-1$, with respect to the jets of order q and class $n-1$, and so on till we get equations of class 1, the remaining equations being of order $\leq q-1$. For each equation of class i we shall call the variables $x^1, ..., x^i$ (respectively $x^{i+1}, ..., x^n$) *multiplicative* (*non-multiplicative*) while all variables will be non-multiplicative for the equations of order $\leq q-1$. The following definition is essential and studies are in progress in order to test it through computer algebra packages:

Definition 5.31. *A system of PD equations is said to be involutive if its first prolongation can be achieved by prolonging its equations only with respect to the corresponding multiplicative variables.*

It can be proved that *this definition is intrinsic though it must be checked in a particular system of coordinates* and this point has not been yet overcome by symbolic computations. An involutive system is formally integrable. Also, one can prove that, *in this case*, the maximum number α of dependent variables that can be given arbitrarily (that are differentially independent) is equal to $m-\beta$. Homogeneous systems with constant coefficients are automatically formally integrable. We shall prove that:

Proposition 5.14. *When R_q is involutive, we say that $\mathcal{D}$ is involutive and, in this case only, we get a finite chain of 1st order involutive operators $\mathcal{D}_1, ..., \mathcal{D}_n$ making up a canonical differential sequence called Janet sequence.*

Proof. Setting $F_0 = J_q(E)/R_q$, let us define the vector bundle F_1 by the *ker/coker* exact sequence of vector bundles over X:

$$0 \longrightarrow g_{q+1} \longrightarrow S_{q+1}T^* \otimes E \stackrel{\sigma_1(\Phi)}{\longrightarrow} T^* \otimes F_0 \longrightarrow F_1 \longrightarrow 0$$

where $\sigma_1(\Phi)$ is the restriction of the first prolongation $\rho_1(\Phi) : J_{q+1}(E) \rightarrow J_1(F_0)$ of the epimorphism $\Phi : J_q(E) \rightarrow F_0$. As g_q is involutive and thus at least 2-acyclic, it follows from Theorem 5.5 by induction on $r \geq 1$ that we have the following commutative and exact diagram depicted in Fig. 5.1(a) where the vertical sequences are δ-sequences.

Using the exactness of the top rows of the preceding diagrams and the assumption of formal integrability, it now follows by induction on $r \geq 0$ that we have the commutative and exact diagram depicted in Fig. 5.1(b).

Accordingly, the compatibility conditions of order $r+1$ are nothing else than the r-prolongation $\rho_r(\Psi_1)$ of the compatibility conditions of order 1, namely $\Psi_1 : J_1(F_0) \rightarrow F_1$ and we may introduce the first order operator $\mathcal{D}_1 = \Psi_1 \circ j_1$. We let the reader check as an exercise that $\mathcal{D}_1$ is again involutive and we may successively construct similarly the first order involutive operators $\mathcal{D}_2, ..., \mathcal{D}_n$.

$$
\begin{array}{ccccccccc}
 & & 0 & & 0 & & 0 & & 0 \\
 & & \downarrow & & \downarrow & & \downarrow & & \downarrow \\
0 \to & & g_{q+r+1} & \to & S_{q+r+1}T^* \otimes E & \to & S_{r+1}T^* \otimes F_0 & \to & S_rT^* \otimes F_1 \\
 & & \downarrow & & \downarrow & & \downarrow & & \downarrow \\
0 \to & & T^* \otimes g_{q+r} & \to & T^* \otimes S_{q+r}T^* \otimes E & \to & T^* \otimes S_rT^* \otimes F_0 & \to & T^* \otimes S_{r-1}T^* \otimes F_1 \\
 & & \downarrow & & \downarrow & & \downarrow & & \\
0 \to & & \Lambda^2T^* \otimes g_{q+r-1} & \to & \Lambda^2T^* \otimes S_{q+r-1}T^* \otimes E & \to & \Lambda^2T^* \otimes S_{r-1}T^* \otimes F_0 & & \\
 & & \downarrow & & \downarrow & & & & \\
0 \to & & \Lambda^3T^* \otimes S_{q+r-2}T^* \otimes E & = & \Lambda^3T^* \otimes S_{q+r-2}T^* \otimes E & \to & 0 & &
\end{array}
$$

(a)

$$
\begin{array}{ccccccccc}
 & & 0 & & 0 & & 0 & & 0 \\
 & & \downarrow & & \downarrow & & \downarrow & & \downarrow \\
0 & \longrightarrow & g_{q+r+1} & \longrightarrow & S_{q+r+1}T^* \otimes E & \stackrel{\sigma_{r+1}(\Phi)}{\longrightarrow} & S_{r+1}T^* \otimes F_0 & \stackrel{\sigma_r(\Psi_1)}{\longrightarrow} & S_rT^* \otimes F_1 \\
 & & \downarrow & & \downarrow & & \downarrow & & \downarrow \\
0 & \longrightarrow & R_{q+r+1} & \longrightarrow & J_{q+r+1}(E) & \stackrel{\rho_{r+1}(\Phi)}{\longrightarrow} & J_{r+1}(F_0) & \stackrel{\rho_r(\Psi_1)}{\longrightarrow} & J_r(F_1) \\
 & & \downarrow & & \downarrow & & \downarrow & & \downarrow \\
0 & \longrightarrow & R_{q+r} & \longrightarrow & J_{q+r}(E) & \stackrel{\rho_r(\Phi)}{\longrightarrow} & J_r(F_0) & \stackrel{\rho_{r-1}(\Psi_1)}{\longrightarrow} & J_{r-1}(F_1) \\
 & & \downarrow & & \downarrow & & \downarrow & & \downarrow \\
 & & 0 & & 0 & & 0 & & 0
\end{array}
$$

(b)

Fig. 5.1. Exact commutative diagrams.

Finally, cutting the first of the preceding diagrams as we did in the proof of Theorem 5.5 and setting $h_1 = im(\sigma_1(\Phi)) \subset T^* \otimes F_0$, we obtain (exercise) the crucial canonical formula $F_r = \Lambda^r T^* \otimes F_0/\delta(\Lambda^{r-1}T^* \otimes h_1)$ showing that $\mathcal{D}_n$ is always formally surjective. ■

Example 5.16. With $n = 4, m = 1, q = 1$, the system R_1 defined by the two PD equations $\Phi^2 \equiv y_4 - x^3 y_2 - y = 0, \Phi^3 \equiv y_3 - x^4 y_1 = 0$ is not formally integrable as one can easily check $\Phi^1 \equiv d_4\Phi^2 - d_3\Phi^3 - x^3 d_2\Phi^2 + x^4 d_1\Phi^3 - \Phi^2 \equiv y_2 - y_1 = 0$. However, the system $R_1' \subset R_1$ defined by the three PD equations $\Phi^1 = 0, \Phi^2 = 0, \Phi^3 = 0$ is involutive with 1 equation of class 4, 1 equation of class 3, 1 equation of class 2 and one checks the 3 first order compatibility conditions:

$$\Psi^3 \equiv d_4\Phi^2 - d_3\Phi^3 - x^3 d_2\Phi^2 + x^4 d_1\Phi^3 - \Phi^2 - \Phi^1 = 0$$

$$\Psi^2 \equiv d_4\Phi^1 - d_2\Phi^3 + d_1\Phi^3 - x^3\Phi^1 - \Phi^1 = 0$$

$$\Psi^1 \equiv d_3\Phi^1 - d_2\Phi^2 + d_1\Phi^2 - x^4 d_1\Phi^1 = 0$$

This is again an involutive system with 2 equations of class 4, 1 equation of class 3 and 1 single compatibility condition of class 4, namely:

$$(d_4 - x^3 d_2 - 1)\Psi^1 + (x^4 d_1 - d_3)\Psi^2 + (d_2 - d_1)\Psi^3 = 0$$

ending the construction of the Janet sequence.

Example 5.17. With $n = 3, m = 1, q = 2$, let us consider the homogeneous second order system $\Phi^1 \equiv y_{22} - y_{11} = 0, \Phi^2 \equiv y_{23} = 0, \Phi^3 \equiv y_{33} - y_{11} = 0$. This system is formally integrable though one needs 2 prolongations to get involution (all jets of order 4 are nul !). There are 3 compatibility conditions of order two, namely $d_{33}\Phi^2 - d_{23}\Phi^3 - d_{11}\Phi^2 = 0, d_{33}\Phi^1 - d_{22}\Phi^3 + d_{11}\Phi^3 - d_{11}\Phi^1 = 0, d_{23}\Phi^1 - d_{22}\Phi^2 + d_{11}\Phi^2 = 0$ and this homogeneous system is again formally integrable though not involutive.

We shall now prove that the "*Kalman form*" is nothing else but a particular case of the so-called "*Spencer form*" existing in the formal theory of systems of PD equations, as *no reference to control theory is needed.*

For simplicity and in a sketchy way, we may say that the Spencer method amounts to use the canonical inclusion $J_{q+1}(E) \subset J_1(J_q(E))$ through the identification $y^k_{\mu,i} = y^k_{\mu+1_i}$ in order to induce a canonical inclusion $R_{q+1} \subset J_1(R_q)$ allowing to consider R_{q+1} as a first order system on R_q, *whenever R_{q+1} is a vector bundle.* When R_q is involutive, this is the *Spencer form* and the corresponding canonical Janet sequence is called *second Spencer sequence* (cf. [18, 19, 24]). It is not so well known that such a method, which allows to bring a system of order q to a system of order 1, is only truly useful when R_q is formally integrable. Indeed, *in this case and only in this case,* R_{q+1} can be considered as a first order system over R_q, without equations of order zero. The following example clarifies this delicate point.

Example 5.18. Looking back to Example 5.16 we notice that $R_2 \longrightarrow R_1$ is *not* surjective and, introducing the parametric jets (derivatives) $z^1 = y, z^2 = y_1, z^3 = y_2$ for R_1, then $R_2 \subset J_1(R_1)$ is defined by first order PD equations *and* by the zero order equation $z^3 - z^2 = 0$. On the contrary, if we introduce the new formally integrable system $R_1' = R_1^{(1)} \subset R_1$, projection of R_2 in R_1, the parametric jets of R_1' are now only $z^1 = y, z^2 = y_1$. *Though this is not evident at first sight,* they are linked by the 7 first order PD equations:

$$\begin{aligned}
&z_4^1 - x^3 z^2 - z^1 = 0,\\
&z_4^2 - x^3 z_1^2 - z^2 = 0,\\
&z_3^1 - x^4 z^2 = 0,\\
&z_3^2 - x^4 z_1^2 = 0,\\
&z_2^1 - z^2 = 0,\\
&z_2^2 - z_1^2 = 0,\\
&z_1^1 - z^2 = 0
\end{aligned}$$

because $dim(R_2') = 10 - 9 = 4 \times 2 - 7 = 1$.

According to the preceding comments and with no loss of generality, we consider from now on a first order system $R_1 \subset J_1(E)$ over E with $R_1 \longrightarrow E$ surjective and we shall bring such a system to a canonical form, noticing that similar methods can be used for nonlinear systems too as in [19]. For this, *changing linearly the system of coordinates if necessary,* we may manage to have a maximum number of PD equations solved with respect to the jets $y_n^1, ..., y_n^{\beta_1^n}$, called equations of *class n*. The remaining equations only contain jets with respect to $x^1, ..., x^{n-1}$ and, *changing again linearly the system of coordinates if necessary,* we may manage to have a maximum number of PD equations solved with respect to $y_{n-1}^1, ..., y_{n-1}^{\beta_1^{n-1}}$ with $\beta_1^{n-1} \leq \beta_1^n$, called equations of *class* $n-1$ and so on, till we get the equations of *class* 1 if any. The complements to m of the β-integers are the *characters* (exercise).

Definition 5.32. *A first order system is said to be involutive if it can be brought to such a canonical form and all the equations of second order can be obtained by differentiating/prolonging the equations of class i with respect to* $x^1, ..., x^i$ *only, for* $i = 1, ..., n$.

In this situation $x^1, ..., x^i$ are called *multiplicative variables* and $x^{i+1}, ..., x^n$ are called *non-multiplicative variables* for the equations of class i.

Example 5.19. The system R_1' defined in the preceding example is easily seen to be in canonical form and involutive with characters $(1, 1, 1, 0)$. In particular, the equation $y_2 - y_1 = 0$ is of class 2 with multiplicative variables x^1, x^2 and non-multiplicative variables x^3, x^4. One has $d_3(y_2 - y_1) \equiv d_2(y_3 - x^4 y_1) - d_1(y_3 - x^4 y_1) + x^4 d_1(y_2 - y_1)$

Denoting by z the variables $y^{\beta_1^{n+1}}, ..., y^m$ and using elementary properties of the *characteristic variety* to be found in section 6, one can prove that the z can be chosen arbitrarily (differential transcendence basis) [10, 18, 19]. More generally, if R_q is formally integrable with an involutive symbol, that is to say a symbol such that all the groups of Spencer δ-cohomology vanish, the corresponding first order system $R_{q+1} \subset J_1(R_q)$ is involutive in the above sense.

It now remains to introduce the linear combinations $\bar{y}^k = y^k - term(z)$ for $k = 1, ..., \beta_1^n$ that will allow to "*absorb*" the z_n in the y_n. The generalization of the Kalman form finally depends on the following striking though technical result on involutive systems that does not seem to have been previously known.

Proposition 5.15. *The new equations of class n only contain $z, z_1, ..., z_{n-1}$ while the equations of class $1, ..., n-1$ no more contain z or its jets.*

Proof. x^n is a non-multiplicative variable for the equations of class $1, ..., n-1$. Hence, if z or any jet of order one should appear in one of the latter equations, by prolonging this equation with respect to x^n and using the involutiveness property, one should get a linear combination of equations of various classes, prolonged with respect to $x^1, ..., x^{n-1}$ and this is not possible as only $z, z_1, ..., z_{n-1}$ do appear. ∎

Example 5.20. With $n = 2, m = 3, q = 1$, let us consider the following linear involutive first order system in solved form:

$$\Phi^3 \equiv y_2^2 - y_1^2 + z_2 - z_1 - z = 0, \Phi^2 \equiv y_2^1 - y_1^2 - z_2 - z_1 - z = 0, \Phi^1 \equiv y_1^1 - y_1^2 - 2z_1 = 0$$

In this system, Φ^2 and Φ^3 are of class 2 while Φ^1 is of class 1. Setting $\bar{y}^1 = y^1 - z, \bar{y}^2 = y^2 + z$, we get the new system:

$$\Phi^3 \equiv \bar{y}_2^2 - \bar{y}_1^2 - z = 0, \Phi^2 \equiv \bar{y}_2^1 - \bar{y}_1^2 - z = 0, \Phi^1 \equiv \bar{y}_1^1 - \bar{y}_1^2 = 0$$

and z or its jets no longer appear in Φ^1.

In order to study differential modules, for simplicity we shall forget about changes of coordinates and consider trivial bundles. Let K be a *differential field* with n commuting derivations $\partial_1, ..., \partial_n$, that is to say K is a field and $\partial_i : K \to K$ satisfies $\partial_i(a+b) = \partial_i a + \partial_i b, \partial_i(ab) = (\partial_i a)b + a\partial_i b, \forall a, b \in K, \forall i = 1, ..., n$ (say $\mathbb{Q}, \mathbb{Q}(x^1, ..., x^n)$ or $\mathbb{Q}(a)$ in the previous examples). If $d_1, ..., d_n$ are *formal derivatives* (pure symbols in computer algebra packages!) which are only supposed to satisfy $d_i a = a d_i + \partial_i a$ in the operator sense for any $a \in K$, we may consider the (non-commutative) ring $D = K[d_1, ..., d_n]$ of differential operators with coefficients in K. If now $y = (y^1, ..., y^m)$ is a set of differential indeterminates, we let D act formally on y by setting $d_\mu y^k = y_\mu^k$ and set $Dy = Dy^1 + ... + Dy^m$. Denoting simply by $D\mathcal{D}y$ the subdifferential module generated by all the given OD or PD equations and all their formal derivatives, we may finally introduce the D-module $M = Dy/D\mathcal{D}y$.

Example 5.21. In Example 5.4 with $K = \mathbb{Q}(x^1, x^2)$ and in Example 5.5 with $K = \mathbb{Q}$, we get $M = 0$ while in Example 5.6 with $K = \mathbb{Q}(x^1, x^2, x^3)$, we get $dim_K(M) = 12$.

More generally, with similar definitions, if A is a differential ring, we can consider $D = A[d]$ to be the ring of differential operators with coefficients in A and D is noetherian (is an integral domain) when A is noetherian (is an integral domain).

It just remains to prove that even in this non-commutative case, one can also define localization and the torsion submodule. We achieve this in the previous situation of differential operators over a differential field K by means of the following proposition.

Proposition 5.16. *$D = K[d]$ is an Ore ring, that is to say, given any nonzero $P, Q \in D$, then one can find $A, B \in D$ such that $AP = BQ$.*

Proof. Let us consider the system with second members $Py = u, Qy = v$ and its prolongations as a linear system over K for the jets of y, u, v. If $deg(P) = p, deg(Q) = q$, we may suppose that $p \leq q$ and, if we prolong r times, the number of jets of y of order $q + r$ is equal to $(n + q + r)!/n!(q + r)! = (q + r + 1)...(q + r + n)/n! = \frac{1}{n!}r^n + ...$ Meanwhile, the number of jets of order r of u and v is equal to $2(n + r)!/n!r! = \frac{2}{n!}r^n + ...$ Hence, when r is large enough, the second number is greater than the first and we can eliminate the jets of y by using Cramer's rule over K. Accordingly, one can find at least one compatibility condition of the form $Au - Bv = 0$ and thus $AP = BQ$. ∎

The application of the preceding results on localization to D and $S = D - \{0\}$ is immediate and we refer the reader to [19] for more general situations.

5.4 Problem Formulation

Though it seems that we are very far from any possible application, let us now present three problems which, both with the previous examples, look like unrelated with what we already said and between themselves.

Problem 5.1. Let a rigid bar of length L be able to slide horizontally and attach at the end of abcissa x (resp. $x + L$) a pendulum of length l_1 (resp. l_2) with mass m_1 (resp. m_2), making an angle θ_1 (resp. θ_2) with the downwards vertical axis. Projecting the dynamical equations on the perpendicular to each pendulum in order to eliminate the respective tension, we get:

$$m_1(\ddot{x}cos\theta_1 + l_1\ddot{\theta}_1) + m_1 g sin\theta_1 = 0$$

where g is the gravity. When θ_1 and θ_2 are small, we get the following two OD equations that only depend on l_1 and l_2 but no longer on m_1 and m_2:

$$\begin{cases} \ddot{x} + l_1\ddot{\theta}_1 + g\theta_1 = 0 \\ \ddot{x} + l_2\ddot{\theta}_2 + g\theta_2 = 0 \end{cases}$$

Now it is easy to check *experimentally* that, when $l_1 \neq l_2$, it is possible to bring any small amplitude motion $\theta_1 = \theta_1(t), \theta_2 = \theta_2(t)$ of the two pendula back to equilibrium $\theta_1 = 0, \theta_2 = 0$, just by choosing a convenient $x = x(t)$ and *the system is said to be controllable.* On the contrary, if $l_1 = l_2$ and unless $\theta_1(t) = \theta_2(t)$, then it is impossible to bring the pendula back to equilibrium and *the system is said to be uncontrollable.* A similar question can be asked when $l_1 = l_1(t), l_2 = l_2(t)$ are given, the variation of length being produced by two small engines hidden in the bar.

Hence, a much more general question concerns the controllability of control systems defined by systems of OD or PD equations as well, like in gasdynamics or magnetohydrodynamics.
In our case, setting $x_1 = x + l_1\theta_1, x_2 = x + l_2\theta_2$, we get:

$$\begin{cases} \ddot{x}_1 + (g/l_1)x_1 - (g/l_1)x = 0 \\ \ddot{x}_2 + (g/l_2)x_2 - (g/l_2)x = 0 \end{cases}$$

and may set $\dot{x}_1 = x_3, \dot{x}_2 = x_4$ in order to bring the preceding system to Kalman form with 4 first order OD equations. The controllability condition is then easily seen to be $l_1 \neq l_2$ but such a result not only seems to depend on the choice of input and output but cannot be extended to PD equations.

Problem 5.2. Any engineer knows about the first set of Maxwell equations:

$$\mathbf{\nabla}.\mathbf{B} = 0, \qquad \mathbf{\nabla} \wedge \mathbf{E} + \frac{\partial \mathbf{B}}{\partial t} = 0$$

and the fact that any solution can be written in the form:

$$\mathbf{B} = \mathbf{\nabla} \wedge \mathbf{A}, \qquad \mathbf{E} = -\mathbf{\nabla}.V - \frac{\partial \mathbf{A}}{\partial t}$$

for an arbitrary vector $\mathbf{A}$ and an arbitrary function V.
According to special relativity, these equations can be condensed on space-time by introducing a 1-form A for the *potential* and a 2-form F for the *field* in order to discover that the above Maxwell equations can be written in the form $dF = 0$ and admit the "generic" solution $dA = F$ where d is the exterior

derivative. Hence, we have "*parametrized*" the field equations by means of a "*potential*", that is the field equations generate the compatibility conditions of the inhomogeneous system allowing to express the field (right member) by means of the potential (left member).

Similarly, in 2-dimensional elasticity theory, if we want to solve the stress equations with no density of force, namely:

$$\partial_1\sigma^{11} + \partial_2\sigma^{21} = 0 \quad , \quad \partial_1\sigma^{12} + \partial_2\sigma^{22} = 0$$

we may use the first PD equation to get:

$$\exists\varphi \ , \ \sigma^{11} = \partial_2\varphi \quad , \quad \sigma^{21} = -\partial_1\varphi$$

and the second PD equation to get:

$$\exists\psi \ , \ \sigma^{12} = -\partial_2\psi \quad , \quad \sigma^{22} = \partial_1\psi$$

Now, we have:

$$\sigma^{12} = \sigma^{21} \Rightarrow \exists\phi \ , \ \varphi = \partial_2\phi \quad , \quad \psi = \partial_1\phi$$

and we finally get the generic parametrization by the Airy function:

$$\sigma^{11} = \partial_{22}\phi \quad , \quad \sigma^{12} = \sigma^{21} = -\partial_{12}\phi \quad , \quad \sigma^{22} = \partial_{11}\phi$$

The reader will have noticed that such a specific computation cannot be extended in general, even to 3-dimensional elasticity theory.

In 1970 J. Wheeler asked a similar question for Einstein equations in vacuum and we present the linearized version of this problem.
Indeed, if $\omega = (dx^1)^2 + (dx^2)^2 + (dx^3)^2 - (dx^4)^2$ with $x^4 = ct$, where c is the speed of light, is the Minkowski metric of space-time, we may consider a perturbation Ω of ω and the linearized Einstein equations in vacuum become equivalent to the following second order system with 10 equations for 10 unknowns:

$$\begin{aligned}\omega^{rs}(d_{ij}\Omega_{rs} + d_{rs}\Omega_{ij} - d_{ri}\Omega_{sj} - d_{sj}\Omega_{ri})\\ -\omega_{ij}(\omega^{rs}\omega^{uv}d_{rs}\Omega_{uv} - \omega^{ru}\omega^{sv}d_{rs}\Omega_{uv}) = 0\end{aligned}$$

Surprisingly, till we gave the (negative) answer in 1995 [17], such a problem had never been solved before.
More generally, if one considers a system of the form $\mathcal{D}_1\eta = 0$, the question is to know whether one can parametrize or not the solution space by $\mathcal{D}\xi = \eta$ with arbitrary potential-like functions ξ, in such a way that $\mathcal{D}_1\eta = 0$ just generates the compatibility conditions of the parametrization. The problem of multiple parameterizations may also be considered, as an inverse to the construction of differential sequences. For example, in vector calculus, the *div* operator is parametrized by the *curl* operator which is itself parametrized by the *grad* operator (cf. [19, 21, 22] for more details).

Problem 5.3. When M is an A-module, there is a canonical morphism $\epsilon = \epsilon_M : M \longrightarrow M^{**}$ given by $\epsilon(m)(f) = f(m), \forall m \in M, \forall f \in M^*$ and M is said to be *torsionless* if ϵ is injective and *reflexive* if ϵ is bijective. Any finitely projective module is reflexive but a reflexive module may not be projective. We have $t(M) \subseteq ker(\epsilon)$ because, if $m \in M$ is a torsion element for $a \neq 0$, then $af(m) = f(am) = f(0) = 0 \Rightarrow f(m) = 0, \forall f \in M^*$ as before and ϵ fails to be injective. Hence, it just remains to study whether this inclusion is strict or not.

The striking result of this lecture is to prove that THESE THREE PROBLEMS ARE IDENTICAL!

5.5 Problem Solution

The *main but highly not evident* trick will be to introduce the *adjoint operator* $\tilde{\mathcal{D}} = ad(\mathcal{D})$ by the formula of integration by part:

$$< \lambda, \mathcal{D}\xi > = < \tilde{\mathcal{D}}\lambda, \xi > + div(\quad)$$

where λ is a test row vector and $<>$ denotes the usual contraction. The adjoint can also be defined formally, as in computer algebra packages, by setting $ad(a) = a, \forall a \in K, ad(d_i) = -d_i, ad(PQ) = ad(Q)ad(P), \forall P, Q \in D$. Another way is to define the adjoint of an operator directly on D by setting $P = \sum_{0 \leq |\mu| \leq p} a^\mu d_\mu \longrightarrow ad(P) = \sum_{0 \leq |\mu| \leq p} (-1)^{|\mu|} d_\mu a^\mu$ for any $P \in D$ with $ord(P) = p$ and to extend such a definition by linearity.
We shall denote by N the differential module defined from $ad(\mathcal{D})$ exactly like M was defined from $\mathcal{D}$ and we have [7, 13, 14, 19].

Theorem 5.7. *The following statements are equivalent:*
- *A control system is controllable.*
- *The corresponding operator is simply (doubly) parameterizable.*
- *The corresponding module is torsion-free (reflexive).*

Proof. Let us start with a free presentation of M:

$$F_1 \xrightarrow{d_1} F_0 \longrightarrow M \longrightarrow 0$$

By definition, we have $M = coker(d_1) \Longrightarrow N = coker(d_1^*)$ and we may exhibit the following free resolution of N:

$$0 \longleftarrow N \longleftarrow F_1^* \xleftarrow{d_1^*} F_0^* \xleftarrow{d_0^*} F_{-1}^* \xleftarrow{d_{-1}^*} F_{-2}^*$$

where $M^* = ker(d_1^*) = im(d_0^*) \simeq coker(d_{-1}^*)$. The deleted sequence is:

$$0 \longleftarrow F_1^* \overset{d_1^*}{\longleftarrow} F_0^* \overset{d_0^*}{\longleftarrow} F_{-1}^* \overset{d_{-1}^*}{\longleftarrow} F_{-2}^*$$

Applying $hom_A(\bullet, A)$ and using the canonical isomorphism $F^{**} \simeq F$ for any free module F, we get the sequence:

$$\begin{array}{ccccccc} 0 \longrightarrow F_1 & \overset{d_1}{\longrightarrow} & F_0 & \overset{d_0}{\longrightarrow} & F_{-1} & \overset{d_{-1}}{\longrightarrow} & F_{-2} \\ & & \downarrow & & \uparrow & & \\ & & M & \overset{\epsilon}{\longrightarrow} & M^{**} & & \\ & & \downarrow & & \uparrow & & \\ & & 0 & & 0 & & \end{array}$$

Denoting as usual a coboundary space by B, a cocycle space by Z and the corresponding cohomology by $H = Z/B$, we get the commutative and exact diagram:

$$\begin{array}{ccccccccc} 0 & \longrightarrow & B_0 & \longrightarrow & F_0 & \longrightarrow & M & \longrightarrow & 0 \\ & & \downarrow & & \| & & \downarrow \epsilon & & \\ 0 & \longrightarrow & Z_0 & \longrightarrow & F_0 & \longrightarrow & M^{**} & & \end{array}$$

An easy chase provides at once $H_0 = Z_0/B_0 = ext_A^1(N, A) \simeq ker(\epsilon)$. It follows that $ker(\epsilon)$ is a torsion module and, as we already know that $t(M) \subseteq ker(\epsilon) \subseteq M$, we finally obtain $t(M) = ker(\epsilon)$. Also, as $B_{-1} = im(\epsilon)$ and $Z_{-1} \simeq M^{**}$, we obtain $H_{-1} = Z_{-1}/B_{-1} = ext_A^2(N, A) \simeq coker(\epsilon)$. Accordingly, a torsion-free (reflexive) module is described by an operator that admits a single (double) step parametrization. ■

This proof also provides an effective test for applications by using D and ad instead of A and $*$ in the differential framework. In particular, a control system is controllable if it does not admit any "*autonomous element*", that is to say any finite linear combination of the control variables and their derivatives that satisfies, *for itself*, at least one OD or PD equation. More precisely, starting with the control system described by an operator $\mathcal{D}_1$, one MUST construct $\tilde{\mathcal{D}}_1$ and then $\mathcal{D}$ such that $\tilde{\mathcal{D}}$ generates all the compatibility conditions of $\tilde{\mathcal{D}}_1$. Finally, M is torsion-free if and only if $\mathcal{D}_1$ generates all the compatibility conditions of $\mathcal{D}$. Though striking it could be, *this is the true generalization of the standard Kalman test.*

Example 5.22. If $\mathcal{D}_1 : (\sigma^{11}, \sigma^{12} = \sigma^{21}, \sigma^{22}) \rightarrow (\partial_1\sigma^{11} + \partial_2\sigma^{21}, \partial_1\sigma^{12} + \partial_2\sigma^{22})$ is the stress operator, then $\tilde{\mathcal{D}}_1 : (\xi^1, \xi^2) \rightarrow (\partial_1\xi^1 = \epsilon_{11}, \frac{1}{2}(\partial_1\xi^2 + \partial_2\xi^1) = \epsilon_{12} = \epsilon_{21}, \partial_2\xi^2 = \epsilon_{22})$ is half of the Killing operator. The only compatibility condition for the strain tensor ϵ is $\tilde{\mathcal{D}}\epsilon = 0 \Leftrightarrow \partial_{11}\epsilon_{22} + \partial_{22}\epsilon_{11} - 2\partial_{12}\epsilon_{12} = 0$ and $\mathcal{D}$ describes the Airy parametrization.

Now, in order to have a full picture of the correspondence existing between differential modules and differential operators, it just remains to explain why and how we can pass from left to right modules and conversely. By this way, we shall be able to take into account the behaviour of the adjoint of an operator under changes of coordinates. We start with a technical lemma (exercise):

Lemma 5.9. *If $f \in aut(X)$ is a local diffeomorphism of X, we may set $y = f(x) \Rightarrow x = f^{-1}(y) = g(y)$ and introduce the Jacobian $\Delta(x) = det(\partial_i f^k(x)) \neq 0$. Then, we have the identity:*

$$\frac{\partial}{\partial y^k}\left(\frac{1}{\Delta(g(y))}\partial_i f^k(g(y))\right) \equiv 0.$$

Accordingly, we notice that, if $\mathcal{D} : E \to F$ is an operator, the way to obtain the adjoint through an integration by parts proves that the test function is indeed a section of the *adjoint bundle* $\tilde{F} = F^* \otimes \Lambda^n T^*$ and that we get an operator $ad(\mathcal{D}) : \tilde{F} \to \tilde{E}$. This is in particular the reason why, in elasticity, the deformation is a covariant tensor but the stress is a contravariant tensor density and, in electromagnetism, the EM field is a covariant tensor (in fact a 2-form) but the induction is a contravariant tensor density.
Also, if we define the adjoint formally, we get, in the operator sense:

$$ad\left(\frac{1}{\Delta}\partial_i f^k \frac{\partial}{\partial y^k}\right) = -\frac{\partial}{\partial y^k} \circ \left(\frac{1}{\Delta}\partial_i f^k\right) = -\frac{1}{\Delta}\partial_i f^k \frac{\partial}{\partial y^k} = -\frac{1}{\Delta}\frac{\partial}{\partial x^i}$$

and obtain therefore:

$$\frac{\partial}{\partial x^i} = \partial_i f^k(x)\frac{\partial}{\partial y^k} \Rightarrow ad\left(\frac{\partial}{\partial x^i}\right) = -\frac{\partial}{\partial x^i} = \Delta ad\left(\frac{1}{\Delta}\partial_i f^k(x)\frac{\partial}{\partial y^k}\right)$$

a result showing that the adjoint of the gradient operator $d : \Lambda^0 T^* \to \Lambda^1 T^*$ is minus the exterior derivative $d : \Lambda^{n-1} T^* \to \Lambda^n T^*$.

If A is a differential ring and $D = A[d]$ as usual, we may introduce the ideal $I = \{P \in D | P(1) = 0\}$ and obtain $A \simeq D/I$ both with the direct sum decomposition $D \simeq A \oplus I$. In fact, denoting by D_q the submodule over A of operators of order q, A can be identified with the subring $D_0 \subset D$ of zero order operators and we may consider any differential module over D as a module over A, just "*forgetting*" about its differential structure. Caring about the notation, we shall set $T = D_1/D_0 = \{\xi = a^i d_i | a^i \in A\}$ with $\xi(a) = \xi^i \partial_i a, \forall a \in A$, so that D can be generated by A and T.

The module counterpart is much more tricky and is based on the following theorem [19]:

Theorem 5.8. *If M and N are right D-modules, then $hom_A(M, N)$ becomes a left D-module.*

Proof. We just need to define the action of $\xi \in T$ by the formula:

$$(\xi f)(m) = f(m\xi) - f(m)\xi, \qquad \forall m \in M$$

Indeed, setting $(af)(m) = f(m)a = f(ma)$ and introducing the bracket $(\xi, \eta) \to [\xi, \eta]$ of vector fields, we let the reader check that $a(bf) = (ab)f, \forall a, b \in A$ and that we have the formulas:

$$\xi(af) = (\xi(a) + a\xi)f, \qquad (\xi\eta - \eta\xi)f = [\xi, \eta]f, \forall a \in A, \forall \xi, \eta \in T$$

in the operator sense. ∎

Finally, if M is a left D-module, according to the comment following Lemma 5.2, then $M^* = hom_D(M, D)$ is a right D-module and thus $N = N_r$ is a right D-module. However, we have the following technical proposition:

Proposition 5.17. *$\Lambda^n T^*$ has a natural right module structure over D.*

Proof. If $\alpha = adx^1 \wedge ... \wedge dx^n \in \Lambda^n T^*$ is a volume form with coefficient $a \in A$, we may set $\alpha.P = ad(P)(a)dx^1 \wedge ... \wedge dx^n$. As D is generated by A and T, we just need to check that the above formula has an intrinsic meaning for any $\xi \in T$. In that case, we check at once:

$$\alpha.\xi = -\partial_i(a\xi^i)dx^1 \wedge ... \wedge dx^n = -\mathcal{L}(\xi)\alpha$$

by introducing the Lie derivative of α with respect to ξ, along the intrinsic formula $\mathcal{L}(\xi) = i(\xi)d + di(\xi)$ where $i()$ is the interior multiplication and d is the exterior derivative on exterior forms. According to well known properties of the Lie derivative, we get:

$$\alpha.(a\xi) = (\alpha.\xi).a - \alpha.\xi(a),$$
$$\alpha.(\xi\eta - \eta\xi) = -[\mathcal{L}(\xi), \mathcal{L}(\eta)]\alpha = -\mathcal{L}([\xi, \eta])\alpha = \alpha.[\xi, \eta].$$

∎

According to the preceding theorem and proposition, the left differential module corresponding to $ad(\mathcal{D})$ is not N_r but rather $N_l = hom_A(\Lambda^n T^*, N_r)$. When D is a commutative ring, this side changing procedure is no longer needed.

Of course, keeping the same module M but changing its presentation or even using an isomorphic module M' (two OD equations of second order or four OD equations of first order as in the case of the double pendulum), then N may change to N'. The following result, *totally unaccessible to intuition*, justifies "*a posteriori*" the use of the extension functor by proving that the above results are unchanged and are thus "*intrinsic*" [19, 22].

Theorem 5.9. *N and N' are projectively equivalent, that is to say one can find projective modules P and P' such that $N \oplus P \simeq N' \oplus P'$.*

Proof. According to Schanuel lemma, we can always suppose, with no loss of generality, that the resolution of M projects onto the resolution of M'. The kernel sequence is a splitting sequence made up with projective modules because the kernel of the projection of F_i onto F'_i is a projective module P_i for $i = 0, 1$. Such a property still holds when applying duality. Hence, if C is the kernel of the epimorphism from P_1 to P_0 induced by d_1, then C is a projective module and the top short exact sequence splits in the following commutative and exact diagram:

$$
\begin{array}{ccccccccccc}
 & & 0 & & 0 & & 0 & & & & \\
 & & \downarrow & & \downarrow & & \downarrow & & & & \\
0 & \to & C & \to & P_1 & \to & P_0 & \to & 0 & & \\
 & & \downarrow & & \downarrow & & \downarrow & & \downarrow & & \\
0 & \to & K & \to & F_1 & \stackrel{d_1}{\to} & F_0 & \to & M & \to & 0 \\
 & & \downarrow & & \downarrow & & \downarrow & & \downarrow & & \\
0 & \to & K' & \to & F'_1 & \stackrel{d'_1}{\to} & F'_0 & \to & M' & \to & 0 \\
 & & \downarrow & & \downarrow & & \downarrow & & \downarrow & & \\
 & & 0 & & 0 & & 0 & & 0 & &
\end{array}
$$

Applying $hom_A(\bullet, A)$ to this diagram while taking into account Corollary 5.1, we get the following commutative and exact diagram:

$$
\begin{array}{ccccccccccc}
 & & 0 & & 0 & & 0 & & & & \\
 & & \uparrow & & \uparrow & & \uparrow & & & & \\
0 & \leftarrow & C^* & \leftarrow & P_1^* & \leftarrow & P_0^* & \leftarrow & 0 & & \\
 & & \uparrow & & \uparrow & & \uparrow & & \uparrow & & \\
0 & \leftarrow & N & \leftarrow & F_1^* & \stackrel{d_1^*}{\leftarrow} & F_0^* & \leftarrow & M^* & \leftarrow & 0 \\
 & & \uparrow & & \uparrow & & \uparrow & & \uparrow & & \\
0 & \leftarrow & N' & \leftarrow & F_1'^* & \stackrel{d_1'^*}{\leftarrow} & F_0'^* & \leftarrow & M'^* & \leftarrow & 0 \\
 & & \uparrow & & \uparrow & & \uparrow & & \uparrow & & \\
 & & 0 & & 0 & & 0 & & 0 & &
\end{array}
$$

In this diagram C^* is also a projective module, the upper and left short exact sequences split and we obtain $N \simeq N' \oplus C^*$. ■

Remark 5.3. When A is a principal ideal ring, it is well known (see [8, 23] for more details) that any torsion-free module over A is free and thus projective. Accordingly, the kernel of the projection of F_0 onto M is free and we can always suppose, with no loss of generality, that d_1 and d'_1 are monomorphisms. In that case, there is an isomorphism $P_0 \simeq P_1$ in the proof of the preceding theorem and $C = 0 \Rightarrow C^* = 0$, that is to say $N \simeq N'$. This is the very specific situation only considered by OD control theory where the OD equations defining the control systems are always supposed to be differentially independent (linearly independent over D).

Accordingly, using the properties of the extension functor, we get:

Corollary 5.6. $ext_A^i(N, A) \simeq ext_A^i(N', A) \qquad \forall i \geq 1.$

We finally apply these results in order to solve the three preceding problems.

Solution 5.1. As the operator $\mathcal{D}$ of the control system is surjective, it follows that the map d_1 of the presentation is injective. When $K = \mathbb{R}$ and $n = 1$, then D can be identified with a polynomial ring in one indeterminate and is therefore a principal ideal domain (any ideal can be generated by a single polynomial). In this case, it is well known, as we just said, that any torsion-free module is indeed free and thus projective. The short exact sequence of the presentation splits, with a similar comment for its dual sequence. Accordingly, *M is torsion-free if and only if $N = 0$* and it just remains to prove that $\tilde{\mathcal{D}}$ is injective. We have to solve the system:

$$\begin{cases} x \longrightarrow \ddot{\lambda}_1 + \ddot{\lambda}_2 \quad = 0 \\ \theta_1 \longrightarrow l_1\ddot{\lambda}_1 + g\lambda_1 = 0 \\ \theta_2 \longrightarrow l_2\ddot{\lambda}_2 + g\lambda_2 = 0 \end{cases}$$

Multiplying the second OD equation by l_2, the third by l_1 and adding them while taking into account the first OD equation, we get:

$$l_2\lambda_1 + l_1\lambda_2 = 0$$

Differentiating this OD equation twice while using the second and third OD equations, we get:

$$(l_2/l_1)\lambda_1 + (l_1/l_2)\lambda_2 + 0$$

The determinant of this linear system for λ_1 and λ_2 is just $l_1 - l_2$, hence the system is controllable if and only if $l_1 \neq l_2$.
Conversely, if $l_1 = l_2 = l$, the corresponding module has torsion elements. In particular, setting $\theta = \theta_1 - \theta_2$ and subtracting the second dynamic equation from the first, we get $l\ddot{\theta} + g\theta = 0$. Hence θ is a torsion element which is solution of an *autonomous* OD equation, that is an OD equation *for itself* which cannot therefore be "controlled" by any means.
We now study the problem of a double-pendulum when one of the pendula has a variable length, namely , setting $l_1 = l(t), l_2 = 1, g = 1$, we study the system:

$$\ddot{x} + l(t)\ddot{\theta}_1 + \theta_1 = 0, \qquad \ddot{x} + \ddot{\theta}_2 + \theta_2 = 0$$

Multiplying these two OD equations by test functions λ^1, λ^2 and integrating by part, we obtain for the kernel of the adjoint operator:

$$\ddot{\lambda}^1 + \ddot{\lambda}_2 = 0, \qquad l\ddot{\lambda}^1 + 2\dot{l}\dot{\lambda}^1 + (\ddot{l}+1)\lambda^1 = 0, \qquad \ddot{\lambda}^2 + \lambda^2 = 0$$

Eliminating λ^2 among the first and third OD equations, we get at once for $\lambda = \lambda^1$ the system:

$$\lambda^{(4)} + \ddot{\lambda} = 0, \qquad l\ddot{\lambda} + 2\dot{l}\dot{\lambda} + (\ddot{l}+1)\lambda = 0$$

which is surely not formally integrable.
Differentiating twice the second OD equation, we successively get:

$$l\lambda^{(3)} + 3\dot{l}\ddot{\lambda} + (3\ddot{l} + 1)\dot{\lambda} + l^{(3)}\lambda = 0$$

$$l\lambda^{(4)} + 4\dot{l}\lambda^{(3)} + (6\ddot{l} + 1)\ddot{\lambda} + 4l^{(3)}\dot{\lambda} + l^{(4)}\lambda = 0$$

Using the first OD equation of order 4, we get:

$$4\dot{l}\lambda^{(3)} + (6\ddot{l} - l + 1)\ddot{\lambda} + 4l^{(3)}\dot{\lambda} + l^{(4)}\lambda = 0$$

Using the previous OD equation of order 3, we get:

$$(6l\ddot{l} - 12\dot{l}^2 - l^2 + l)\ddot{\lambda} + (4ll^{(3)} - 12\dot{l}\ddot{l} - 4\dot{l})\dot{\lambda} + (ll^{(4)} - 4\dot{l}l^{(3)})\lambda = 0$$

Using the previous OD equation of order 2, we get:

$$(4l^2l^{(3)} - 24l\dot{l}\ddot{l} - 6l\dot{l} + 2l^2\dot{l} + 24\dot{l}^3)\dot{\lambda} + (l^2l^{(4)} - 4l\dot{l}l^{(3)} - (\ddot{l}+1)(6l\ddot{l} - 4\dot{l}^2 - l^2 + l))\lambda = 0$$

It remains to differentiate this OD equation, substitute the $\ddot{\lambda}$ obtained from the first OD equation of order 2 found for λ and eliminate $\dot{\lambda}, \lambda$ in the resulting linear system of two OD equations by equating to zero the corresponding determinant. We should obtain for $l(t)$ *a highly nonlinear OD equation of order 5*. The MAPLE output has been produced by Daniel Robertz (daniel@momo.math.rwth-aachen.de) and covers half a page!
Of course, no computer algebra technique may work in order to look for all the solutions of this OD equation or even have any idea of possible solutions. By chance, in this specific case, exactly as in [19, Example 1.104] where one should point out a confusion as the adjoint operator corresponds to the reduced Kalman type system with variable length and not to the bipendulum, though the conceptual approach is similar, there is a way to find out directly the general solution by integrating the system for $\lambda^1 = \lambda, \lambda^2$ and l. From the third OD equation, we get:

$$\lambda^2 = \alpha cost + \beta sint$$

where α and β are arbitrary constants and thus, from the first OD equation, we therefore obtain:

$$\lambda = at + b - \alpha cost - \beta sint$$

where a and b are again arbitrary constants. Accordingly, we get:

$$(d^2/dt^2)(l\lambda) = -\lambda = -at - b + \alpha cost + \beta sint$$

and thus, with two new arbitrary constants c and d, we get:

$$l\lambda = -\frac{1}{6}at^3 - \frac{1}{2}bt^2 + ct + d - \alpha cost - \beta sint$$

We finally get the general solution:

$$l(t) = \frac{-\frac{1}{6}at^3 - \frac{1}{2}bt^2 + ct + d - \alpha cost - \beta sint}{at + b - \alpha cost - \beta sint}$$

For example, if $a = b = 0$, we get:

$$l(t) = 1 - \frac{ct + d}{\alpha cost + \beta sint}$$

but other possibilities may be of interest and could be tested or simulated, provided $l(t) \geq 0$. Of course, if $l(t) = 1$, we get $a = b = c = d = 0 \Rightarrow \lambda^1 + \lambda^2 = 0$ while, if $l(t) = cst \neq 1$, we also get $\alpha = \beta = 0 \Rightarrow \lambda^1 = \lambda^2 = 0$ in a coherent way with the constant coefficient case.

Solution 5.2. After a short computation left to the reader as an exercise, one checks easily that *the Einstein operator is self-adjoint* because the 6 terms are just exchanged between themselves. Then, it is well known that the compatibility condition is made by the standard divergence operator and its adjoint is the Killing operator (Lie derivative of the Minkowski metric) which admits the linearized Riemann curvature (20 PD equations) as compatibility conditions and not the Einstein equations (10 PD equations only). Hence, the Einstein operator cannot be parameterizable and it follows that *Einstein equations cannot be any longer considered as field equations* (For a computer algebra solution, see [27]).

Solution 5.3. It has already been provided by the preceding theorems.

Remark 5.4. Writing a Kalman type system in the form $-\dot{x} + Ax + Bu = 0$ and multiplying on the left by a test row vector λ, the kernel of the adjoint operator is defined by the system:

$$\dot{\lambda} + \lambda A = 0, \qquad \lambda B = 0\,.$$

Differentiating the second equations, we get:

$$\dot{\lambda} B = 0 \Longrightarrow \lambda AB = 0 \Longrightarrow \lambda A^2 B = 0 \Longrightarrow \ldots$$

and we discover that *the Kalman criterion just amounts to the injectivity of the adjoint operator*. Hence, in any case, *controllability only depends on formal integrability*. Comparing with Examples 5.4, 5.5, 5.6 we notice that, when a constant coefficient operator is injective, the fact that we can find differentially independent compatibility conditions is equivalent to the Quillen-Suslin theorem saying roughly that a projective module over a polynomial ring is indeed free (see [9, 23] for details). More generally, by duality we obtain at once $t(M) \simeq ext^1_A(N, A) \Leftrightarrow t(N) \simeq ext^1_A(M, A)$ and this result is coherent with the introduction of this lecture provided we say that a control system is "*observable*" or rather "*extendable*" (cf. [14, Chap. VIII, Par. 14] and [23]) if $ext^1_A(M, A) = 0$. Finally, in the case of the linearized system with variable coefficients provided in Example 5.11 at the end of Section 2, multiplying by a test function λ and integrating by parts, we have to study the system :

$$U \longrightarrow \dot{\lambda} + \dot{y}\lambda = 0, \qquad Y \longrightarrow u\dot{\lambda} + \dot{u}\lambda = 0.$$

Multiplying the first OD equation by u and subtracting the second, we get $a\lambda = 0$ and the linearized system is controllable if and only if $a \neq 0$.

We now explain the link existing between localization and parametrization (cf. [21]). This result will prove at once that localization is the good substitute for superseding the transfer matrix approach and can even be used for systems with variable coefficients. The main idea is contained in the following technical proposition.

Proposition 5.18. *If A is an integral domain with quotient field $K = Q(A)$ and M is a finitely generated torsion-free module over A, then M can be embedded into a free module of the same rank.*

Proof. As $t(M) = 0$, the module $M = (x_1, ..., x_n)$ can be identified with a submodule of $K \otimes_A M$. Also, as $K \otimes_A M$ is a vector space over K, the $x_1, ..., x_n$ may not be linearly independent over K and we may select a basis $(e_1, ..., e_m)$ with $rk_A(M) = m \leq n$. We have therefore $x_i = \sum_{j=1}^{m} b_i^j e_j$ with $b_i^j \in K$. We may reduce all these coefficients to the same denominator $s \in S = A - \{0\}$ and write $x_i = \sum \frac{a_i^j}{s} e_i = \sum a_i^j \bar{e}_j$ with $a_i^j \in A$. Accordingly, M becomes a submodule of the free module generated over A by the $\bar{e}$ which are linearly independent over K and thus over A. ∎

When D is used in place of A with $S = D - \{0\}$, the expression $x_i = \sum_{j=1}^{m} a_i^j \bar{e}_j$ just describes the parametrization of the corresponding system by m *potentials*. In the non-commutative case, the technique relies on the following lemma that essentially uses the adjoint operator.

Lemma 5.10. $S^{-1}D \simeq DS^{-1}$.

Proof. Let $U \in S$ and $P \in D$. We may consider the adjoint operators $\tilde{U} = ad(U) \in S$ and $\tilde{P} = ad(P) \in D$. Taking into account the Ore property of $S \subset D$, we may find $\tilde{V} = ad(V) \in S$ and $\tilde{Q} = ad(Q) \in D$ such that $\tilde{V}\tilde{P} = \tilde{Q}\tilde{U}$ and obtain therefore $PV = UQ$, that is to say $U^{-1}P = QV^{-1}$. ∎

Example 5.23. With $n = 2, m = 2, K = \mathbb{Q}(x^1, x^2)$, let us consider the system $d_1y^1 - d_2y^2 - x^2y^1 = 0$. We have $d_2y^2 = d_1y^1 - x^2y^1$ and thus $y^2 = d_2^{-1}(d_1 - x^2)y^1$. Using the above lemma, we obtain (exercise) the identity:

$$(d_1 - x^2)d_{22} = d_2(d_{12} - x^2 d_2 + 1)$$

We have therefore $y^2 = (d_{12} - x^2d_2 + 1)d_{22}^{-1}y^1$. Hence, setting $d_{22}^{-1}y^1 = z$, we get the parametrization:

$$y^1 = d_{22}z, \qquad y^2 = (d_{12} - x^2d_2 + 1)z$$

This parametrization is of course not unique and one could exhibit the other (very different) parametrization:

$$y^1 = (d_{12} - x^2 d_2 - 2)z, \qquad y^2 = (d_{11} - 2x^2 d_1 + (x^2)^2)z$$

In both cases, we check that the corresponding compatibility conditions are indeed generated by the given system.

We have therefore exhibited a kind of non-commutative SISO transfer function for a system with variable coefficients. In actual practice, unless we know by means of other techniques that M is torsion-free, the only way to look for a parametrization is to proceed as above for exhibiting one parametrization and to check that it is indeed generic. Of course, many other different parameterizations may exist as in the above example.

At the end of this section, we shall present a new constructive way to study the torsion submodule $t(M)$ of a differential module M having torsion. For this, setting $D' = K[d_1, ..., d_{n-1}] \subset D$, let us now introduce the nested chain:

$$M_{(0)} \subseteq M_{(1)} \subseteq ... \subseteq M_{(\infty)} = M$$

of D'-submodules of the D-module M by defining $M_{(r)}$ as the residue of the set of all parametric jets for which y^k_μ is such that $1 \leq k \leq \beta^n_1, \mu_n = 0$ and z^l_ν is such that $1 \leq l \leq m - \beta^n_1, \nu_n \leq r$.

Definition 5.33. *We shall introduce the chain of D'-submodules:*

$$M_{(-\infty)} \subseteq ... \subseteq M_{(-1)} \subseteq M_{(0)}$$

by setting $M_{(-r-1)} = \{m \in M_{(-r)} \mid d_n m \in M_{(-r)}\}$.

Our main result is the following theorem (compare to [1]):

Theorem 5.10. $M_{(-\infty)} = t(M)$.

Proof. By analogy with the method used for the Kalman form, a torsion element cannot contain any z_ν with $\nu_n \geq 1$ and thus $t(M) \subset M_{(0)}$. As $t(M)$ is a D-module, it is stable by d_n and $t(M) \subseteq M_{(-\infty)}$. Let now $w \in M_{(-\infty)}$ be an element of order q. The number of derivatives of w of order r is a polynomial in r with leading term $\frac{1}{n!}r^n$. Let us now call *jet of class* i any jet with $\mu_1 = ... = \mu_{i-1} = 0$ and *equation of class* i any equation solved with respect to a leading jet of class i, called *principal.* We notice that the prolongations of an involutive solved form are also in involutive solved form because the prolongation of an equation of class i with respect to a multiplicative variable x^j becomes an equation of class j, as $j \leq i$. Accordingly, one can express all the principal jets as linear combinations of the parametric jets and the number of such jets of order $q+r$ for each dependent variable is a polynomial in r with leading term not greater than $\frac{1}{(n-1)!}r^{n-1}$. Hence, when r is large enough, one can eliminate the parametric jets among the derivatives of w that must therefore satisfy at least one OD or PD equation for itself, that is to say $M_{(-\infty)} \subseteq t(M)$. ■

Remark 5.5. Using $(\bar{y}, z)$ in place of (y, z) as we did for the Kalman form, we discover that a torsion element cannot contain anymore z or its jets and only depends on $\bar{y}$ and its jets with respect to $x^1, ..., x^{n-1}$. *The Kalman form is thus only the particular case $n = 1$ of the modified Spencer form.* In this case, $M_{(-1)} = \{\Sigma\lambda x\} \subset M_{(0)} = \{\Sigma\lambda x + \Sigma\mu u\}$ and $M_{(-r-2)}$ can be identified with the orthogonal space to S_r in $M_{(-1)}$ with $dim(M_{(-r-2)}) = n - dim(S_r)$ for $r \geq 0$.

Example 5.24. Looking back at Examples 5.3 and 5.20, we notice that $M_{(0)}$ is generated by $(\bar{y}^1, \bar{y}^2, z)$ and their jets in x^1, modulo the only equation $\bar{y}^1_1 - \bar{y}^2_1 = 0$ of class 1. However, $d_2 z \notin M_{(0)}$ and thus $M_{(-1)}$ is only generated by $(\bar{y}^1, \bar{y}^2)$. Finally, $t(M)$ is generated by $w = \bar{y}^1 - \bar{y}^2 = y^1 - y^2 - 2z$ which satisfies $d_2 w = 0$ but *also* $d_1 w = 0$ and we get $t(M) = M_{(-2)}$.

The effectivity of this recursive algorithm lays on the following non-trivial corollary which constitutes the core of the procedure and brings a quite new understanding of the controllability indices (compare to [1]).

Corollary 5.7. *$t(M) = M_{(-r)}$ for a certain $r < \infty$.*

Proof. The previous definitions amount to the exact sequences:

$$0 \longrightarrow M_{(-r-1)} \longrightarrow M_{(-r)} \xrightarrow{d_n} M_{(-r+1)}/M_{(-r)}$$

and we obtain therefore the induced monomorphisms:

$$0 \longrightarrow M_{(-r)}/M_{(-r-1)} \xrightarrow{d_n} M_{(-r+1)}/M_{(-r)}$$

of D'-modules. Composing these monomorphisms, we deduce that all these quotient modules are torsion-free D'-modules because they are submodules of $M_{(0)}/M_{(-1)}$ which is only generated by the z, exactly like in classical control theory.

Now, if $F \simeq D^r$ is a free D-module, we define its *differential rank* to be $rk_D(F) = r$ and, more generally, if M is a D-module, we define its *differential rank* $rk_D(M)$ to be the differential rank of a maximum free differential submodule $F \subseteq M$, and, in this case, $T = M/F$ is a torsion module. In our situation, setting $rk_{D'}(M_{(-r-1)}/M_{(-r-2)}) = l_r$, we notice that, if $l_r = 0$, then $M_{(-r-1)}/M_{(-r-2)}$ is a torsion D'-module and get a contradiction unless $M_{(-r-1)} = M_{(-r-2)}$. Otherwise, if $l_r > 0$ *strictly*, using the additivity property of the differential rank, we obtain the *strict* inequalities $rk_{D'}(M_{(-r-2)}) < rk_{D'}(M_{(-r-1)}) < rk_{D'}(M_{(-r)}) < ... < rk_{D'}(M_{(-1)}) =$ $nb\, of\, \bar{y} - nb\, of\, equations\, of\, class\, (n-1) < \infty$ and the inductive procedure necessarily stops after a finite number of steps. When $n = 1$, the l_r are nothing else but the controllability indices and $\bar{y}$ is just the output/state. ■

5.6 Poles and Zeros

In order to explain our aim in this section, let us start with a simple motivating example and consider the SISO system:

$$\ddot{y} - 3\dot{y} + 2y = \ddot{u} - \dot{u}$$

Using a Laplace transform, we should get:

$$(s-1)(s-2)\hat{y} = s(s-1)\hat{u} \Rightarrow \hat{y} = \frac{s}{s-2}\hat{u}$$

as we can divide both members by $s-1$. In the rational transfer function thus obtained, we may pay attention *separately* to the numerator s or to the denominator $s-2$, which are both polynomials in s, and look for their roots $s = 0$ or $s = 2$ respectively called *zeros* and *poles*. Of course, in the present traditional setting, we may obtain $\hat{y}$ from $\hat{u}$ and thus y from u but we could also, asking for the inverse problem, try to obtain $\hat{u}$ from $\hat{y}$ and thus u from y. Meanwhile, the common factor $s-1$ with root $s = 1$ just "*disappeared*", though, in view of the way to obtain the transfer function, it could be interpreted either as a zero or as a pole.

It becomes clear that the challenge is now to extend these concepts of poles and zeros to MIMO systems defined by PD equations in a way that should not depend on any multivariable Laplace transform and could be applied to systems with variable coefficients. Also, even if we understand at once that poles and zeros do not any longer, as in the previous sections, depend on a structural property of the control system but rather on an input/output property, we should like to define poles and zeros in a unique intrinsic way related to module theory.

For this, we need a few more definitions and results on modules that were not needed in the previous sections.

If M is a module over a ring A and $ann(M) = ann_A(M)$ is the annihilator of M in A, it follows from its definition that, in any case, $ann(M) = 0$ if M is not a torsion module. Also, if M' is a submodule of M, then $ann(M) \subseteq ann(M')$. The proof of the following lemma is left to the reader as an exercise:

Lemma 5.11. *If M' and M'' are submodules of M, we have:*

- $ann(M' + M'') = ann(M') \cap ann(M'')$
- $ann(M' \oplus M'') = ann(M') \cap ann(M'')$
- $ann(M' \cap M'') \supseteq ann(M') + ann(M'')$.

The key result is the following proposition:

Proposition 5.19. *For any short exact sequence:*

$$0 \to M' \xrightarrow{f} M \xrightarrow{g} M'' \to 0$$

one has the relations:

$$ann(M) \subseteq ann(M') \cap ann(M'')$$
$$\Longrightarrow rad(ann(M)) = rad(ann(M')) \cap rad(ann(M''))$$

Proof. If $a \in ann(M)$, then $f(ax') = af(x') = 0, \forall x' \in M'$ because $x = f(x') \in M$ and thus $ax' = 0$ because f is a monomorphism. It follows that $ann(M) \subset ann(M')$. Now, if $x'' \in M''$, we have $x'' = g(x)$ for some $x \in M$ because g is an epimorphism and we get $ax'' = ag(x) = g(ax) = g(0) = 0$, that is $ann(M) \subseteq ann(M'')$. It follows that $ann(M) \subseteq ann(M') \cap ann(M'')$ and thus $rad(ann(M)) \subseteq rad(ann(M')) \cap rad(ann(M''))$.
Conversely, let $a \in rad(ann(M')) \cap rad(ann(M''))$. As $a \in rad(ann(M''))$, we may find $x \in M$ such that $g(x) = x''$ and we have $a^r x'' = 0$ for a certain $r \in \mathbb{N}$, that is $a^r g(x) = g(a^r x) = 0$. Hence we may find $x' \in M'$ such that $a^r x = f(x')$ because the sequence is exact. As $a \in rad(ann(M'))$, we have $a^s x' = 0$ for a certain $s \in \mathbb{N}$ and we get $a^{r+s}x = a^s a^r x = a^s f(x') = f(a^s x') = 0$, that is $a \in rad(ann(M))$, a result leading to the converse inclusion $rad(ann(M')) \cap rad(ann(M'')) \subseteq rad(ann(M))$. ■

Definition 5.34. *For any ideal $\mathfrak{a} \in A$, we may introduce the zero of $\mathfrak{a}$ to be the family $Z(\mathfrak{a}) = \{\mathfrak{p} \in spec(A) | \mathfrak{p} \supseteq rad(\mathfrak{a}) \supseteq \mathfrak{a}\}$ of proper prime ideals of A.*

Definition 5.35. *If $\mathfrak{p} \in spec(A)$, the localization with respect to the multiplicatively closed set $S = A - \mathfrak{p}$ is denoted by $M_\mathfrak{p}$ and the support of M is the family $supp(M)$ of proper prime ideals $\mathfrak{p} \in spec(A)$ such that $M_\mathfrak{p} \neq 0$.*

Proposition 5.20. *If M is a finitely generated A-module, then $supp(M) = Z(ann(M))$.*

Proof. If $\mathfrak{p} \in spec(A), \mathfrak{p} \not\supseteq ann(M)$, then we can find $s \in ann(M) \subset A, s \notin \mathfrak{p}$ and $M_\mathfrak{p} = 0$, that is to say $supp(M) \subseteq Z(ann(M))$.
Conversely, if $M = Ax_1 + ... + Ax_r$ is a finitely generated A-module and $\mathfrak{p} \in spec(A)$ is such that $M_\mathfrak{p} = 0$, then we can find $s_i \in A - \mathfrak{p}$ such that

$s_i x_i = 0$ for $i = 1, ..., r$. We have $s = s_1 ... s_r \in ann(M)$ but $s \notin \mathfrak{p}$ and therefore $\mathfrak{p} \not\supseteq ann(M)$, that is to say $Z(ann(M)) \subseteq supp(M)$. ∎

We obtain at once the following corollary from the two preceding propositions:

Corollary 5.8. *For any short exact sequence:*

$$0 \to M' \xrightarrow{f} M \xrightarrow{g} M'' \to 0$$

one has the relation:

$$supp(M) = supp(M') \cup supp(M'').$$

Having in mind the preceding example, we now define poles and zeros for systems with coefficients in a field k of constants.
With $D = k[d] = k[d_1, ..., d_n]$ as usual, separating the control variables into *inputs* u and *outputs* y, we may use the canonical injection $Du \subset Du + Dy$, take the intersection of Du with the submodule of equations in $Du + Dy$ allowing to define M and define by residue the differential *input module* M_{in}. A similar procedure may be applied with y in place of u and leads to the differential *output module* M_{out}. We may then introduce the differential modules $M'_{in} = M_{in} + t(M), M'_{out} = M_{out} + t(M)$ and obtain the following commutative diagram of inclusions:

$$\begin{array}{ccccc} & & M & & \\ & \nearrow & & \nwarrow & \\ M'_{in} & & & & M'_{out} \\ & \nwarrow & & \nearrow & \\ \uparrow & & t(M) & & \uparrow \\ & & & & \\ M_{in} & & \uparrow & & M_{out} \\ & \nwarrow & & \nearrow & \\ & & 0 & & \end{array}$$

We now prove that all the known results on poles and zeros just amount to apply the preceding corollary to the various modules we have introduced, both with their sums and quotient whenever they can be defined.
For example, in order to define $supp(M/M_{in})$, that provides the so-called *system poles*, we just need to add $u = 0$ to the control OD equations and look for the annihilator of the differential module M/M_{in} which is a torsion module by tradition in OD control. In the preceding example, we get the ideal $((s-1)(s-2))$ and the only two prime ideals containing it are $(s-1)$ and $(s-2)$. Now, the torsion submodule $t(M)$ is easily seen to be generated by $z = \dot{y} - 2y - \dot{u}$ satisfying $\dot{z} - z = 0$. Hence, in order to define $supp(M/M'_{in})$, that provides the so-called *controllable poles*, we just need to

add $u = 0, \dot{y} - 2y - \dot{u} = 0$ to the control OD equations and look for the annihilator of the differential module M/M'_{in} which is generated by $(s-2)$and is already prime. We have thus recovered the " poles" and could similarly recover the " zeros" by using now $supp(M/M_{out})$ and $supp(M/M'_{out})$. Finally, in order to define $supp(t(M))$, that provides the so-called *input decoupling zeros*, we have to look for the annihilator of $t(M)$ which is generated by $(s-1)$ and thus prime.

The only difficulty left is to define the corresponding concepts in the non-commutative case $D = K[d]$ when K is an arbitrary differential field. For this, using the inclusion $D_q = \{P \in D | ord(P) \leq q\} \subset D$, we obtain the inclusion $D_q y \subset Dy$ inducing by residue, as above, a submodule $M_q \subset M$ *which is not a differential module* with an action $D_r M_q \subseteq M_{q+r}, \forall q, r \geq 0$ providing an equality for q large enough. Looking at the composition $P, Q \in D \to P \circ Q \neq Q \circ P \in D$, we notice (exercise) that $gr(D) = \oplus_{q=0}^{\infty} D_q/D_{q-1}$ is isomorphic to the commutative ring $K[\chi] = K[\chi_1, ..., \chi_n]$ of polynomials in n indeterminates with coefficients in K. Introducing $G_q = M_q/M_{q-1}$ and setting $G = gr(M) = \oplus_{q=0}^{\infty} G_q$, we notice that G is a module over $gr(D)$ and we are brought back to the commutative case with $gr(D)$ and G in place of D and M. As a byproduct, we may state:

Definition 5.36. *$char(M) = supp(G)$ is called the characteristic set of M while $ann(G)$ is called the characteristic ideal.*

Remark 5.6. According to the last corollary, one must use $rad(ann(G))$ in place of $ann(G)$. For more details, see [10, 19]. More specific results on poles and zeros for the particular case $n = 1$ can be found in Chap. 6 of this textbook.

Remark 5.7. The above technique may also be used in order to define poles and zeros for non-linear systems through a generic linearization as will become clear from the following example.

Example 5.25. With $n = 2, M = 1, q = 2$, let us consider the following non-linear system:

$$y_{22} - \frac{1}{3}(y_{11})^3 = 0, \qquad y_{12} - \frac{1}{2}(y_{11})^2 = 0$$

We let the reader prove that this system is involutive with respective multiplicative variables (x^1, x^2) and (x^1). The generic linearization:

$$Y_{22} - (y_{11})^2 Y_{11} = 0, \qquad Y_{12} - y_{11} Y_{11} = 0$$

is defined (exercise) over the differential field $K = \mathbb{Q}(y, y_1, y_2, y_{11}, y_{111}, ...)$ and the characteristic ideal in $K[\chi_1, \chi_2]$ becomes:

$$((\chi_2)^2 - (y_{11})^2(\chi_1)^2, \chi_1(\chi_2 - y_{11}\chi_1))$$

Its radical is the prime ideal $(\chi_2 - y_{11}\chi_1)$.

When k is a ring of constants, the commutative ring $D = k[d_1, ..., d_n]$ of linear differential operators in n commuting formal derivatives with constant coefficients in k can be identified with the polynomial integral domain $A = k[\chi_1, ..., \chi_n]$ in n indeterminates. In 1955, J.P. Serre conjectured that any projective module over A was free, that is to say isomorphic to A^r for some positive integer r (cf. [8, 9, 23]). This result has been proved in 1976, independently by D. Quillen and A.A. Suslin through very technical arguments and a constructive proof has only been given less than ten years ago in [15]. As a generalization of the above results, N.K. Bose and Z. Lin proposed in 1998 a conjecture on the possibility to factorize a certain type of full rank polynomial matrices through zero prime polynomial matrices in order to factor out the P.G.C.D. of the major minors (cf. [3]). Our purpose is to use *algebraic analysis* in order to solve positively this conjecture while giving its intrinsic module theoretic meaning.

Any linear operator $\mathcal{D}$ of order q with constant coefficients acting on m differential indeterminates $y^1, ..., y^m$ can be written as $\mathcal{D} = \sum_{0\leq|\mu|\leq q} a^\mu d_\mu$ when using a multi-index $\mu = (\mu_1, ..., \mu_n)$ with $|\mu| = \mu_1 + ... + \mu_n$ and $a^\mu \in k^{l\times m}$ for some l while setting for short $d_\mu = (d_1)^{\mu_1}...(d_n)^{\mu_n}$. According to what we said, $\mathcal{D}$ can be identified with the polynomial matrix $\varphi = (\sum a^\mu \chi_\mu) \in A^{l\times m}$. From now on, we shall suppose that $l \leq m$ and that φ has l rows, m columns and maximal rank equal to l. In general, any operator $\mathcal{D}$ may have (generating) *compatibility conditions* expressed by another operator $\mathcal{D}_1$ such that $\mathcal{D}_1 \circ \mathcal{D} \equiv 0$ and the above assumption just amounts to the fact that $\mathcal{D}_1$ is the zero operator and we shall say in this case that $\mathcal{D}$ is (formally) *surjective.*

Similarly, we may introduce the (formal) *adjoint operator* $\tilde{\mathcal{D}} = ad(\mathcal{D})$ as usual. In actual practice, for a constant coefficient operator $\mathcal{D}$ with matrix φ, we shall identify the matrix of $\tilde{\mathcal{D}}$ with the simple transposed $\tilde{\varphi}$ of φ, forgetting about the need to also change χ to $-\chi$.

Example 5.26. When $n = 3$, the *curl* operator is not surjective, is equal to its own adjoint and the action of the corresponding polynomial matrix just amounts to exterior product by the covector $\chi = (\chi_1, ..., \chi_n)$, up to sign. The *div* operator is surjective and its adjoint is the *grad* operator.

Let us introduce the module M over A defined by the finite presentation $A^l \xrightarrow{\varphi} A^m \to M \to 0$ where φ acts on the right on row vectors. Similarly, let N be the module defined by the cokernel of $\tilde{\varphi}$. Following [4], we set:

Definition 5.37. *We denote by $I_i(\varphi)$ the ideal of A generated by the determinants of the $i \times i$ submatrices (minors of size i) of φ, with the convention*

$I_0(\varphi) = A$. One can prove that the ideals $Fitt_r(M) = I_{m-r}(\varphi)$ of A, called r-Fitting ideal of M, only depend on M and not on its presentation. Moreover we have $Fitt_r(M) = Fitt_{r+l-m}(N)$.

If M is any module over A, we denote by $ann_A(M)$ or simply $ann(M)$ the *annihilator* of M in A and we shall set as usual $Z = zero$ and $M^\star = hom_A(M, A)$. We quote the two main (*very technical*) results on Fitting ideals that will be crucially used in the sequel with $l \leq m$ and $r = m - l$.

Theorem 5.11. *$Fitt_0(N) \subseteq ann(N)$ and $rad(Fitt_0(N)) = rad(ann(N))$.*

Theorem 5.12. *M is a projective (thus free) module of rank r if and only if $Fitt_r(M) = Fitt_0(N) = A$.*

We now recall from section 3 the following three numbers:
- The last character $\alpha_q^n = rk_D(M) \leq m$.
- The number of nonzero characters $d(M) = d_M = d \leq n$.
- The smallest nonzero character α.

These numbers are intrinsic because, when the corresponding system R_q is involutive in the differential geometric framework, the Hilbert polynomial of M is by definition:

$$P_M(r) = dim_k(R_{q+r}) = \frac{\alpha}{d!} r^d + ...$$

It follows from the delicate Hilbert-Serre theorem (cf. [10, 19]) that d is also equal to the maximum dimension of the irreducible components of the characteristic set, this number being equal, *in the present situation*, to the maximum dimension of the irreducible components of the algebraic set defined by $ann_A(M)$ as we are dealing with systems having constant coefficients. In particular, we have that $d = n \Longleftrightarrow \alpha^n q = \alpha^n \neq 0$ We shall define the *codimension* $cd(M)$ of M to be $n - d$ and we may introduce the two following concepts that coincide when $n = 1$ (cf. [20]).

Definition 5.38. *φ is said to be zero prime if $Fitt_r(M) = Fitt_0(N) = A$ or, equivalently, if $d(N) = -1$ (convention).*

Definition 5.39. *φ is said to be minor prime if the elements of $Fitt_0(N)$ have no common divisor in $A \backslash k$ or, equivalently, if $d(N) \leq n - 2$.*

We are now ready to state the conjecture we want to prove:

Conjecture 5.1. Let us suppose that the greatest common divisor $c \in A$ of the $m!/(m-l)!l!$ minors $a_i = ca'_i$ of φ is such that $(a'_1, ..., a'_{m!/(m-l)!l!}) = A$, then one can find $\varphi' \in A^{l\times m}$ and $\psi \in A^{l\times l}$ such that $\varphi = \psi \circ \varphi'$ and $det(\psi) = c$.

Surprisingly, in order to understand the intrinsic meaning of this conjecture, we shall need many more (delicate) facts from algebraic analysis [7, 10, 19, 20]. In particular, if $\ldots \xrightarrow{d_2} F_1 \xrightarrow{d_1} F_0 \xrightarrow{p} M \rightarrow 0$ is a free resolution of the A-module M, we recall that the groups of cohomology of the dual complex $\ldots \xleftarrow{d_2^\star} F_1^\star \xleftarrow{d_1^\star} F_0^\star \leftarrow 0$, where $d^\star(f) = f \circ d, \forall f \in hom_A(F, A) = F^\star$, do not depend on the resolution and will be denoted by $ext_A^i(M, A) = ker(d_{i+1}^\star)/im(d_i^\star)$ or simply by $ext^i(M)$ with $ext^0(M) = M^\star$. The proof of the three following results is quite delicate but valid for an arbitrary differential module M.

Theorem 5.13. $cd(ext^i(M)) \geq i$.

Theorem 5.14. $cd(M) \geq r \Longleftrightarrow ext^i(M) = 0, \forall i < r$.

Setting $char(M) = Z(ann(M)) = \{\mathfrak{p} \in spec(A) | \mathfrak{p} \supseteq ann(M)\}$, we have the following result

Theorem 5.15. $char(M) = \cup_{i=0}^n char(ext^i(M))$.

We shall now use these results in order to give an intrinsic interpretation and solution of the previous conjecture.
Introducing the *torsion submodule* $t(M) = \{x \in M | \exists 0 \neq a \in A, ax = 0\}$ and the torsion-free module $M' = M/t(M)$, the main trick will be to use N and N' in the preceding theorems, in order to study $t(M), M$ and M'.
First of all, if $c \in k$, then $c \neq 0$ because of the maximum rank assumption on φ and the conjecture is trivial, that is M is projective, thus free. In particular, if $l = m$, then $M = 0$ (see the Janet conjecture in [20]). Hence, we may suppose that $c \in A \backslash k$. Surprisingly, in this case, we shall need *all* the previous results in order to prove that the quoted conjecture is equivalent to the following theorem:

Theorem 5.16. *M' is a projective module if and only if $char(N)$ is the union of irreducible varieties of the same dimension $n - 1$.*

Proof. If M' is projective, the kernel K' of the composition $A^m \rightarrow M \rightarrow M'$ of epimorphisms is a projective module because the short exact sequence $0 \rightarrow K' \rightarrow A^m \rightarrow M' \rightarrow 0$ splits. It is thus a free module of rank l because of the additivity property of the rank and the fact that, if $Q(A)$ is the quotient field of A, we have an isomorphism $Q(A) \otimes_A M \simeq Q(A) \otimes_A M'$ of vector spaces over $Q(A)$. We obtain therefore a commutative and exact diagram:

$$\begin{array}{ccccccccc} 0 & \longrightarrow & A^l & \xrightarrow{\varphi} & A^m & \longrightarrow & M & \longrightarrow & 0 \\ & & \downarrow & & \| & & \downarrow & & \\ 0 & \longrightarrow & A^l & \xrightarrow{\varphi'} & A^m & \longrightarrow & M' & \longrightarrow & 0 \end{array}$$

inducing the matrix $\psi : A^l \to A^l$ on the left, with $\varphi = \psi \circ \varphi'$ acting on the right on row vectors. According to Theorem 5.12, we get $a_i = det(\psi)a'_i$ for the corresponding minors and obtain the assumption of the conjecture with $c = det(\psi)$.

The hard step is the converse way. First of all, if $char(N)$ is $(n-1)$-equidimensional, we find the assumption of the conjecture as we have indeed $ann(N) = (c)$, though this monogenic ideal needs not be equal to its radical. Now, using Theorem 5.15 for N, we get:

$$char(N) = char(ext^0(N)) \cup char(ext^1(N)) \cup char(ext^2(N)) \cup ...$$

Applying $hom_A(\bullet, A)$ to the *ker/coker* exact sequence:

$$0 \longleftarrow N \longleftarrow A^l \xleftarrow{\tilde{\varphi}} A^m \longleftarrow M^\star \longleftarrow 0$$

and using the fact that $\mathcal{D}$ surjective $\Leftrightarrow \varphi$ injective, we get $N^\star = ext^0(N) = 0$ and $char(0) = \emptyset$.
It then follows from Section 5 that we have the *ker/coker* exact sequence:

$$0 \longrightarrow ext^1(N) \longrightarrow M \xrightarrow{\epsilon} M^{\star\star} \longrightarrow ext^2(N) \longrightarrow 0$$

where $\epsilon(x)(f) = f(x), \forall x \in M, \forall f \in M^\star$ and $ext^1(N) = t(M)$, with $d(t(M)) = n-1$ *strictly* in our case.
Using Theorem 5.13, we have $d(ext^i(N)) < n-1$ when $i = 2, 3, ..., n$ and, from the assumption of the theorem, this is impossible unless $ext^i(N) = 0, \forall i = 2, 3, ..., n$.
Then, applying $hom_A(\bullet, A)$ to the short exact sequence $0 \to t(M) \to M \to M' \to 0$, we get $M^\star = M'^\star$ because, if $x \in t(M)$ is such that $ax = 0$ with $a \neq 0$, we get $af(x) = f(ax) = f(0) = 0 \Rightarrow f(x) = 0$ as A is an integral domain and thus $t(M)^\star = 0$.
Using the commutative and exact diagram:

$$\begin{array}{ccccccccccc} 0 & \longrightarrow & t(M) & \longrightarrow & M & \xrightarrow{\epsilon} & M^{\star\star} & \longrightarrow & ext^2(N) & \longrightarrow & 0 \\ & & \downarrow & & \downarrow & & \downarrow & & \downarrow & & \\ 0 & \longrightarrow & t(M') & \longrightarrow & M' & \xrightarrow{\epsilon'} & M'^{\star\star} & \longrightarrow & ext^2(N') & \longrightarrow & 0 \end{array}$$

where $t(M') = ext^1(N') = 0$ because M' is torsion-free, we obtain from an easy chase the isomorphism $ext^2(N') \simeq ext^2(N) = 0$.
Finally, from the previous *ker/coker* exact sequences defining N and N', using twice the connecting sequence for the ext, we get:

$$ext^i(N') \simeq ext^{i-2}(M'^{\star}) \simeq ext^{i-2}(M^{\star}) \simeq ext^i(N) = 0, \forall i = 3, ..., n.$$

It follows that $ext^i(N') = 0, \forall i \geq 1$ and M' is projective according to [20, Corollary 4]. ∎

Remark 5.8. This result is coherent with the fact that N is defined up to an isomorphism (exercise) while N' is only defined up to a projective equivalence and $ext^i(P) = 0, \forall i \geq 1$ for any projective module P. It is also coherent with the long exact connecting sequence for ext :

$$0 \to M'^{\star} \to M^{\star} \to 0 \to ext^1(M') \to ext^1(M) \to ext^1(t(M)) \to ...$$

as we deduce from it the monomorphism $0 \to N \to ext^1(ext^1(N))$ showing that N is 1-*pure* [19].

Remark 5.9. In a more specific way, taking into account Theorem 5.14 and applying it to N, we notice that:

$$N^* = hom_D(N, D) = ext^0_D(N, D) = 0 \Leftrightarrow N torsion \Leftrightarrow \mathcal{D}\, formally\, surjective.$$

We obtain therefore the following recapitulating tabular for systems made up by differentially independent PD equations [22]:

Module M	$\mathbf{ext}^i_D(N, D)$	***d(N)***	**Primeness**
with torsion	$\mathrm{ext}^0_D(N, D) = 0$	$\leq n-1$	$\emptyset$
torsion-free	$\mathrm{ext}^i_D(N, D) = 0,\ 0 \leq i \leq 1$	$\leq n-2$	minor prime
reflexive	$\mathrm{ext}^i_D(N, D) = 0,\ 0 \leq i \leq 2$	$\leq n-3$	
.	.	.	.
.	.	.	.
projective	$\mathrm{ext}^i_D(N, D) = 0,\ 0 \leq i \leq n$	≤ -1	zero prime

where we set $d(N) = -1$ when $char(N) = \emptyset$. For example, the divergence operator for $n = 3$ provides a good example of a differential module which is reflexive but not projective because $ext^3_D(N, D) \neq 0$.

Example 5.27. With $n = 3$ and $k = \mathbb{Q}$, let M be defined by the two independent PD equations $y^3_{12} - y^2_3 - y^3 = 0, y^3_{22} - y^1_3 = 0$. We have $a_1 = (\chi_3)^2, a_2 = \chi_3(\chi_1\chi_2 - 1), a_3 = \chi_3(\chi_2)^2$ and $b_1 = \chi_3, b_2 = \chi_1\chi_2 - 1, b_3 = (\chi_2)^2$ with $(\chi_1)^2 b_3 - (\chi_1\chi_2 + 1)b_2 = 1$, a result leading to $ann(N) = rad(ann(N)) = (\chi_3)$. The generating torsion element $z = y^2_{22} - y^1_{12} + y^1$ satisfies $z_3 = 0$. Though this is *not evident at all*, we may define M' by the two independent PD equations $y^2_{22} - y^1_{12} + y^1 = 0, y^2_{123} - y^1_{113} + y^2_3 + y^3 = 0$ and we have the injective parametrization $u_{22} = y^1, u_{12} - u = y^2, u_3 = y^3$ showing that $M' \simeq A$ is free.

5.7 Conclusion

We hope to have convinced the reader that the results presented are striking enough to open a wide future for applications of computer algebra. The systematic use of the adjoint operator has allowed to relate together results as far from each other as the Quillen-Suslin theorem in module theory and the controllability criterion in control theory. A similar criterion for projective modules does exist and relies on the possibility to have finite length differential sequences [19, 20]. We believe that the corresponding symbolic packages will be available in a short time. It will thus become possible to classify (differential) modules, having in mind that such a classification always describes hidden but deep concepts in the range of applications.

5.8 Exercises

We provide below four exercises which can help the reader recapitulating the various concepts introduced through this chapter.

Exercise 1: In the motivating examples of Section 2, we have seen that the system:

$$Py \equiv d_{22}y = u, \qquad Qy \equiv d_{12}y - y = v$$

where $P, Q \in D = \mathbb{Q}[d_1, d_2]$, admits the two generating compatibility conditions of order 4:

$$A \equiv d_{1122}u - u - d_{1222}v - d_{22}v = 0, \qquad B \equiv d_{1112}u - d_{11}u - d_{1122}v = 0$$

or the single compatibility condition of order 2:

$$C \equiv d_{12}u - u - d_{22}v = 0$$

1) Prove that the two systems $A = 0, B = 0$ on one side and $C = 0$ on the other side for (u, v) determine the same differential module $M \simeq Dy = D$.
2) Determine the unique generating compatibility condition of order 2 satisfied by (A, B).
3) Exhibit the corresponding resolution (') of M:

$$0 \longrightarrow D \longrightarrow D^2 \longrightarrow D^2 \longrightarrow M \longrightarrow 0$$

with central morphism $DA + DB \to Du + Dv$.
4) Exhibit similarly a resolution (") of M:

$$0 \longrightarrow D \longrightarrow D^2 \longrightarrow M \longrightarrow 0$$

with morphism $DC \to Du + dv$.
5) Prove that (') projects onto (") by exhibiting an epimorphism $D^2 \to D$: $DA + DB \to DC$.

6) Prove that the kernel of the preceding epimorphism is a free module by finding out an injective parametrization of the second order operator just obtained.
7) Find a lift for the preceding epimorphism and exhibit a short split exact sequence:

$$0 \longrightarrow D \longrightarrow D^2 \longrightarrow D \longrightarrow 0$$

with morphism $DA + DB \to DC$ on the right.
8) Applying $hom(\bullet, D)$ to the commutative diagram just obtained, prove that, if N' and N'' are the modules "N" that can be constructed from the two preceding resolutions of M, then one has $N' \simeq N'' \oplus D$ and thus $ext_D(N', D) \simeq ext_D(N'', D) = 0, \forall i \geq 1$.

Exercise 2: With $m = 2, n = 2, q = 1$, let us consider the following first order system with two variable coefficients, namely one arbitrary constant α and one arbitrary function $a(x) = a(x^1, x^2)$ (for example $\alpha \in \mathbb{Q} = k, a \in K = \mathbb{Q}(x^1, x^2)$):

$$d_2 y^1 - \alpha d_1 y^1 - d_2 y^2 + a(x) y^2 = 0$$

Study how the corresponding module M does depend on the coefficients by exhibiting a "tree" of 4 possibilities, from the less generic to the most generic one (successively with torsion, free, torsion-free, projective).

1) In the case M has torsion, exhibit a generator for the torsion submodule $t(M)$.
2) In the case M is free, prove that $M \simeq D$.
3) In the case M is torsion-free but not free, prove that $M \subset D$ with strict inclusion and prove that M is not projective.
4) In the case M is projective, exhibit a lift for the presentation $0 \to D \to D^2 \to M \to 0$ which becomes therefore a split short exact sequence.

Exercise 3: With $n = 1, \chi = \frac{d}{dt}$ and $D = \mathbb{Q}[d] \simeq \mathbb{Q}[\chi]$, we shall revisit the following example of a third order OD control system given in 1980 by Blomberg and Ylinen in [2]:

$$\begin{pmatrix} 2\chi^3 + \chi^2 - 8\chi - 4 & 2\chi + 1 & 4\chi^2 + 2\chi \\ \chi^2 - 4 & \chi^3 + 5\chi^2 + 8\chi + 5 & 3\chi + 2 \end{pmatrix} \begin{pmatrix} y^1 \\ y^2 \\ u \end{pmatrix} = \begin{pmatrix} 0 \\ 0 \end{pmatrix}$$

with one input u and two outputs (y^1, y^2).
1) Check that the above operator matrix $\mathcal{D}$ can be factorized through $\mathcal{D} = \mathcal{P} \circ \mathcal{D}'$ with $\mathcal{P}$ square and $\mathcal{D}'$ the operator matrix:

$$\begin{pmatrix} \chi^2 - 4 & 1 & 2\chi \\ 0 & \chi^2 + 3\chi + 2 & 1 \end{pmatrix}$$

Provide the corresponding commutative diagram interpretation, both in terms of operators and in terms of the differential modules M and M' respectively determined by $\mathcal{D}$ and $\mathcal{D}'$.
2) Prove that M' is torsion-free.
3) Check that the ideal generated in $\mathbb{Q}[\chi]$ by the 3 determinants of the 2×2 minors of $\mathcal{D}'$ is equal to $\mathbb{Q}[\chi]$, that is this ideal contains 1. Use this result in order to find out a lift for $\mathcal{D}'$.
4) Prove that $M' \simeq M/t(M)$.
5) Denoting by M_{in} the submodule of M induced by the inclusion $Du \subset Du + Dy^1 + Dy^2$ and introducing ${M_{in}}' = M_{in} + t(M) \subset M$, work out the ideals $\mathfrak{a} = ann(M/{M_{in}}')$, $\mathfrak{a}' = ann({M_{in}}'/M_{in})$, $\mathfrak{a}'' = ann(M/{M_{in}}')$ and prove that $rad(\mathfrak{a}) = rad(\mathfrak{a}') \cap rad(\mathfrak{a}'')$ though $\mathfrak{a} \neq \mathfrak{a}' \cap \mathfrak{a}''$.
For this last question, one may refer to the general theory by showing that $t(M) \cap M_{in} = 0 \Rightarrow {M_{in}}'/M_{in} = t(M)$ and exhibit the short exact sequence:

$$0 \longrightarrow {M_{in}}'/M_{in} \longrightarrow M/M_{in} \longrightarrow M/{M_{in}}' \longrightarrow 0$$

and finally find:

$$\mathfrak{a} \subset rad(\mathfrak{a}) = ((2\chi + 1)(\chi + 1)(\chi + 2)(\chi - 2)),$$

$$\mathfrak{a}' = rad(\mathfrak{a}') = ((2\chi + 1)(\chi + 2)),$$

$$\mathfrak{a}'' \subset rad(\mathfrak{a}'') = ((\chi + 1)(\chi + 2)(\chi - 2)).$$

Exercise 4: We want to study a few points concerning Example 6.
1) Considering the compatibility condition $A = 0$ as a SISO control system for (u, v), use the general test in order to check whether this system is controllable or not. Find a torsion element.
2) Similarly prove that the new SISO control system $A = 0$, $B = 0$ for (u, v) defines a reflexive module which is, nevertheless, not projective.

References

1. E.Aranda-Bricaire, C.H. Moog, J.-B. Pommet, (1995) A Linear Algebraic Framework for Dynamics Feedback Linearization, *IEEE Transactions on Automatic Control*, 40, 1, 127-132.
2. H. Blomberg, Y. Ylinen, (1983) *Algebraic Theory for Multivariable Linear Systems*, Academic Press.
3. N.K. Bose, Z. Lin (2001) A generalization of Serre's conjecture and related issues, *Linear Algebra and its Applications*, 338, 125-138.
4. D. Eisenbud (1995) *Commutative Algebra with a view towards Algebraic Geometry*, Graduate Texts in Mathematics 150, Springer Verlag.

5. M. Janet (1920) Sur les Systèmes aux dérivées partielles, *Journal de Math.*, 8, 3, 65-151.
6. E.R. Kalman, Y.C. Yo, K.S. Narenda (1963) Controllability of Linear Dynamical Systems, *Contrib. Diff. Equations*, 1, 2, 189-213.
7. M. Kashiwara (1995) *Algebraic Study of Systems of Partial Differential Equations*, Mémoires de la Société Mathématique de France 63, (Transl. from Japanese of his 1970 Master's Thesis).
8. E. Kunz (1985) *Introduction to Commutative Algebra and Algebraic Geometry*, Birkhäuser.
9. T.Y. Lam (1978) *Serre's Conjecture*, Lecture Notes in Mathezmatics 635, Springer Verlag.
10. P. Maisonobe, C. Sabbah (1993) *D-Modules Cohérents et Holonomes*, Travaux en Cours, 45, Hermann, Paris.
11. B. Malgrange (1966) *Cohomologie de Spencer (d'après Quillen)*, Sém. Math. Orsay.
12. H. Matsumura (1986) *Commutative Ring Theory*, Cambridge Studies in Advanced Mathematics 8, Cambridge University Press.
13. U. Oberst (1990) Multidimensional Constant Linear Systems, *Acta Appl. Math.*, 20, 1-175.
14. V.P. Palamodov (1970) *Linear Differential Operators with Constant Coefficients*, Grundlehren der Mathematischen Wissenschaften 168, Springer Verlag.
15. H. Park, C. Woodburn (a995) An Algorithmic Proof of Suslin's Stability Theorem for Polynomial Rings, *J. Algebra*, 178, 277-298.
16. J.F. Pommaret (1986) Géométrie Différentielle Algébrique et Théorie du Contrôle, *C. R. Acad. Sci. Paris*, 302, série I, 547-550.
17. J.F. Pommaret (1995) Dualité Différentielle et Applications, *C. R. Acad. Sci. Paris*, 320, Série I, 1225-1230.
18. J.F. Pommaret (1994) *Partial Differential Equations and Group Theory: New Perspectives for Applications*, Kluwer.
19. J.F. Pommaret (2001) *Partial Differential Control Theory*, Kluwer. (http://cermics.enpc.fr/~pommaret/home.html)
20. J.F. Pommaret, A. Quadrat (1999) Algebraic Analysis of Linear Multidimensional Control Systems, *IMA Journal of Mathematical Control and Informations*, 16, 275-297.
21. J.F. Pommaret, A. Quadrat (1999) Localization and Parametrization of Linear Multidimensional Control Systems, *Systems and Control Letters*, 37, 247-260.
22. A. Quadrat (1999) *Analyse Algébrique des Systèmes de Contrôle Linéaires Multidimensionnels*, Thèse de Docteur de l'Ecole Nationale des Ponts et Chaussées, (http://www-sop.inria.fr/cafe/Alban.Quadrat/ index.html).
23. J.J. Rotman (1979) *An Introduction to Homological Algebra*, Pure and Applied Mathematics, Academic Press.
24. D.C. Spencer (1965) Overdetermined Systems of Partial Differential Equations, *Bull. Amer. Math. Soc.*, 75, 1-114.
25. W.M. Wonham (1985) *Linear Multivariable Control: a Geometric Approach*, Springer Verlag.
26. J. Wood (2000) Modules and behaviours in nD systems theory, *Multidimensional Systems and Signal Processing*, 11, 11-48.
27. E. Zerz (2000) *Topics in Multidimensional Linear Systems Theory*, Lecture Notes in Control and Information Sciences 256, Springer Verlag.

6

Structural Properties of Discrete and Continuous Linear Time-Varying Systems: A Unified Approach

Henri Bourlès

SATIE, ENS de Cachan et CNAM, 61 Ave du Président Wilson, 94230 Cachan, France. E-mail: henri.bourles@satie.ens-cachan.fr

6.1 Introduction

The aim of this chapter is the study of structural properties of linear systems. Systems with time-varying coefficients are considered, both in the continuous- and discrete-time cases. These two cases are merged into a general framework. This study is based on *module theory.* Let us briefly explain what a module is, in connection with the theory of differential and difference equations.

The notion of vector space over a field $\mathbf{k}$ is classic. Let V be a $\mathbf{k}$-vector space, $\lambda \neq 0$ be a scalar (i.e. an element of $\mathbf{k}$) and v be a vector (i.e. an element of V). If $\lambda\, v = 0$, then $v = 0$.

A module M is a set similar to a vector space, but defined over a *ring* $\mathbf{R}$ of scalars. The axioms of a module are the same as those of a vector space (and, as a field is a ring, a vector space is a module of a special kind). The point is that a nonzero element of a ring is not necessarily invertible, therefore the above property does not hold. Let m be an element of an $\mathbf{R}$-module M, and $\lambda \neq 0$ be a scalar (i.e. an element of $\mathbf{R}$). The equality $\lambda\, m = 0$ does not imply $m = 0$: such an element m is called a *torsion element* of M. The only torsion element of a vector space is 0.

Consider the differential equation

$$\frac{dm}{dt} = 0. \tag{6.1}$$

Let $\mathbf{R} = \Re\,[\partial]$ be the ring of polynomials with real coefficients and indeterminate $\partial = \frac{d}{dt}$. Equation (6.1) can be written $\lambda\, m = 0$, with $\lambda = \partial \in \mathbf{R}$. This scalar λ is nonzero, and obviously (6.1) does not imply $m = 0$; from the above, m is a torsion element of an $\mathbf{R}$-module M. This elementary example explains why module theory is a suitable framework for the study of linear differential equations –and, more generally, of linear dynamical *systems.*

F. Lamnabhi-Lagarrigue et al. (Eds.): Adv. Top. in Cntrl. Sys. Theory, LNCIS 311, pp. 225–280, 2005

In the discrete-time case, let us consider the difference equation

$$(\mathbf{q}-1)\,m=0 \tag{6.2}$$

where $\mathbf{q}$ is the usual shift forward operator $m\,(t) \longrightarrow m\,(t+1)$. Equation (6.2) can be put in the form (6.1), setting $\partial = \mathbf{q}-1$. In everything that follows, this operator ∂ is called the *discrete-time derivative.* For our purpose, it is more suitable to use the operator ∂, rather than $\mathbf{q}$. Using ∂, indeed, (6.2) becomes a "discrete-time differential equation", similar to the "continuous-time differential equation" (6.1).

Module theory has been developing in Mathematics since the end of the 19th century (the notion of module is due to Dedekind: see [24]), and its use for solving systems of differential equations with *constant* coefficients is explained in classical textbooks (e.g., [7], Chap. IV, Sect. 2, n°9).

In the control community, Kalman was the first who realized that the concept of module can be used to study some aspects of systems theory [28]. However, his approach was quite different from the one explained in the sequel, which was introduced by Fliess [16]. The fact that the set of solutions (called the *behavior* in what follows) of a linear system with *constant* coefficients is a module over $\mathbf{R} = \Re\,[\partial]$ had been noticed before by several authors (e.g., [2]) but not exploited.

Module theory was also used to study *multidimensional* linear systems with constant coefficients [42]; the case of multidimensional systems with time-varying coefficients (which can be viewed as systems governed by *partial differential equations*) was recently considered [46]. To our knowledge, these extensions were not applied to real control problems; nevertheless, they are closely connected with the theory of "$\mathcal{D}$-modules", a subject which is very topical in Mathematics (see [35] and the references therein).

In this chapter, we focus on "monodimensional systems", i.e. systems governed by *ordinary* differential equations.

When discovering the subject, the reader may assume that the systems under consideration have constant coefficients, since the theory is then much easier. However, the present chapter is self-contained for any reader having a basic background in Automatic control and Algebra (notions such as *group, ring, field, vector space*, etc., are assumed to be known). The references indicated in the text should enable the reader to fill possible deficiencies. Many exercises are proposed; for the most difficult ones, a hint, or a reference where the solution can be found, are given.

The chapter is organized as follows: differential polynomials are presented in Sect. 2. This part is classical in Mathematics and the main reference is [12] (for the case of constant coefficients, [40] is a good reference). Modules are studied in connection with linear differential equations in Sect. 3. A good introduction to modules (and in particular to modules over commutative principal ideal domains) can be found in [40]. The reader may also have a look on ([7], Chap. IV, Sect. 2). In the general case, the main reference is once more [12]. Linear time-varying systems are studied in Sect. 4 in a module-theoretic setting. This approach was revealed by Fliess in the seminal papers [15] and especially [16]. The "behavioral approach", developed by Willems [55] (and extensively expounded in [45]), is succinctly presented in Sect. 5, using the theory explained in Sect. 4 and the theory of "duality". This "duality" refers to the theory of categories and functors [39], whose essential points are explained. In the context of linear systems, one of the first contributions to duality is due to Oberst [42] (following a remark of Malgrange [36]). An up-to-date presentation of duality (from the mathematical point of view) can be found in [30]; [48] is also a good reference. Last, Sect. 5 is devoted to concluding remarks.

6.2 Differential Polynomials

6.2.1 Differential Fields

The Coefficient Field

In what follows, **K** denotes the set to which the coefficients of the system under study belong; this set is assumed to be a commutative field[1], and is called the "coefficient field", for short. As these coefficients are (possibly) time-varying, they are functions of time t. It is assumed that the system coefficients are indefinitely differentiable (in the sense specified in Sect. 6.1), which implies that the commutative field **K** is equipped with a derivation denoted by δ. This derivation δ is the same as the indeterminate ∂ introduced in Sect. 6.1, with the only difference that δ operates on the system coefficients and ∂ operates on the system variables. As we will see, the derivation δ satisfies the Leibniz rule or a suitable generalization of this rule. The field **K**, equipped with the derivation δ, is called a "differential field". This is specified below.

Examples

Let us take some examples.

(I) $\mathbf{K} = \Re$, the field of real numbers.

(II) $\mathbf{K} = \Re(t)$, the field of rational functions with real coefficients.

(III) $\mathbf{K} = \Re((t))$, the field of formal Laurent series with real coefficients. An element of this field is of the form

[1] The more general case where **K** is a ring is considered in [23]. Unfortunately, systems with coefficients belonging to a ring, not a field, have poor structural properties.

$$\sum_{i \geq \nu} a_i \, t^i \quad , a_i \in \Re.$$

(IV) $\mathbf{K} = \Re\{\{t\}\}$, the field of *convergent* Laurent series with real coefficients.

(V) $\mathbf{K} = \mathbf{M}(\Re)$, the field of meromorphic functions on the real line (i.e. of restrictions to the real line of meromorphic functions on the complex plane).

Continuous-Time Case

If the continuous-time case, all above examples are valid. Let us consider two elements a and b of $\mathbf{K}$. By the Leibniz rule:

$$\frac{d}{dt}(a \, b) = a \, \frac{db}{dt} + \frac{da}{dt} \, b$$

and as $\frac{d}{dt}(.) = (.)^{\delta}$, this yields

$$(a \, b)^{\delta} = a \, b^{\delta} + a^{\delta} \, b \tag{6.3}$$

Discrete-Time Case

In the discrete-time case, let us consider two elements a and b of $\mathbf{K}$. As $\delta = \mathbf{q} - 1$, one obtains:

$$\begin{aligned}(a \, b)^{\delta}(t) &= a(t+1) \; b(t+1) - a(t) \; b(t) \\ &= a(t+1) \; [b(t+1) - b(t)] + [a(t+1) - a(t)] \; b(t)\end{aligned}$$

i.e.

$$(a \, b)^{\delta} = a^{\mathbf{q}} \, b^{\delta} + a^{\delta} \, b. \tag{6.4}$$

In examples (I), (II) and (V) above, $\mathbf{q}$ is a mapping from $\mathbf{K}$ to $\mathbf{K}$, and is then an *endomorphism* of $\mathbf{K}$ (i.e. $\mathbf{q}$ is $\mathbf{K}$-linear).

A General Framework

The continuous- and discrete-time cases can be merged into a general framework. Set $\alpha = 1$ (resp. $\alpha = \mathbf{q}$) in the continuous- (resp. discrete-) time case. The rules (6.3) and (6.4) become

$$(a \, b)^{\delta} = a^{\alpha} \, b^{\delta} + a^{\delta} \, b. \tag{6.5}$$

A derivation δ satisfying the rule (6.5) is called an "α-derivation" ([12], Sect. 0.8); δ is an endomorphism of the *abelian group* $\mathbf{K}$. In both the continuous- and discrete-time cases, α is an *automorphism* of $\mathbf{K}$ (since α is surjective and has an inverse $\beta = \alpha^{-1}$).

A *constant* of $\mathbf{K}$ is an element a such that $a^{\delta} = 0$. This implies $a^{\alpha} = a$ in the discrete-time case and this equality is always satisfied in the continuous-time case. In addition, $\alpha\,\delta = \delta\,\alpha$. Therefore:

In everything that follows, a *differential field* is a field equipped with an α-derivation δ such that: α is an *automorphism* of $\mathbf{K}$, $a^{\delta} = 0$ implies $a^{\alpha} = a$, and $\alpha\,\delta = \delta\,\alpha$.[2]

The set of all constants of $\mathbf{K}$ is a subfield of $\mathbf{K}$, called the *subfield of constants* of $\mathbf{K}$, and denoted by $\mathbf{k}$. The ring $\mathbf{K}$ is said to be a *field of constants* if all its elements are constants (such as in example (I)).

In the *continuous-time case*, the subfield of constants of $\mathbf{K}$ is the field $\Re$ in all above examples.

Let us consider the *discrete-time case.*

Exercise 6.1. Prove that in example (II), the subfield of constants of $\mathbf{K}$ is the field $\Re$. (Hint: a nonzero polynomial has a finite number of roots.)

In example (V), a constant is a meromorphic periodic function whose period is of the form $1/n$ for some positive integer n, e.g. $a(t) = A \tan n\,\pi\,(t + \varphi)$, $A \in \Re$.

6.2.2 Rings of Differential Polynomials

Let $\mathbf{K}$ be a differential field. A *left* differential polynomial is an element which can be uniquely written in the form

$$a(\partial) = \sum_{i=0}^{n} a_i\, \partial^{n-i}, \quad a_i \in \mathbf{K} \tag{6.6}$$

where a_0 is assumed to be nonzero whenever $a(\partial) \neq 0$ ([29], Sect. I.1(6)). The natural integer n is the *degree* of the left polynomial $a(\partial)$.

A *right* differential polynomial is an element which can be uniquely written in the form

$$b(\partial) = \sum_{i=0}^{m} \partial^{n-i}\, b_i, \quad b_i \in \mathbf{K}$$

[2] In the *general theory*, as it is presented in [12], α is only an *endomorphism* of $\mathbf{K}$, $a^{\delta} = 0$ does not imply $a^{\alpha} = a$, and the equality $\alpha\,\delta = \delta\,\alpha$ does not necessarily hold. The case we are considering is much simpler.

where its leading coefficient b_0 is assumed to be nonzero whenever $b(\partial) \neq 0$. The natural integer m is the *degree* of the right polynomial $b(\partial)$.

A left differential polynomial such as (6.6) is considered as an *operator* on system variables. Let w be such a variable and $a \in \mathbf{K}$. By the same rationale as in Sect. 6.2.1, $\partial(a\,w) = a^{\alpha}\,\partial w + a^{\delta}\,w$. As this equality is valid for any system variable w, one obtains the *commutation rule*

$$\partial a = a^{\alpha}\partial + a^{\delta}. \tag{6.7}$$

Therefore, a right differential polynomial of degree 1 can be put in the form of a left differential polynomial of degree 1. By induction, a right differential polynomial of any degree can be put in the form of a left differential polynomial of the same degree.

One obtains from (6.7), with $\beta = \alpha^{-1}$:

$$a\partial = \partial a^{\beta} - a^{\beta\,\delta}. \tag{6.8}$$

Therefore, by the same rationale as above, a left differential polynomial of any degree can be put in the form of a right differential polynomial of the same degree.

In the sequel, the set of all (left or right) differential polynomials, equipped with the commutation rule (6.7), is denoted as $\mathbf{R} = \mathbf{K}[\partial; \alpha, \delta]$. The reader can check that $\mathbf{R}$ is a ring. The degree of any nonzero element $a = a(\partial)$ of $\mathbf{R}$ is a well-defined natural integer denoted by $d^{\circ}(a)$. In addition, we set $d^{\circ}(0) = -\infty$.

A left (resp. right) differential polynomial is said to be *monic* if its leading coefficient is 1.

6.2.3 Properties of General Rings

Integral Domains

In this chapter, a ring $\mathbf{D}$ is called an *integral domain*, or a *domain*, if it is integral, i.e. without zero divisors[3]. This means that if a and b are two elements of $\mathbf{D}$ such that $a\,b = 0$, then $a = 0$ or $b = 0$. A *unit* of $\mathbf{D}$ is an invertible element.

Two nonzero elements a and b of an integral domain $\mathbf{D}$ are said to be *associated* if there there exist units υ and ϑ such that $a = \upsilon b \vartheta$. If $\upsilon = 1$ (resp. $\vartheta = 1$), they are said to be *right* (resp. *left*) *associated.*

[3] For many authors, a *domain* is a *commutative* integral ring. We follow the terminology of Cohn [12]. Similarly, in this chapter, a *field* means a *possibly noncommutative field* (called a *skew field* or a *division ring* by many authors).

An *atom* of $\mathbf{D}$ is a nonzero element which is not the product of two nonunits ([12], p. xix)[4].

The proof of the following proposition is easy and left to the reader:

Proposition 6.1. *(i) The ring* $\mathbf{R} = \mathbf{K}[\partial; \alpha, \delta]$ *is an integral domain.*

(ii) If a *and* b *are nonzero differential polynomials, then* $d^\circ(a\,b) = d^\circ(a) + d^\circ(b)$*. (iii) For any differential polynomials* a *and* b, $d^\circ(a - b) \leq \max\{d^\circ(a), d^\circ(b)\}$*, with equality holding whenever* $d^\circ(a) \neq d^\circ(b)$ *(iv) The ring* $\mathbf{R}$ *is commutative if, and only if* $\mathbf{K}$ *is a field of constants.*

Ideals

Most of important properties of rings are connected to properties of their ideals.

Let $\mathbf{A}$ be a ring and $\mathbf{a}$ be a subset of $\mathbf{A}$. The set $\mathbf{a}$ is said to be a *left* (resp. *right*) *ideal* of $\mathbf{A}$ if it is a subgroup of the abelian group $\mathbf{A}$ and if $a\,b \in \mathbf{a}$ whenever $a \in \mathbf{A}$ and $b \in \mathbf{a}$ (resp. $a \in \mathbf{a}$ and $b \in \mathbf{A}$). If $\mathbf{a}$ is both a left and right ideal, it is called a *two-sided ideal.* If $\mathbf{A}$ is a commutative ring, any left or right ideal of $\mathbf{A}$ is two-sided. Let $\mathbf{A}$ be any ring; 0 (i.e. the subring of $\mathbf{A}$ whose only element is 0) and $\mathbf{A}$ are two-sided ideals of $\mathbf{A}$. A left or right ideal of $\mathbf{A}$ different from 0 and from $\mathbf{A}$ is said to be *proper.* A field has no proper left or right ideals.

Let $(a_j)_{j \in J}$ be a family of elements of the ring $\mathbf{A}$. The *left* ideal *generated* by this family is $\sum_{j \in J} \mathbf{A}\,a_j$, i.e. the subgroup of $\mathbf{A}$ consisting of all elements of the form $\sum_{j \in J} b_j\,a_j$, where $b_j \in \mathbf{A}$ and all but a finite number of b_j are zero. Similarly, the right ideal generated by the family $(a_j)_{j \in J}$ is $\sum_{j \in J} a_j\,\mathbf{A}$. The left (resp. right) ideal generated by a single element $a \in \mathbf{A}$ is called the *principal left* (resp. *right*) *ideal* generated by a; it is equal to $\mathbf{A}\,a$ (resp. $a\,\mathbf{A}$).

Let $\mathbf{a}$ and $\mathbf{b}$ be two left (resp. right) ideals. Then $\mathbf{a}+\mathbf{b}$ is the set consisting of all elements of the form $a + b$, $a \in \mathbf{a}$, $b \in \mathbf{b}$. The product $\mathbf{a}\,\mathbf{b}$ is the set consisting of all elements of the form $\sum_{finite} a_i\,b_i$, $a_i \in \mathbf{a}$, $b_i \in \mathbf{b}$. Both sets $\mathbf{a}+\mathbf{b}$ and $\mathbf{a}\,\mathbf{b}$ are left (resp. right) ideals ([32], Chap. II, Sect. 1).

The proof of the following result is easy and left to the reader (or see [12], Sect. 6.1, prop. 1.4):

Proposition 6.2. *Let* $\mathbf{a}$ *be a nonzero ideal of an integral domain* $\mathbf{D}$*. If* $\mathbf{a}$ *is principal both as left and as right ideal, there exists* $0 \neq a \in \mathbf{D}$ *such that* $\mathbf{D}\,a = a\,\mathbf{D}\,a = a\,\mathbf{D}$*. Such an element* a *is called an* invariant element *of* $\mathbf{D}$.

[4] This terminology is used only in the context of noncommutative rings. In the commutative case, it is replaced by the term "prime" ([32], Chap. II, Sect. 5) or "extremal element" ([4], Chap. VI, Sect. 1, n°13); the necessity of introducing this somewhat unusual expression is explained below.

Ore Domains

A *commutative ring* $\mathbf{D}$ can be embedded in a field of quotients (also called a field of fractions) if, and only if $\mathbf{D}$ is an *integral domain* (this field is then unique, and consists of all elements of the form b/a, $b \in \mathbf{D}$, $0 \neq a \in \mathbf{D}$).

In the *noncommutative case*, this condition is still necessary, but no longer sufficient. Left fractions, of the form $a^{-1}\,b$, must be distinguished from right fractions, of the form $b\,a^{-1}$.

An integral domain $\mathbf{D}$ is said to be a *left Ore domain* if for any nonzero elements a and b of $\mathbf{D}$, $\mathbf{D}\,a \cap \mathbf{D}\,b \neq 0$. A right Ore domain is defined similarly, replacing principal left ideals by right ones.

An integral domain $\mathbf{D}$ can be embedded in a (unique) field of left (resp. right) fractions if, and only if it is a left (resp. right) Ore domain. If $\mathbf{D}$ is a two-sided Ore domain (i.e. a left Ore domain which is a right Ore domain), its fields of left and right fractions coincide ([12], Sect. 0.8).

Principal Ideal Domains and Euclidean Domains

Principal One-Sided Ideal Domains and Principal Ideal Domains

A *principal left* (resp. *right*) *ideal domain* is an integral domain whose all left (resp. right) ideals are principal. A *principal ideal domain* is a principal left ideal domain which is a principal right ideal domain ([38], Sect. 0.1.8).

By Proposition 6.2, *the two-sided ideals of a principal ideal domain are generated by an invariant element.* In a noncommutative principal ideal domain, all left or right ideals are principal but *not necessarily two-sided.*

A principal left (resp. right) ideal domain is a left (resp. right) noetherian domain[5], therefore it is a left (resp. right) Ore domain ([12], Sect. 0.8, Corollary 8.10).

One-sided euclidean domains and euclidean domains

A *left* (resp. *right*) *euclidean domain* is an integral domain $\mathbf{D}$ equipped with a function $\theta : \mathbf{D} \longrightarrow \mathbb{N} \cup \{-\infty\}$ (where $\mathbb{N}$ is the set of natural integers), called a *left* (resp. *right*) *degree function*, satisfying the 4 following conditions ([12], Sect. 2.1):

(E1) $\theta(0) = -\infty; \quad \theta(1) = 0.$

(E2) For any elements a and b of $\mathbf{D}$, $\theta(a - b) \leq \max\{\theta(a), \theta(b)\}$;

[5] A ring $\mathbf{A}$ is left (resp. right) *noetherian* if every left (resp. right) ideal of $\mathbf{A}$ is finitely generated ([38], 0.1.2).

(E3) If a and b are nonzero elements of $\mathbf{D}$, then $\theta(a\,b) = \theta(a) + \theta(b)$;

(E4) For any elements a and b of $\mathbf{D}$ such that $b \neq 0$, there exists q and r in $\mathbf{D}$ such that

$$a = q\,b + r \text{ (resp. } a = b\,q + r), \quad \theta(r) < \theta(b).$$

Condition (E4) is called the *left* (resp. *right*) *division algorithm.* Notice that by the left division algorithm, one "right-divides" a by b.

An *euclidean domain* is a domain equipped with a left and right degree function[6].

Lemma 6.1. *(i) Condition (E4) is equivalent to the following condition:*

(E5) For any elements a and b of $\mathbf{D}$ such that $b \neq 0$ and $\theta(a) \geq \theta(b)$, there exists $c \in \mathbf{D}$ such that

$$\theta(a - c\,b) < \theta(a) \quad (\text{resp. } \theta(a - b\,c) < \theta(a)).$$

(ii) In a left or right euclidean domain, $\theta(v) = 0$ if, and only if v is a unit.

Proof. (i) The proof is given for the right division algorithm. If (E4 right) holds, then (E5 right) follows by taking $c = q$. Conversely, if (E5 right) holds and $a, b \in \mathbf{D}$, $b \neq 0$, choose $q \in \mathbf{D}$ such that $\theta(a - b\,q)$ takes its least value. If $\theta(a - b\,q) \geq \theta(b)$, then by (E5 right) there exists $c \in \mathbf{D}$ such that $\theta(a - b\,q - b\,c) < \theta(a - b\,q)$, a contradiction. Therefore, $\theta(a - b\,q) < \theta(b)$ and (E4 right) is satisfied with $r = a - b\,q$. (ii) If v is a unit, $0 = \theta\left(v\,v^{-1}\right) = \theta(v) + \theta\left(v^{-1}\right)$, thus $\theta(v) = \theta\left(v^{-1}\right) = 0$. Conversely, let v be such that $\theta(v) = 0$. Right dividing 1 by v, one obtains $1 = q\,v + r$ where $\theta(r) < 0$, thus $r = 0$. Therefore, $1 = q\,v$, hence q and v are units. ■

Proposition 6.3. *(i) A left (resp. right) euclidean domain is a principal left (resp. right) ideal domain.*

(ii) An euclidean domain is a principal ideal domain.

Proof. (i): Let $\mathbf{a} \neq 0$ be a left ideal of a left euclidean domain $\mathbf{D}$. Let b be an element of $\mathbf{a}$ of smallest left degree ≥ 0. Let $a \neq 0$ be an element of $\mathbf{a}$. By the left division algorithm (E4), we find $q, r \in \mathbf{D}$ such that $a = q\,b + r$ and $\theta(r) < \theta(b)$. But $r = a - q\,b$, whence $r \in \mathbf{a}$. Since b has minimal left degree ≥ 0 in $\mathbf{a}$, it follows that $r = 0$, therefore $\mathbf{a} = \mathbf{D}\,b$ and $\mathbf{D}$ is a principal left ideal domain. (ii) is then an obvious consequence of the definitions. ■

[6] Our terminology is different from the one chosen in [12], where an euclidean domain is a left *or* right euclidean domain. Our terminology is justified by Proposition 6.3(ii).

Proposition 6.4. *The ring* $\mathbf{R} = \mathbf{K}[\partial; \alpha, \delta]$ *is an euclidean domain.*

Proof. The "degree" d° of a differential polynomial, as defined in Sect. 6.2.2, satisfies conditions (E1), (E2) and (E3), by Proposition 6.1. Let us show that d° satisfies (E5) (left and right). Consider the nonzero left differential polynomial a defined by (6.6), and let $b = \sum_{i=0}^{m} b_i\, \partial^{m-i}$ be another left differential polynomial, such that $a_0 \neq 0, b_0 \neq 0$ and $m \leq n$. Then $d^\circ\left(a - a_0\, b_0^{-1}\, b\, \partial^{n-m}\right) < d^\circ(a)$. Thus the left condition (E5) is satisfied. For the right condition, the rationale is similar, considering right differential polynomials. ■

Simple Rings

A ring $\mathbf{A}$ is said to be *simple* if it has no proper two-sided ideals.

A commutative simple ring is a commutative field ([29], Sect. I.1). In the noncommutative case, a simple ring may have proper left or right ideals.

Let $\mathbf{D}$ be a principal ideal domain. By Proposition 6.2, $\mathbf{D}$ is simple if, and only if its only invariant elements are *units*.

Proposition 6.5. *Consider a ring* $\mathbf{R} = \mathbf{K}[\partial; \alpha, \delta]$ *of differential polynomials. Let* $f = \sum_{i=1}^{n} a_i\, \partial^i$, $a_0 = 1$. *This element* f *is invariant if, and only if conditions (i) and (ii) below are satisfied:*

(i) $f\, c = c^{\alpha^n} f$ *for all* $c \in \mathbf{K}$

(ii) $f\, \partial = (\partial + a_1 - a_1^\alpha)\, f$.

Proof. (a) Let us prove that conditions (i) and (ii) are necessary for f to be "left invariant", i.e. for the left ideal $\mathbf{R}f$ to be two-sided. This means $f\mathbf{R} \subset \mathbf{R}f$, i.e. that for each $g \in \mathbf{R}$, there exists $g_1 \in \mathbf{R}$ such that $g_1\, f = f\, g$. In particular, let $g = c \in \mathbf{K}$. There should exist elements c_1, u and v such that $c_1\, f = f\, c$ and $(\partial u + v)\, f = f\, \partial$. By comparison of degrees, c_1, u and v belong to $\mathbf{K}$. The first equality implies

$$c_1\, (\partial^n + ...) = (\partial^n + ...)\, c,$$

therefore $c_1 = c^{\alpha^n}$ by (6.7), which proves (i). Regarding the second equality,

$$f\, \partial = \partial^{n+1} + a_1\, \partial^n + ...,$$

$$(\partial u + v)\, f = u\, \partial^{n+1} + (v + u\, a_1^\alpha)\, \partial^n + ...$$

therefore, $u = 1$ and $v = a_1 - a_1^\alpha$, which proves (ii). (b) Conversely, assuming that conditions (i) and (ii) are satisfied, one can show by induction on the degree that f is left invariant. (c) As α is an automorphism with inverse β, (i) and (ii) may be restated as $c\, f = f\, c^{\beta^n}$ and $\partial\, f = f\, \left(\partial - a^{\beta^n}\right)$ with $a = a - a_1^\alpha$, which is the necessary and sufficient condition for f to be "right invariant". ■

Proposition 6.6. *Let* $\mathbf{R} = \mathbf{K}[\partial; \alpha, \delta]$ *be a ring of differential polynomials. If* $\mathbf{K}$ *is not a field of constants, the only invariant elements of* $\mathbf{R}$ *are the nonzero constants of* $\mathbf{K}$.

Proof. Let f be a nonzero element of $\mathbf{K}$. (1) Conditions (i) and (ii) of Proposition 6.5 are satisfied if f is a constant. (2) Let $f = \sum_{i=1}^{n} a_i \, \partial^i$ be a differential polynomial of degree $n \geq 1$, with $a_0 = 1$. Let $c \in \mathbf{K}$. As $\alpha \, \delta = \delta \, \alpha$, it is easy to prove by induction that

$$\partial^n c = \sum_{i=0}^{n} \binom{n}{i} c^{\alpha^i \delta^{n-i}} \partial^i. \tag{6.9}$$

Therefore,

$$f \, c = c^{\alpha_n} \, \partial^n + \left(n \, c^{\alpha^{n-1} \delta} + a_1 \, c \right) \partial^{n-1} + \ldots$$

For condition (i) of Proposition 6.5 to be satisfied, one must have

$$n \, c^{\alpha^{n-1} \delta} = a_1 \left(c^{\alpha^n} - c \right)$$

This equality is satisfied for every $c \in \mathbf{K}$ if, and only if $\mathbf{K}$ is a field of constants. ∎

Let us summarize the main results:

Theorem 6.1. *Let* $\mathbf{R} = \mathbf{K}[\partial; \alpha, \delta]$ *be a ring of differential polynomials. This ring is an euclidean domain, thus a principal ideal domain, and it is simple if* $\mathbf{K}$ *is not a field of constants.*

Divisibility

One-Sided Divisibility

Let us study *right divisibility.*

Let $\mathbf{D}$ be an *integral domain* and $a, c \in \mathbf{D}$, $a \neq 0$; a is said to be a *right divisor* of c (and c is said to be a *left multiple* of a) if there exists $q \in \mathbf{D}$ such that $c = q \, a$. This property holds if, and only if $\mathbf{D} \, c \subset \mathbf{D} \, a$.

Assuming that $\mathbf{D}$ is a *principal left ideal domain,* let $a, b \in \mathbf{D}$. The set $\mathbf{D} \, a \cap \mathbf{D} \, b$ is a left ideal, thus there exists $c \in \mathbf{D}$ such that $\mathbf{D} \, c = \mathbf{D} \, a \cap \mathbf{D} \, b$. Such an element c is said to be a *least common left multiple* (l.c.l.m.) of a and b, since any common left multiple of a and b is a left multiple of c. An l.c.l.m. of two elements is unique up to left associates. An element $c \neq 0$ is a right divisor of two elements a and b such that $a \, b \neq 0$ if, and only if

$\mathbf{D}\,a+\mathbf{D}\,b\subset\mathbf{D}\,c$. The left ideal $\mathbf{D}\,a+\mathbf{D}\,b$ is principal, thus there exists $d\in\mathbf{D}$ such that $\mathbf{D}\,a+\mathbf{D}\,b=\mathbf{D}\,d$; d is a *greatest common right divisor* (g.c.r.d.) of a and b, since every common right divisor of a and b is a right divisor of d. A g.c.r.d. of two elements a and b (such that $a\,b\neq 0$) is unique up to left associates. The elements a and b are said to be *right coprime* if 1 is a g.c.r.d. of a and b.

Left divisibility can be studied similarly.

Total divisibility

Let $\mathbf{D}$ be an *integral domain* and c be an invariant element of $\mathbf{D}$ (if any: see Proposition 6.2). An element a is a left multiple of c if, and only if $a\in\mathbf{D}\,c=c\,\mathbf{D}$, thus the sets of left and right multiples of c coincide.

An element $b\neq 0$ is a right divisor of c if, and only if there exists $d\neq 0$ such that $c=d\,b$. As $\mathbf{D}\,c=c\,\mathbf{D}$, there exists $q\neq 0$ such that $d\,c=c\,q$, thus $d\,(d\,b-b\,q)=0$, whence $d\,b=b\,q$. Therefore, b is a left divisor of c, and the sets of left and right divisors of c coincide. As a result, the notations $b\mid c$ and $c\mid a$ are not ambiguous when c is invariant.

Let a and b be two nonzero elements of $\mathbf{D}$; b is said to be a *total divisor* of a (written $b\parallel a$) if there exists an invariant element c such that $b\mid c\mid a$.

An *invariant non unit* p of $\mathbf{D}$ is said to be a *prime* if: $p\mid a\,b$ implies $p\mid a$ or $p\mid b$.

In a *commutative principal ideal domain* (or, more generally, in a commutative unique factorization domain), the notions of *prime* and of *atom* coincide ([32], Chap. II, Sect. 5).

Let $\mathbf{D}$ be a *principal ideal domain* and c be an invariant element of $\mathbf{D}$ (if any). The set C of all two-sided ideals containing c is inductively ordered by inclusion; therefore, by Zorn's lemma, C has a maximal element $\mathbf{D}\,p\,\mathbf{D}$ ([32], Appendix 2, Sect. 2). The invariant element p is a prime which divides c.

6.3 Modules and Systems of Linear Differential Equations

6.3.1 Modules

Some Definitions

Let $\mathbf{A}$ be a ring. A set M is said to be a left $\mathbf{A}$-module if for any $m_1,m_2\in M$ and any $\lambda\in\mathbf{A}$, $m_1+m_2\in M$ and $\lambda m_1\in M$. All modules are *left modules* in the sequel (except when explicitly stated), i.e. in the product of a scalar and of an element of a module, the scalar appears on the left-hand side. An **A**-module M is said to be *generated* by a family $(w_i)_{i\in I}$ (written $M = \left[(w_i)_{i\in I}\right]_{\mathbf{A}}$) if every element $m \in M$ is an **A**-linear combination of the elements w_i, i.e. if there exists a family $(\lambda_i)_{i\in I}$ of elements of **A** such that $m = \sum_{i\in I} \lambda_i\, w_i$ where all but a finite number of λ_i are zero. A module is said to be *finitely generated* (f.g.) if it is generated by a finite sequence of its elements. A left ideal of **A** (and in particular **A** itself) is an **A**-module.

Let M, M' be **A**-modules. A mapping $f : M \to M'$ is said to be **A**-linear (or an **A**-morphism) if for any $m_1, m_2 \in M$ and any $\lambda \in \mathbf{A}$, $f(m_1 + m_2) = f(m_1) + f(m_2)$ and $f(\lambda\, m_1) = \lambda\, f(m_1)$. Such a mapping f is called a *monomorphism* (or a *monic* map) if $\ker f = 0$, an *epimorphism* (or an *epic* map) if $\operatorname{Im} f = M'$, and an *isomorphism* if it is both monic and epic. The notation $M \cong M'$ means that there exists an isomorphism $M \xrightarrow{\sim} M'$. A submodule N of M is a subset of M which is a module; the mapping $N \ni n \to n \in M$ is called the *canonical monomorphism.*

Let M be an **A**-module and S be a subset of M. The *annihilator* of S, written $\operatorname{Ann}(S)$, is the set of all $\lambda \in \mathbf{A}$ such that $\lambda\, S = 0$; $\operatorname{Ann}(S)$ is a *left ideal.* Assuming that **A** is an integral domain, an **A**-module M is said to be *bounded* if $\operatorname{Ann}(M) \neq 0$.

Products, Coproducts and Sums

Let $(M_j)_{j\in J}$ be a family of **A**-modules.

The *product* $\prod_{j\in J} M_j$ is the set of all sequences $(m_j)_{j\in J}$ where $m_j \in M_j$ for each index $j \in J$.

The *coproduct* $\coprod_{j\in J} M_j$ is the subset of $\prod_{j\in J} M_j$ consisting of those elements $(m_j)_{j\in J}$ such that $m_j = 0$ for all but a finite number of indexes $j \in J$ ([48], Chap. 2).

Let $(M_j)_{j\in J}$ be a family of *submodules* of an **A**-module M. The *sum* $\sum_{j\in I} M_j$ is the module generated by $\bigcup_{j\in J} M_j$.

This sum is said to be *direct* (and then written $\bigoplus_{j\in J} M_j$) if for every $k \in J, M_k \bigcap \left(\sum_{j\neq k} M_j\right) = 0$.

Exercise 6.2. (i) Show that $\prod_{j\in J} M_j$, $\coprod_{j\in J} M_j$ and $\bigoplus_{j\in J} M_j$ have a natural structure of **A**-module.

(ii) Prove that there exists an isomorphism $\coprod_{j\in J} M_j \cong \bigoplus_{j\in J} M_j$. (For (ii) see, e.g., ([3], Sect. 1, n°8).)

Remark 6.1. Due to isomorphism to be proved in exercise 6.2(ii), coproducts and direct sums can be identified.

Free Modules

An $\mathbf{A}$-module M is said to be *free* if it has a basis, i.e. if there exists a family $(e_i)_{i\in I}$ such that: M is generated by this family, and the elements e_i are $\mathbf{A}$-linearly independent. This means that any element m of M can be written, *in a unique manner*, in the form: $m = \sum_{i\in I} \lambda_i \, e_i$.

Let I be a nonempty set and $\mathbf{A}^{(I)} = \coprod_{i\in I} \mathbf{A}$.

Exercise 6.3. (i) Show that $\mathbf{A}^{(I)}$ is an $\mathbf{A}$-module.

(ii) Let $c_i : I \to \mathbf{A}$ be the mapping defined as: $c_i(j) = 1$ if $i = j$ and $c_i(j) = 0$ otherwise. Show that $(c_i)_{i\in I}$ is a basis of $\mathbf{A}^{(I)}$ (called the *canonical basis*), which implies that $\mathbf{A}^{(I)}$ is free. (iii) Let M be a free $\mathbf{A}$-module; show that there exists a set I and an isomorphism $M \cong \mathbf{A}^{(I)}$. (Hint: to clarify ideas, first consider the case where I is finite; see ([3], Chap. II, Sect. 1, n°11).)

The *rank* of the free module M is the cardinal of I such that $M \cong \mathbf{A}^{(I)}$. Note that the rank of a free module is well-defined if, and only if the ring $\mathbf{A}$ has the following property: $\mathbf{A}^{(I)} \cong \mathbf{A}^{(J)}$ if, and only if $\operatorname{card} I = \operatorname{card} J$, i.e. if $\mathbf{A}$ has "invariant basis number"; a left or right noetherian domain has invariant basis number ([12], Sect. 0.2).

Quotient Modules

Let N be a submodule of an $\mathbf{A}$-module M and $x \in M$. The *class* of x (mod N) is the set of all $y \in M$ such that $x - y \in N$; this class is denoted by $\bar{x} = x + N$. The set of all classes $\bar{x}$, $x \in M$, is easily shown to be an $\mathbf{A}$-module, called the *quotient module* M/N. The mapping $M \ni x \to \bar{x} \in M/N$ is $\mathbf{A}$-linear, and is called the *canonical epimorphism*. Denoting by φ the canonical epimorphism, $\varphi(x) = 0$ if, and only if $x \in N$. As a consequence, $\mathbf{A}/\mathbf{A} = 0$ and $\mathbf{A}/0 = \mathbf{A}$.

The following facts are useful ([48], Chap. 2, 12-15):

Proposition 6.7. *(i) Let M, N be two $\mathbf{A}$-modules and $f : M \to N$ be an $\mathbf{A}$-linear morphism. Then, $\operatorname{Im} f \cong M/\ker f$.*

(ii) If M_1 and M_2 are submodules of M, then

$$M_1/(M_1 \cap M_2) \cong (M_1 + M_2)/M_2 .$$

(iii) If $M_2 \subset M_1$ are submodules of M, then $(M/M_2)/(M_1/M_2) \cong M/M_1$.

(iv) If M' is a submodule of M, there is a one-to-one correspondence between the submodules S of M/M' and the intermediate submodules of M containing M', given by $S \to \varphi^{-1}(S)$, where $\varphi : M \to M/M'$ is the canonical epimorphism.

Exercise 6.4. Let N be a submodule of an $\mathbf{A}$-module M and $\varphi : M \to M/N$ be the canonical epimorphism. An $\mathbf{A}$-morphism $f : M \to M'$ has the form $\bar{f} \circ \varphi$, where $\bar{f} : M/N \to M'$ is $\mathbf{A}$-linear, if, and only if $N \subset \ker f$. This map $\bar{f}$ is then unique (and is called the "induced map").

Lemma 6.2. *Every module is isomorphic to a quotient of a free module.*

Proof. Let $M = \left[(w_i)_{i \in I}\right]_{\mathbf{A}}$, $(c_i)_{i \in I}$ be the canonical basis of $\mathbf{A}^{(I)}$, and $\varphi : \mathbf{A}^{(I)} \to M$ be the $\mathbf{A}$-morphism defined by

$$\varphi(c_i) = w_i. \tag{6.10}$$

The morphism φ is epic, thus $M \cong \mathbf{A}^{(I)}/\ker \varphi$ by Proposition 6.7(i). ■

Exact Sequences

Two $\mathbf{A}$-morphisms

$$M' \xrightarrow{f} M \xrightarrow{g} M"$$

are *exact at* M if $\operatorname{Im} f = \ker g$. A sequence of $\mathbf{A}$-morphisms

$$\ldots \to M_{n+1} \xrightarrow{f_{n+1}} M_n \xrightarrow{f_n} M_{n-1} \to \ldots$$

is *exact* if each adjacent pair of $\mathbf{A}$-morphisms is exact.

Exercise 6.5. (i) The sequence $0 \to M' \xrightarrow{f} M$ is exact if, and only if f is monic.

(ii) The sequence $M \xrightarrow{g} M" \to 0$ is exact if, and only if g is epic.

(iii) If $f : M \to N$ is an $\mathbf{A}$-morphism, there is an exact sequence $0 \to \ker f \to M \xrightarrow{f} N \to \operatorname{coker} f \to 0$, where $\operatorname{coker} f$ (the "cokernel" of f) is defined as $N/\operatorname{Im} f$. (iv) The sequence $0 \to M' \xrightarrow{f} M \xrightarrow{\varphi} M" \to 0$ is exact if, and only $M' \cong f(M')$ and $M" \cong M/f(M')$.

(Remark: exact sequences such as in (iv) are called *short exact sequences.* Hint: for (iii) and (iv), use Proposition 6.7(i).)

Presentation of a Module

Let $M = \left[(w_j)_{j \in J}\right]_{\mathbf{A}}$ be an $\mathbf{A}$-module. As shown by Lemma 6.2, there exists an exact sequence $\mathbf{A}^{(J)} \xrightarrow{\varphi} M \to 0$. Let $F = \ker \varphi \subset \mathbf{A}^{(J)}$. Similarly, there exists an exact sequence $\mathbf{A}^{(I)} \xrightarrow{g} F \to 0$. Let $f : \mathbf{A}^{(I)} \to \mathbf{A}^{(J)}$ be defined by $f(x) = g(x)$ for every $x \in \mathbf{A}^{(I)}$. The following sequence is exact:

$$\mathbf{A}^{(I)} \xrightarrow{f} \mathbf{A}^{(J)} \xrightarrow{\varphi} M \to 0. \tag{6.11}$$

Definition 6.1. *The exact sequence* (6.11) *is called a* presentation *of the module M. The free module* $\mathbf{A}^{(J)}$ *is the* module of generators *of M, and the module $Im\, f = ker\, \varphi$ is the* module of relations.

If M is *finitely generated*, the cardinal of J is finite, e.g. $\operatorname{card} J = k$, thus $\mathbf{A}^{(J)} = \mathbf{A}^k$. If, in addition, $\mathbf{A}$ is *noetherian* (and, in particular, if $\mathbf{A}$ is a *principal ideal domain*), the submodule $F = \ker \varphi$ of $\mathbf{A}^k$ is finitely generated ([48], Theorem 4.1), thus the cardinal of I is finite, e.g. $\operatorname{card} I = q$. In this case, the module M is said to be *finitely presented*, since it is presented by the following exact sequence:

$$\mathbf{A}^q \xrightarrow{f} \mathbf{A}^k \xrightarrow{\varphi} M \to 0. \tag{6.12}$$

Assuming that M is *finitely presented*, let $(a_i)_{1 \le i \le q}$ and $(c_j)_{1 \le j \le k}$ be the canonical bases of $\mathbf{A}^q$ and $\mathbf{A}^k$, respectively, set

$$f(a_i) = \sum_{j=1}^{k} b_{ij}\, c_j, \quad 1 \le i \le q, \quad b_{ij} \in \mathbf{A}, \tag{6.13}$$

and let B be the matrix with entries b_{ij}. Representing the elements of $\mathbf{A}^q$ and $\mathbf{A}^k$ by *rows* in the canonical bases, B is the matrix of f in these bases; moreover, the $\mathbf{A}$-morphism f is identified (in these bases) by the right multiplication by B, written $f = \bullet B$. The image of a_i by f is the submodule of $\mathbf{A}^k$ generated by the i-th row of B, and

$$M = \varphi\left(\mathbf{A}^k\right) \cong \mathbf{A}^k / \mathbf{A}^q \bullet B = \operatorname{coker} \bullet B. \tag{6.14}$$

By (6.10) and (6.13),

$$\sum_{j=1}^{k} b_{ij}\, w_j = 0, \quad 1 \le i \le q$$

i.e., setting $w = \begin{bmatrix} w_1 & \cdots & w_k \end{bmatrix}^T$,[7]

$$B\, w = 0. \tag{6.15}$$

Equation (6.15) and the matrix B are called, respectively, the *equation* and the *matrix of definition* of the module $M = [w]_{\mathbf{A}}$ (in the canonical bases); M is said to be *defined by generators* (the elements c_j such that $\varphi(c_j) = w_j$) *and relations* (the rows of (6.15)) ([48], Chap. 3).

[7] In the sequel, a finite sequence $\lambda = (\lambda_j)_{1 \le j \le k}$ (resp. $w = (w_j)_{1 \le j \le k}$) of elements of a ring $\mathbf{A}$ (resp. of an $\mathbf{A}$-module M) is identified with the row $\begin{bmatrix} \lambda_1 & \cdots & \lambda_k \end{bmatrix}$ (resp. with the column $\begin{bmatrix} w_1 & \cdots & w_k \end{bmatrix}^T$). Therefore, the linear combination $\sum_{1 \le j \le k} \lambda_j\, w_j$ can be written $\lambda\, w$.

Torsion Modules

Let $\mathbf{D}$ be an integral domain and M a $\mathbf{D}$-module. An element $m \in M$ is called a *torsion element* if there exists $0 \neq \lambda \in \mathbf{D}$ such that $\lambda m = 0$ (see Sect. 6.1).

Proposition 6.8. *Let* $\mathbf{D}$ *be a left Ore domain. The set of all torsion elements of a* $\mathbf{D}$*-module* M *is a* submodule $\mathcal{T}(M)$.

Proof. (a) Let $m_1, m_2 \in M$ and λ_1, λ_2 be nonzero elements of $\mathbf{D}$ such that $\lambda_1 m_1 = \lambda_2 m_2 = 0$. As $\mathbf{D}$ is left Ore, $\mathbf{D}\lambda_1 \cap \mathbf{D}\lambda_2 \neq 0$, i.e. there exist $\lambda_1', \lambda_2' \in \mathbf{D}$ such that $\lambda_2' \lambda_1 = \lambda_1' \lambda_2 \neq 0$. Therefore, $\lambda_2' \lambda_1 (m_1 + m_2) = \lambda_2' \lambda_1 m_2 = \lambda_1' \lambda_2 m_2 = 0$, thus $m_1 + m_2$ is torsion. (b) Similarly, the reader may check that if $m \in M$ is a torsion element, λm is torsion for every $\lambda \in \mathbf{D}$ ([12], Sect. 0.9). ■

A $\mathbf{D}$-module is said to be *torsion-free* if $\mathcal{T}(M) = 0$. The quotient module $M/\mathcal{T}(M)$ is torsion-free. A module M is said to be *torsion* if $M = \mathcal{T}(M)$.

Let $\mathbf{D}$ be a left Ore domain, $\mathbf{Q}$ be its field of left fractions, and M be a $\mathbf{D}$-module. The *tensor product* $\mathbf{Q} \otimes M$ is the set of all elements

$$\sum_{finite} \lambda_i \otimes m_i, \quad \lambda_i \in \mathbf{Q},$$

where $\lambda_i \otimes m_i$ is the formal product $\lambda_i m_i$.

Exercise 6.6. ([12], Sect. 0.9, Prop. 9.1). (i) $\hat{M} = \mathbf{Q} \otimes M$ is a $\mathbf{Q}$-vector space.

(ii) Every element of $\hat{M}$ has the form $a^{-1} m, m \in M, 0 \neq a \in \mathbf{Q}$.

(iii) Let $\hat{\varphi}$ be the canonical mapping $M \to \hat{M}$ defined by $\hat{\varphi}(m) = \frac{1}{1} m$. This map is $\mathbf{D}$-linear and $\ker \hat{\varphi} = \mathcal{T}(M)$.

(iv) The $\mathbf{D}$-morphism $\hat{\varphi}$ is an embedding (i.e. a monomorphism) if, and only if $\mathcal{T}(M) = 0$.

(v) Any submodule of a free module is torsion-free.

Let $\mathbf{D}$ be a left Ore domain with field of left fractions $\mathbf{Q}$, and M be a $\mathbf{D}$-module. The *rank* of M (written $\operatorname{rk} M$) is defined as the dimension of $\mathbf{Q} \otimes M$ as $\mathbf{Q}$-vector space.

Exercise 6.7. Let $\mathbf{D}$ be a left Ore domain. (i) If M is a free $\mathbf{D}$-module, the above definition of $\operatorname{rk} M$ agrees with the one given in the paragraph "Free modules" above.

(ii) A $\mathbf{D}$-module T is torsion if, and only if $\operatorname{rk} T = 0$.

The following result will be useful ([12], Prop. 9.1):

Proposition 6.9. *Let N be a submodule of a* $\mathbf{D}$*-module M; then $rk\,M = rk\,N + rk\,M/N$.*

Cyclic Modules

An $\mathbf{A}$-module M is said to be *cyclic* if it is generated by a single element. Let M be a cyclic $\mathbf{A}$-module, w be its generator, and $\mathbf{a}$ =Ann(w). Let $\varphi : \mathbf{A} \to \mathbf{A}/\mathbf{a}$ be the canonical epimorphism. As $\ker \varphi = M$, there exists an isomorphism $M \cong \mathbf{A}/\mathbf{a}$ by Proposition 6.7(i). Conversely, a quotient $\mathbf{A}/\mathbf{a}$ is cyclic, generated by $\varphi(1)$, since $\mathbf{A}$ is generated by 1.

An $\mathbf{A}$-module M is said to be *decomposable* if there exist two *proper* submodules M_1 and M_2 of M (i.e. submodules of M different from 0 and M) such that $M = M_1 \oplus M_2$ (these modules M_1 and M_2 are then called *direct summands* of M); otherwise, M is said to be *indecomposable.*

Let $\mathbf{D}$ be an integral domain; an element $a \in \mathbf{D}$ is said to be *bounded* if the left $\mathbf{D}$-module $\mathbf{D}/\mathbf{D}\,a$ is bounded (see above in this section). An element u is said to be *totally unbounded* if it contains no bounded nonunit factors ([12], Sect. 6.4).

Proposition 6.10. *(i) Let* $\mathbf{a}$ *be a* two-sided *ideal. The cyclic module* $\mathbf{A}/\mathbf{a}$ *is decomposable* if, and only if *there exist two* two-sided *ideals* $\mathbf{b}$ *and* $\mathbf{c}$*, different from* 0 *and* $\mathbf{A}$*, such that* $\mathbf{A} = \mathbf{b}+\mathbf{c}$ *and* $\mathbf{a} = \mathbf{b} \cap \mathbf{c}$ *; then,* $\mathbf{A}/\mathbf{a} \cong \mathbf{A}/\mathbf{b} \oplus \mathbf{A}/\mathbf{c}$.

(ii) Let $\mathbf{b}$ *and* $\mathbf{c}$ *be two* left *ideals such that* $\mathbf{A} = \mathbf{b} + \mathbf{c}$*, and* $\mathbf{a} = \mathbf{b} \cap \mathbf{c}$. *Then,* $\mathbf{A}/\mathbf{a} \cong \mathbf{A}/\mathbf{b} \oplus \mathbf{A}/\mathbf{c}$.

(iii) Let $\mathbf{D}$ *be an integral domain and* $a, b \in \mathbf{D}$, $b \neq 0$. *Then* $\mathbf{D}/\mathbf{D}\,a \cong \mathbf{D}\,b/\mathbf{D}\,a\,b$ *and this is a direct summand of* $\mathbf{D}/\mathbf{D}\,a\,b$ *if, and only if* $\mathbf{D}\,a + b\,\mathbf{D} = \mathbf{D}$*; in that case,* $\mathbf{D}/\mathbf{D}\,a\,b \cong \mathbf{D}/\mathbf{D}\,a \oplus \mathbf{D}/\mathbf{D}\,b$.

Proof. (i): see ([3], Chap. 1, Sect. 8, n°11, Prop. 10). (ii): $\mathbf{A}/\mathbf{a} = (\mathbf{b}+\mathbf{c})/(\mathbf{b} \cap \mathbf{c}) = \mathbf{b}/(\mathbf{b} \cap \mathbf{c}) + \mathbf{c}/(\mathbf{b} \cap \mathbf{c})$, and $(\mathbf{b}/(\mathbf{b} \cap \mathbf{c})) \cap (\mathbf{c}/(\mathbf{b} \cap \mathbf{c})) = 0$. In addition, $\mathbf{b}/(\mathbf{b} \cap \mathbf{c}) \cong \mathbf{A}/\mathbf{c}$ and $\mathbf{c}/(\mathbf{b} \cap \mathbf{c}) \cong \mathbf{A}/\mathbf{b}$ by Proposition 6.7(ii). (iii): the right multiplication by b, written $\bullet b$, is an isomorphism $\mathbf{D}/\mathbf{D}\,a \to \mathbf{D}\,b/\mathbf{D}\,a\,b$; see ([12], Sect. 3.6, Lemma 6.7) for the other statements. ■

Two nonzero elements $a, b \in \mathbf{D}$ are said to be *similar* if $\mathbf{D}/\mathbf{D}\,a \cong \mathbf{D}/\mathbf{D}\,b$. This notion is left-right symmetric: a and b are similar if, and only if $\mathbf{D}/a\,\mathbf{D} \cong \mathbf{D}/b\,\mathbf{D}$ (as right $\mathbf{D}$-modules) ([12], Sect. 0.6, Corollary 6.4).

Simple Modules

An $\mathbf{A}$-module $S \neq 0$ is said to be *simple* if it has no *proper* submodule, i.e. if its only submodules are 0 and S. Let $M \neq 0$ be a f.g. $\mathbf{A}$-module and N be a submodule of M; the set of all submodules of M which contain N is inductively ordered by inclusion ([5], Sect. 3, n°1, prop. 4), therefore (according to Zorn's lemma) this set possesses a maximal element P. The quotient $S = M/P$ is simple. Therefore, we have:

Proposition 6.11. *Every nonzero f.g. module possesses a simple quotient.*

Exercise 6.8. (i) A simple $\mathbf{A}$-module is cyclic.

(ii) A left $\mathbf{A}$-module $\mathbf{A}/\mathbf{a}$ is simple if, and only if $\mathbf{a}$ is a maximal left ideal of $\mathbf{A}$.

(iii) Let $\mathbf{D}$ be a left principal ideal domain and $S \cong \mathbf{D}/\mathbf{D}\,a$ be a cyclic $\mathbf{D}$-module; S is simple if, and only if a is an atom. (iv) Let a be an atom of a left principal ideal domain $\mathbf{D}$, and b be an element of $\mathbf{D}$ similar to a; then, b is an atom. (Hint: the solution of (ii) can be found in ([5], Sect. 3, n°1, prop. 2); for (iii), use Proposition 6.10, (iii).)

Definition 6.2. *Let $(S_i)_{i\in I}$ be a nonempty family of simple $\mathbf{A}$-modules. This family is called a* representative system *of simple $\mathbf{A}$-modules if (i) $S_i \ncong S_j$ for $i \neq j$, and (ii) for any simple $\mathbf{A}$-module S, there exists one index $i \in I$ and an isomorphism $S \cong S_i$.*

According to the choice axiom ([40], Chap. I, Sect. 2, th.1 and exerc. 11), a representative system of simple $\mathbf{A}$-modules exists.

Primary Decomposition of a Cyclic Module

Let $\mathbf{D}$ be a *principal ideal domain* and a be a nonzero element of $\mathbf{D}$. The following proposition is a generalization to the noncommutative case of the classical *unique factorization theorem into irreductible elements* ([32], Chap II, Sect. 5). This generalization is proven in ([12], Sect. 6.4, th. 4.5).

Proposition 6.12. *The left ideal $\mathbf{D}\,a$ has the decomposition*

$$\mathbf{D}\,a = \mathbf{D}\,\pi_1 \cap \ldots \mathbf{D}\,\pi_m \cap \mathbf{D}\,u \tag{6.16}$$

where each π_i is a product of similar bounded atoms, while atoms in different π_i's are dissimilar and u is a unit or is totally unbounded. Moreover the π_i's and u are unique up to left associates.

As a consequence of Proposition 6.12, we have the following theorem:

Theorem 6.2. *(i) The following isomorphism of $\mathbf{D}$-modules holds:*

$$\frac{\mathbf{D}}{\mathbf{D}\,a} \cong \frac{\mathbf{D}}{\mathbf{D}\,\pi_1} \oplus \ldots \oplus \frac{\mathbf{D}}{\mathbf{D}\,\pi_m} \oplus \frac{\mathbf{D}}{\mathbf{D}\,u}. \tag{6.17}$$

(ii) The cyclic modules $\mathbf{D}/\mathbf{D}\,\pi_i$, as well as $\mathbf{D}/\mathbf{D}\,u$ if u is not a unit, are indecomposable, and are called primary modules. *The elements π_i, as well as u if the latter is not a unit, are called the* elementary divisors *of the module $\mathbf{D}/\mathbf{D}\,a$.*

(iii) The 3 following properties are equivalent: (1) *the module $\mathbf{D}/\mathbf{D}\,a$ is bounded;* (2) *the module $\mathbf{D}/\mathbf{D}\,u$ is bounded;* (3) *u is a unit.*

(iv) If the principal ideal domain $\mathbf{D}$ is simple, all nonzero elements of $\mathbf{D}$ which are not units are totally unbounded.

Proof . (i): Let $\pi = \prod_{1 \leq i \leq m} \pi_i$. The elements π and u are left coprime, i.e. $\mathbf{D}\,\pi + \mathbf{D}\,u = \mathbf{D}$. Therefore, by Proposition 6.10(ii), $\mathbf{D}/\left(\mathbf{D}\,\pi \cap \mathbf{D}\,u\right) = \mathbf{D}/\mathbf{D}\,\pi \oplus \mathbf{D}/\mathbf{D}\,u$. This rationale can be repeated, and by induction one obtains (6.17).

(ii): The modules $\mathbf{D}/\mathbf{D}\,\pi_i$ are indecomposable by Proposition 6.10(i). If u is a unit, $\mathbf{D}/\mathbf{D}\,u = 0$. If not, assume that there exists an element u_1 such that $\mathbf{D}/\mathbf{D}\,u_1$ is a direct summand of $\mathbf{D}/\mathbf{D}\,u$. By Proposition 6.10(iii), there exists $u_2 \neq 0$ such that $\mathbf{D}/\mathbf{D}\,u \cong \mathbf{D}/\mathbf{D}\,u_1 \oplus \mathbf{D}/\mathbf{D}\,u_2$ and $\mathbf{D}\,u_1 + u_2\,\mathbf{D} = \mathbf{D}$, thus the ideals generated by u_1 and u_2 are two-sided. Therefore, u_1 is invariant, and there exists a prime which divides u_1 (see Sect. 6.2.3). This prime divides u, a contradiction.

(iii): If u is a unit, $\pi\left(\mathbf{D}/\mathbf{D}\,a\right) = 0$. Conversely, assuming that $\mathbf{D}/\mathbf{D}\,u$ is bounded, there exists $0 \neq \lambda \in \mathbf{D}$ such that $\lambda\left(\mathbf{D}/\mathbf{D}\,u\right) = 0$, i.e. $\lambda\,\mathbf{D} \subset \mathbf{D}\,u$, thus $\mathbf{D}\,\lambda\,\mathbf{D} \subset \mathbf{D}\,u$, i.e. $u \mid \lambda$ where λ is invariant. This is impossible since the invariant λ has a decomposition such as (6.16) where all ideals are two-sided. (iv) is obvious. ∎

6.3.2 Autonomous Linear Differential Equations

Definition

An *autonomous linear differential equation* is an equation

$$a\left(\partial\right)\,w = 0 \tag{6.18}$$

where $a\left(\partial\right)$ is nonzero and belongs to a ring of differential polynomials $\mathbf{R} = \mathbf{K}\left[\partial; \alpha, \delta\right]$. Equation (6.18) is a particular case of (6.15).

Cyclic Module Defined by a Linear Differential Equation

Let $\mathbf{K}$ be any differential field. Consider equation (6.18) and let $\varphi : \mathbf{R} \rightarrow \mathbf{R}/\mathbf{R}\,a$ be the canonical epimorphism (where $a = a\left(\partial\right)$), and $w = \varphi\left(1\right)$.

From Sect. 6.3.1, the module *defined by the autonomous linear differential equation* (6.18) is the cyclic module $M = \mathbf{R}/\mathbf{R}\,a = [w]_{\mathbf{R}}$ where $a(\partial)\,w = 0$. This module M is torsion. If a is a unit, $M = 0$ and the differential equation is trivial. In any case, $a(\partial)$ can be assumed to be a *monic* left differential polynomial without loss of generality, i.e. of the form (6.6) with $a_0 = 1$.

Companion Matrices and Their Conjugates

Companion matrix The cyclic torsion module $\mathbf{R}/\mathbf{R}\,a$ has a structure of $\mathbf{K}$-vector space[8]. Set $x_1 = \partial^{n-1}w$, ..., $x_n = w$, and $x = \begin{bmatrix} x_1 & \cdots & x_n \end{bmatrix}^T$. By (6.18),

$$\partial x = \begin{bmatrix} -a_1 & -a_2 & \cdots & \cdots & -a_n \\ 1 & 0 & & & 0 \\ 0 & \ddots & \ddots & & \vdots \\ \vdots & \ddots & \ddots & \ddots & 0 \\ 0 & \cdots & 0 & 1 & 0 \end{bmatrix} x \tag{6.19}$$

The matrix C_a in the right-hand side of (6.19) is called the[9] *companion matrix* of the polynomial $a(\partial)$. The elements $x_1, ..., x_n$ are $\mathbf{K}$-linearly independent, hence form a *basis* of the $\mathbf{K}$-vector space $\mathbf{R}/\mathbf{R}\,a$. The $\mathbf{K}$-vector space $\mathbf{R}/\mathbf{R}\,a$ and the basis $x = (x_i)_{1 \le i \le n}$ are said to be ∂-cyclic (or (α, δ)-cyclic) since they are generated by the single vector x_n and its images by powers of ∂.

Pseudo-linear mappings and (α, δ)-conjugate matrices. The $\mathbb{Z}$-endomorphism ∂ of $\mathbf{R}/\mathbf{R}\,a$ is not $\mathbf{K}$-linear, due to the commutation rule (6.7), except if $\mathbf{K}$ is a field of constants. It is said to be *pseudo-linear* with respect to (α, δ) ([12], Sect. 8.4). Let $y = \sum_{i=1}^{n} Y_i\, x_i$ be any vector of $\mathbf{R}/\mathbf{R}\,a$, and let $Y = \begin{bmatrix} Y_1 & \ldots & Y_n \end{bmatrix}$ (representing the vector y by the *row* Y in the basis $(x_i)_{1 \le i \le n}$: see footnote 7). The expression $y = Y\,x$ yields $\partial y = Y^{\alpha}\,\partial x + Y^{\delta}\,x = \left(Y^{\alpha}\,C_a + Y^{\delta}\right) x$. Therefore, in the basis $(x_i)_{1 \le i \le n}$, ∂ is the mapping $Y \to Y^{\alpha}\,C_a + Y^{\delta}$, and C_a is called the *matrix of the pseudo-linear mapping* ∂ in the basis $(x_i)_{1 \le i \le n}$.

Exercise 6.9. ([12], Sect. 8.4). Let $\mathbf{GL}_n(\mathbf{K})$ be the group of unimodular (i.e. invertible) $n \times n$ matrices over $\mathbf{K}$ and $P \in \mathbf{GL}_n(\mathbf{K})$. Let $(v_i)_{1 \le i \le n}$ be a basis of the $\mathbf{K}$-vector space $\mathbf{R}/\mathbf{R}\,a$ and set $x = P\,v$ (so that P is the change of basis matrix from $(x_i)_{1 \le i \le n}$ to $(v_i)_{1 \le i \le n}$). (i) Let $Z = \begin{bmatrix} Z_1 & \ldots & Z_n \end{bmatrix}$ be the row representing y in the basis $(v_i)_{1 \le i \le n}$; show that in the basis $(v_i)_{1 \le i \le n}$, ∂ is

[8] This cyclic module is torsion since $a \neq 0$. Assuming that $a = 0$, $\mathbf{R}/\mathbf{R}\,a = \mathbf{R}$ is free of rank 1.

[9] With a slight abuse of language since several companion matrices can be associated with a polynomial, depending, e.g., on the order chosen for the components of x.

the mapping $Z \rightarrow Z^{\alpha} A + Z^{\delta}$ where $C_a P = P^{\alpha} A + P^{\delta}$. (ii) More generally, let $(g_i)_{1 \leq i \leq n}$ be a basis of the **K**-vector space $\mathbf{R}/\mathbf{R}\,a$ and $V \in \mathbf{GL}_n(\mathbf{K})$ be the change of basis matrix from $(g_i)_{1 \leq i \leq n}$ to $(v_i)_{1 \leq i \leq n}$; let B (resp. A) be the matrix of the pseudo-linear mapping $\overline{\partial}$ in the basis $(g_i)_{1 \leq i \leq n}$ (resp. $(v_i)_{1 \leq i \leq n}$); show that

$$B V = V^{\alpha} A + V^{\delta}. \tag{6.20}$$

(iii) Two matrices A, B related by (6.20) are said to be (α, δ)-conjugate (by V); show that (α, δ)-conjugacy is an equivalence relation.

A matrix which is (α, δ)-conjugate to a companion matrix is said to be (α, δ)-cyclic ([12], Sect. 8.4).

If **K** is a field of constants, ∂ becomes linear and (6.20) reduces to the usual formula $B = V A V^{-1}$ (since $\alpha = 1$ and $\delta = 0$) which expresses that the matrices A and B are *similar* (written $A \approx B$). A $(1,0)$-cyclic matrix is simply said to be *cyclic.*

The decomposition (6.17) of $\mathbf{R}/\mathbf{R}\,a$ holds (with $\mathbf{R} = \mathbf{D}$). From Theorems 6.1 and 6.2, one obtains the following result:

Theorem 6.3. *(i) If* **K** *is a field of constants, the companion matrix C_a of the polynomial $a(\partial)$ is similar to the diagonal sum*[10] $C_{\pi_1} \oplus ... \oplus C_{\pi_m}$, *where the polynomials $\pi_j(\partial)$ are the elementary divisors of the module* $\mathbf{R}/\mathbf{R}\,a$; *in addition,* $\mathbf{R}/\mathbf{R}\,u = 0$, *and assuming that the polynomials $a(\partial)$ and $\pi_j(\partial)$ $(1 \leq i \leq m)$ are monic, $u = 1$.*

(ii) If **K** *is not a field of constants, $a(\partial) = u(\partial)$ is a unit or is totally unbounded; the companion matrix C_u is indecomposable, i.e. is not (α, δ)-conjugate to a proper diagonal sum of submatrices.*

Right Roots and Right Eigenvalues

In the scalar case, from (6.20), the class of all (α, δ)-conjugates of an element $r \in \mathbf{K}$ is the set $\Delta^{\alpha,\delta}(r)$ of all elements $v^{\alpha} v^{-1} r + v^{\delta} v^{-1}$ $(0 \neq v \in \mathbf{K})$. The set $\Delta^{\alpha,\delta}(r)$ reduces to $\{r\}$ if **K** is a field of constants and is infinite otherwise ([12], Sect. 8.6, th. 6.2 & lemma 6.3).

An element $r \in \mathbf{K}$ is called a *right root* of the differential polynomial $a(\partial) \neq 0$ if there exists a differential polynomial $b(\partial)$ such that $a(\partial) = b(\partial)(\partial - r)$.

Let $A \in \mathbf{K}^{n \times n}$; an element $\lambda \in \mathbf{K}$ is called a *right eigenvalue* of A if there exists a nonzero row $X \in \mathbf{K}^{1 \times n}$ such that $X A = X \lambda$. We have the following:

Proposition 6.13. *(i) r is a right root of $a(\partial)$ if, and only if $a(r) = 0$, where the "evaluation" $a(r)$ is* defined *as the remainder of the right division of $a(\partial)$ by $\partial - r$, i.e. by the left division algorithm $a(\partial) = q(\partial)(\partial - r) + a(r)$.*

[10] The "diagonal sum" $C_{q_1} \oplus ... \oplus C_{q_m}$ is defined as $diag(C_{q_1}, ..., C_{q_m})$.

(ii) A polynomial $p(\partial)$ is similar to $\partial - r$ if, and only if it is left associated with an element $\partial - \rho$, $\rho \in \Delta^{\alpha,\delta}(r)$.

(iii) The right roots of a differential polynomial of degree n lie in at most n conjugacy classes.

Proof. (i) and (ii): see ([31], Sect. 2). (iii): see ([12], Sect. 8.5, Corollary 5.11). ∎

Exercise 6.10. In the discrete-time case, show that $\Delta^{\alpha,\delta}(z-1)$ is the set of all $r' = z' - 1$ such that $z' = v^{\alpha} v^{-1} z$ $(0 \neq v \in \mathbf{K})$. What is the meaning of z and z'?

A finite subset S of $\mathbf{K}$ is said to be *right algebraic* (or, for short, *algebraic*) if there exists a nonzero polynomial $f(\partial)$ such that $f(s) = 0$ for every $s \in S$. The *minimal polynomial* of S is defined as the unique monic left differential polynomial f_S of the least degree vanishing on S.

Proposition 6.14. *(i) Let S be an algebraic subset of $\mathbf{K}$. The minimal polynomial f_S factors into $(\partial - r_1)\dots(\partial - r_n)$ where each $r_i \in \mathbf{K}$ is (α,δ)-conjugate to some element of S.*

(ii) The right eigenvalues of the companion matrix C_a of the differential polynomial $a(\partial)$ coincide with the (α,δ)-conjugates of the right roots of $a(\partial)$.

Proof. (i): see ([31], Sect. 2). (ii): see ([12], Sect. 8.5, th. 5.10). ∎

Definition 6.3. *The field $\mathbf{K}$ is* right algebraically closed *(or, for short,* algebraically closed*) if every finite subset of $\mathbf{K}$ is right algebraic ([29], Sect. 16, p. 255).*

From the above, $\mathbf{K}$ is algebraically closed if, and only if every differential polynomial $f(\partial) \in \mathbf{R}$ of positive degree splits into a product of linear factors $\partial - r_i$, $r_i \in \mathbf{K}$ ([29], Sect. 16).

Exercise 6.11. Let $\mathbf{K}$ be an algebraically closed field. An element of $\mathbf{R}$ is an atom if, and only if it is a differential polynomial of degree 1. There exists a bounded atom in $\mathbf{R}$ if, and only if $\mathbf{K}$ is a field of constants.

Jordan Canonical Form of a Cyclic Matrix

Let $\mathbf{K} = \mathbf{k}$ be a *field of constants* and $\bar{\mathbf{k}}$ be an algebraic closure of $\mathbf{k}$.

A prime polynomial p is of the form $\partial - r, r \in \bar{\mathbf{k}}$, and an elementary divisor is of the form $\pi = (\partial - r)^k$, $k = d^{\circ}(\pi)$ (assuming that these polynomials are *monic*).

A primary module $\mathbf{R}/\mathbf{R}\,\pi$ has a structure of $\bar{\mathbf{k}}$-vector space, denoted by V_π. Let v be a generator of the module $\mathbf{R}/\mathbf{R}\,\pi$, so that $(\partial - r)^k v = 0$. Set $\eta_{k-i} = (\partial - r)^i v$, $1 \leq i \leq k$, and $\eta = \begin{bmatrix} \eta_1 & \cdots & \eta_k \end{bmatrix}^T$. This yields $(\partial - r)\,\eta_1 = 0$ and $\partial\,\eta_{i+1} = \eta_i + r\,\eta_{i+1}$, $1 \leq i \leq k-1$, thus

$$\partial \eta = \begin{bmatrix} r & 0 & & 0 \\ 1 & \ddots & \ddots & \\ & \ddots & \ddots & 0 \\ 0 & & 1 & r \end{bmatrix} \eta$$

i.e. $\partial\eta = J_{k,r}\,x$, where $J_{k,r}$ is the *Jordan block* of order k associated with r. Therefore, V_π is the $\bar{\mathbf{k}}$-vector space of dimension $k = d^\circ(\pi)$ with basis $(\eta_i)_{1\leq i\leq k}$. The $\mathbf{R}$-endomorphism ∂ of the module $\mathbf{R}/\mathbf{R}\,\pi$ is a $\bar{\mathbf{k}}$-endomorphism A_π of $\bar{V}$. Both the minimal polynomial and the characteristic polynomial of A_π are equal to $\pi(\partial)$, and A_π is represented by the Jordan block $J_{k,r}$ in the basis $(\eta_i)_{1\leq i\leq k}$.

By the primary decomposition (6.17), the cyclic module $\mathbf{R}/\mathbf{R}\,a$ is a $\bar{\mathbf{k}}$-vector space V_a of dimension $\mathrm{d}^\circ(a) = \sum_{j=1}^m d^\circ(\pi_j)$, since $V_a = \bigoplus_{j=1}^m V_{\pi_j}$. Choosing in each ∂-cyclic subspace V_{π_j} the suitable basis, and denoting by X the concatenation of these bases, one obtains $\partial X = J\,X$, where $J = J_{q_1} \oplus \ldots \oplus J_{q_m}$. The $\mathbf{R}$-endomorphism ∂ of the module $\mathbf{R}/\mathbf{R}a$ is a $\bar{\mathbf{k}}$-endomorphism A_a of V_a, represented by the matrix J in the basis X; J is called the *Jordan canonical form* of A_a.[11]

Example 6.1. Consider the equation

$$(\partial - 1)(\partial - 2)^2 w = 0.$$

Setting $a = (\partial - 1)(\partial - 2)^2$, $\pi_1 = \partial - 1$ and $\pi_2 = (\partial - 2)^2$, the following primary decomposition holds:

$$\mathbf{R}/\mathbf{R}\,a \cong \mathbf{R}/\mathbf{R}\,\pi_1 \oplus \mathbf{R}/\mathbf{R}\,\pi_2\,.$$

Therefore, the generator w can be decomposed in a unique manner as $w = w_1 + w_2$ where $(\partial - 1)\,w_1 = 0$ and $(\partial - 2)^2 w_2 = 0$. Setting $X_1 = w_1$, $X_2 = (\partial - 2)\,w_2$ and $X_3 = w_2$, one obtains the canonical representation

$$\partial X = \begin{bmatrix} 1 & 0 & 0 \\ 0 & 2 & 0 \\ 0 & 1 & 2 \end{bmatrix} X.$$

[11] From Sect. 6.3.2, the vector space V_a is cyclic. The endomorphism A_a itself is said to be *cyclic* since it is represented by a cyclic matrix in any basis of V_a.

Remark 6.2. The Jordan canonical form of a square matrix over $\mathbf{R} = \mathbf{K}[\partial; \alpha, \delta]$ can be defined when $\mathbf{K}$ is more general than a field of constants: see [11], and [33] where diagonalization is studied. We will not develop these points.

6.3.3 Systems of Linear Differential Equations

Introduction

Our aim is now to study *systems* of linear differential equations such as

$$B(\partial)\, w = 0 \tag{6.21}$$

where $B(\partial)$ is a $q \times k$ matrix with entries in a ring of differential polynomials $\mathbf{R} = \mathbf{K}[\partial; \alpha, \delta]$ and where $w = \begin{bmatrix} w_1 & \cdots & w_k \end{bmatrix}^T$. Equation (6.21) is of the same form as (6.15) and can be viewed as a generalization of (6.18). To study the properties of (6.21), we will first study the properties of a matrix with entries in a principal ideal domain and the "Smith canonical form" of such a matrix.

Smith Canonical Form

Unimodular matrices Let $\mathbf{D}$ be an *integral domain.* A matrix $U \in \mathbf{D}^{n \times n}$ is said to be *unimodular* if it is invertible over $\mathbf{D}$ (i.e. if U^{-1} exists and belongs to $\mathbf{D}^{n \times n}$). The set of all unimodular matrices belonging to $\mathbf{D}^{n \times n}$ is a group (for the multiplication) denoted by $\mathbf{GL}_n(\mathbf{D})$ and called the "general linear group" of $\mathbf{D}^{n \times n}$.

Remark 6.3. If $\mathbf{D}$ is *commutative*, a matrix $U \in \mathbf{D}^{n \times n}$ is unimodular if, and only if its determinant $|U|$ is a unit of $\mathbf{D}$. If $\mathbf{D}$ is a *noncommutative Ore domain*, this ring can be embedded in its quotient field $\mathbf{Q}$; considering U as element of $\mathbf{Q}^{n \times n}$, the "Dieudonné determinant" $|U|$ of this matrix can be calculated [13]. The necessary and sufficient condition for U to be invertible over $\mathbf{Q}$ is $|U| \neq 0$, but the invertibility of U over $\mathbf{D}$ cannot be easily characterized using determinants, since $|U|$ is a *rational function* of the entries of U.

Equivalence of matrices. Two matrices $B, B' \in \mathbf{D}^{q \times k}$ are said to be *left-* (resp. *right-*) *equivalent* if the exists a unimodular matrix $U \in \mathbf{GL}_q(\mathbf{D})$ (resp. $V \in \mathbf{GL}_k(\mathbf{D})$) such that $UB = B'$ (resp. $BV^{-1} = B'$). These matrices B and B' are said to be *equivalent* if they are both left- and right-equivalent.

Elementary and Secondary Operations

Elementary operations Three elementary row (resp. column) operations are defined as follows:

(i) interchange two rows (resp. columns);

(ii) multiply a row (resp. a column) on the left (resp. on the right) by a unit;

(iii) add a left (resp. right) multiple of one row (resp. column) to another.

Each of (i)-(iii) corresponds to the left (resp. right) multiplication by an "elementary matrix". Elementary matrices are unimodular. To clarify ideas, assume that $q = 2$. Consider the three following unimodular matrices:

$$U_1 = \begin{bmatrix} 0 & 1 \\ 1 & 0 \end{bmatrix}, \quad U_2 = \begin{bmatrix} 1 & 0 \\ 0 & \upsilon \end{bmatrix}, \quad U_3 = \begin{bmatrix} 1 & \lambda \\ 0 & 1 \end{bmatrix},$$

where υ and λ are, respectively, a unit and any element of $\mathbf{D}$. Left multiplications of $B \in \mathbf{D}^{2\times k}$ by U_1, U_2 and U_3 correspond to the elementary row operations (i), (ii) and (iii), respectively.

Secondary operations A secondary row (resp. column) operation is defined as:

(iv) left- (resp. right-) multiply two rows (resp. columns) by an invertible 2×2 matrix.

For example, let us left multiply the first two rows of a matrix B by such a matrix, assuming that the (1,1) entry b_{11} of B is nonzero. This yields

$$\begin{bmatrix} \alpha & \beta \\ * & * \end{bmatrix} \begin{bmatrix} b_{11} & * \\ b_{21} & * \end{bmatrix} = \begin{bmatrix} \alpha\, b_{11} + \beta\, b_{21} & * \\ * & * \end{bmatrix};$$

the elements α and β can be chosen so that $\alpha\, b_{11} + \beta\, b_{21}$ is a g.c.r.d. of b_{11} and b_{21} (see Sect. 6.2.3).

Rank of a Matrix over an Ore Domain

Let $\mathbf{D}$ be an Ore domain and $B \in \mathbf{D}^{q\times k}$. This matrix B can be considered as an element of $\mathbf{Q}^{q\times k}$, where $\mathbf{Q}$ is the quotient field of $\mathbf{D}$. Minors of the matrix $B \in \mathbf{Q}^{q\times k}$ are defined using "Dieudonné determinants", as usual ([3], Chap. III, Sect. 8, n°5). The rank r of B (written $r = \operatorname{rk} B$) is the maximal order of the nonzero minors of B. This integer r is equal to the number of rows or columns of B which are linearly independent over $\mathbf{Q}$, i.e. to $\operatorname{rk} A^q \bullet B$ (see Sect. 6.3.1). If $r = q$ (resp. $r = k$), B is said to be *full row* (resp. *column*) *rank* over $\mathbf{D}$ or $\mathbf{Q}$.

Smith Canonical Form of a Matrix over a Principal Ideal Domain

Let $\mathbf{D}$ be a principal ideal domain (thus an Ore domain from Sect. 6.2.3). The following theorem, which is classical in the commutative case (see, e.g., [40], Chap. X, Sect. 7, th. 15), is proven in the noncommutative case in ([12], Sect. 8.1).

Theorem 6.4. *Let $B \in \mathbf{D}^{q\times k}$ and $r = rk(B) \geq 1$. There exist $U \in \mathbf{GL}_q(\mathbf{D})$ and $V \in \mathbf{GL}_k(\mathbf{D})$ such that $U\,B\,V^{-1} = \Sigma$ where*

$$\Sigma = diag\,(b_1, ..., b_r, 0, ..., 0)\,, \qquad b_i \parallel b_{i+1}, \quad b_r \neq 0.$$

The matrix Σ (which is not necessarily square and may have zeros outside the diagonal: this is implicit in the above equation) is called the Smith canonical form *of B; the nonzero elements b_i $(1 \leq i \leq r)$ are called the* invariant factors *of the matrix B.*

The Smith form of B is obtained using elementary and secondary operations. It is often necessary to also calculate unimodular matrices U and V such that $U\,B\,V^{-1} = \Sigma$. For this, consider the matrix

$$H = \begin{bmatrix} B & I_q \\ I_k & 0 \end{bmatrix};$$

doing the same operations on H, one obtains

$$\begin{bmatrix} U & 0 \\ 0 & I_k \end{bmatrix} H \begin{bmatrix} V^{-1} & 0 \\ 0 & I_q \end{bmatrix} = \begin{bmatrix} \Sigma & U \\ V^{-1} & 0 \end{bmatrix}$$

so that U and V^{-1} are also calculated.

Corollary 6.1. *Let $\mathbf{D}$ be a* simple principal ideal domain *and $B \in \mathbf{D}^{q\times k}$ be a matrix of rank $r \geq 1$; the Smith form of B is the diagonal sum $I_{r-1} \oplus b \oplus 0$, $0 \neq u \in \mathbf{D}$.*

Proof. As $\mathbf{D}$ is simple, its only invariant elements are units. Therefore, $b_i \parallel b_{i+1}$ implies that b_i is a unit. ∎

Remark 6.4. (i) Assuming that the principal ideal domain $\mathbf{D}$ is *commutative*, the invariant factors b_i of a matrix $B = (b_{ij})$ with entries in $\mathbf{D}$ and of rank $r \geq 1$ can be calculated as follows, up to associates ([4], Chap. 7, Sect. 4, n°6, prop. 6): b_1 is a greatest common divisor (g.c.d.) of all elements b_{ij}; for $1 \leq j \leq r$, $\prod_{i=1}^{j} b_i$ is a g.c.d. of all nonzero minors of order j of B. (ii) The term "*canonical* form" suggests that this form is somehow unique. This is specified by Theorem 6.5 below.

Exercise 6.12. Assuming that the matrix B is in Smith canonical form, check that the result in Remark 6.4(i) is correct.

Divisibility of Matrices

Let $\mathbf{D}$ be a principal ideal domain, $D \in \mathbf{D}^{n\times m}, N \in \mathbf{D}^{n\times p}$. A matrix $L \in \mathbf{D}^{n\times n}$ is said to be a *common left divisor* of D and N if there exist matrices $D \in \mathbf{D}^{n\times m}$ and $N \in \mathbf{D}^{n\times p}$ such that

$$\begin{bmatrix} D & \vdots & N \end{bmatrix} = L \begin{bmatrix} D' & \vdots & N' \end{bmatrix}.$$

A matrix $L \in \mathbf{D}^{n\times n}$ is said to be a *greatest common left divisor* (g.c.l.d.) of D and N if (i) it is a common left divisor of D and N and (ii) every left divisor L' of D and N is a left divisor of L.

Exercise 6.13. (i) The matrices D and N have a greatest common left divisor L such that $\begin{bmatrix} D & \vdots & N \end{bmatrix}$ and $\begin{bmatrix} L & \vdots & 0 \end{bmatrix}$ are right-equivalent (see Sect. 6.3.3). (ii) Assuming that $\operatorname{rk} \begin{bmatrix} D & \vdots & N \end{bmatrix} = n$, show that all g.c.l.d.s of D and N are right-equivalent. (Hint: for (i), use the Smith canonical form of $\begin{bmatrix} D & \vdots & N \end{bmatrix}$: there exist unimodular matrices U and V such that $U\,B\,V^{-1} = \Sigma = \begin{bmatrix} \Delta & \vdots & 0 \end{bmatrix}$; check that $L = U^{-1}\,\Delta$.)

The matrices $D \in \mathbf{D}^{n\times m}$ and $N \in \mathbf{D}^{n\times p}$ are said to be *left-coprime* if every g.c.l.d. of D and N is unimodular.

Right divisibility and right-coprimeness are defined and studied analogously.

Application of the Smith Canonical Form to Systems of Linear Differential Equations

Consider the system of linear differential equations (6.21). Let $U(\partial) \in \mathbf{GL}_q(\mathbf{R})$ and $V \in \mathbf{GL}_k(\mathbf{R})$ be such that $U(\partial)\,B(\partial)\,V^{-1}(\partial) = \Sigma(\partial)$ is the Smith canonical form of $B(\partial)$. Setting $v = V(\partial)\,w$, (6.21) reduces to

$$b_i(\partial)\,v_i = 0, \quad 1 \le i \le r. \tag{6.22}$$

Therefore:

- if $q > r$, one has $q - r$ "trivial equations" $0 = 0$;
- if $k > r$, the $k - r$ last components of v are "free variables", i.e. subject to no equation;
- the r first components of v satisfy an autonomous linear differential equations (such as in Sect. 6.3.2).

Using the Smith canonical form, system (6.21) has been completely "decoupled". This decoupling, combined with the theory in Sect. 6.3.2, completely elucidates the structure of this system (at least in principle; some additional theory is still necessary to clarify this structure). This is illustrated by the following exercise:

Exercise 6.14. Consider the system of differential equations (6.21), where

$$B(\partial) = \begin{bmatrix} \partial & \partial & \partial^2 \\ \partial & \partial^2 & \partial^4 \end{bmatrix}.$$

(1) Show that the Smith canonical form of $B(\partial)$ is

$$\Sigma(\partial) = \begin{bmatrix} \partial & 0 & 0 \\ 0 & \partial(\partial-1) & 0 \end{bmatrix}$$

obtained with

$$U(\partial) = \begin{bmatrix} 1 & 0 \\ -1 & 1 \end{bmatrix}, \qquad V^{-1}(\partial) = \begin{bmatrix} 1 & -1 & \partial^2 \\ 0 & 1 & -\partial(\partial+1) \\ 0 & 0 & 1 \end{bmatrix}.$$

(2) Deduce from the above the following parametrization of w:

$w_1 + w_2 + \partial w_3 = v_1$ such that $\partial v_1 = 0$

$w_2 + \partial(\partial+1)\, w_3 = v_2$ such that $\partial(\partial-1)\, v_2 = 0$.

Remark 6.5. If $\mathbf{K}$ is *not a field of constants*, $\mathbf{R}$ is simple (according to Theorem 6.1). Therefore, by Corollary 6.1, (6.22) yields $v_i = 0, 1 \leq i \leq r-1$, and $b(\partial)\, v_r = 0$, where $b(\partial)$ is the only invariant factor of $B(\partial)$ which can be a nonunit. As a consequence, a *system* of linear differential equations reduces to *one* autonomous linear differential equation.

Finitely Generated Modules over Principal Ideal Domains

Let $M = [w]_{\mathbf{D}}$ be a f.g. module over a principal ideal domain $\mathbf{D}$. This module is defined by an equation such as (6.15), where $B \in \mathbf{D}^{q \times k}$. According to Theorem 6.4, there exist unimodular matrices $U \in \mathbf{GL}_q(\mathbf{D})$ and $V \in \mathbf{GL}_k(\mathbf{D})$ such that $U\,B\,V^{-1} = \Sigma = diag(b_1, ..., b_r, 0, ..., 0)$ is the Smith canonical form of Σ. Let $v = V\,w$. As V is unimodular, $M = [v]_{\mathbf{D}}$, and by the same rationale as in Sect. 6.3.3, $M = \bigoplus_{i=1}^{k} [v_i]_{\mathbf{D}}$. For $1 \leq i \leq r$, $[v_i]_{\mathbf{D}} \cong \mathbf{D}/\mathbf{D}\,b_i$. For $r+1 \leq i \leq k$, $[v_i]_{\mathbf{D}} \cong \mathbf{D}$. Therefore, we obtain the following important theorem:

Theorem 6.5. *(i) The following relations hold:*

$$(a)\ M = \mathcal{T}(M) \oplus \Phi, \quad (b)\ \mathcal{T}(M) \cong \bigoplus_{i=1}^{r} \frac{\mathbf{D}}{\mathbf{D}\,b_i},$$
$$(c)\ \Phi \cong M/\mathcal{T}(M) \cong \mathbf{D}^{k-r}$$

(where $\mathcal{T}(M)$ is the torsion submodule of M). The elements b_i $(1 \le i \le r)$, such that $b_i \parallel b_{i+1}$, are called the invariant factors *of M and are* unique up to similarity *(see above in this section).*

(ii) The "trivial equations" of (6.15) *(see Sect. 6.3.3) can be removed. Then, $q = r \le k$ and the module M (of rank $k - r$) is presented by the* short exact sequence

$$0 \longrightarrow \mathbf{D}^r \xrightarrow{\bullet B} \mathbf{D}^k \longrightarrow M \longrightarrow 0. \tag{6.23}$$

Corollary 6.2. *(i) Over a principal ideal domain* $\mathbf{D}$*, a f.g. module is* free *if, and only if it is* torsion-free. *(ii) If* $\mathbf{D}$ *is a* simple *principal ideal domain, every f.g.* torsion $\mathbf{D}$*-module $T \neq 0$ is cyclic, i.e. $T \cong \mathbf{D}/\mathbf{D}\,u$, where u is* totally unbounded.

Proof. (i) is an obvious consequence of Theorem 6.5(i). (ii): By Theorem 6.5(i) and Corollary 6.1, $T \cong \mathbf{D}/\mathbf{D}\,b$, thus T is cyclic. The element b is totally unbounded (thus denoted by u) by Theorem 6.2(iv). ■

Corollary 6.3. *Let $\mathbf{R} = \mathbf{K}[\partial; \alpha, \delta]$ be a ring of differential polynomials and M be a f.g. $\mathbf{R}$-module; M is a finite dimensional $\mathbf{K}$-vector space if, and only if it is a* torsion $\mathbf{R}$*-module.*

Proof. If $M = \mathcal{T}(M)$, M is a direct sum of cyclic modules $\mathbf{R}/\mathbf{R}\,b_i(\partial)$ by Theorem 6.5(i), and $\mathbf{R}/\mathbf{R}\,b_i(\partial)$ is a $\mathbf{K}$-vector space of dimension $d^\circ(b_i(\partial)) < +\infty$ from Sect. 6.3.2. Conversely, if M is not torsion, there exists a free submodule $\Phi \neq 0$ of M. Let $0 \neq m \in \Phi$; the elements $m, \partial m, ..., \partial^n m, ...,$ are $\mathbf{K}$-linearly independent, thus M is an infinite dimensional $\mathbf{K}$-vector space. ■

Jordan Canonical Form and Smith Zeros

Jordan Canonical Form

Let $\mathbf{k}$ be a field of constants, $\bar{\mathbf{k}}$ be an algebraic closure of $\mathbf{k}$, $\mathbf{R} = \mathbf{k}[\partial]$ and T be a f.g. *torsion* $\mathbf{R}$-module. By Corollary 6.3, T is a finite dimensional $\mathbf{k}$-vector space. Let $(x_i)_{1 \le i \le r}$ be a basis of the $\mathbf{k}$-vector space T. For every $i \in \{1, ...r\}$, ∂x_i is a $\mathbf{k}$-linear combination of the elements $x_j, 1 \le j \le r$; therefore, there exists a matrix $A \in \mathbf{k}^{r \times r}$ such that

$$\partial x = A\,x \tag{6.24}$$

where $x = (x_i)_{1 \le i \le r}$. Equation (6.24) defines the $\mathbf{R}$-module T, i.e. $T = [x]_{\mathbf{R}} = [x]_{\mathbf{k}}$. By Theorem 6.5, $T \cong \bigoplus_{i=1}^{r} \mathbf{R}/\mathbf{R}\, b_i$, where the polynomials $b_i = b_i(\partial)$ (assumed to be monic, and then uniquely defined) are the invariant factors of T (or of the matrix $B(\partial) = \partial I_r - A$) and are called the *invariants* of A over $\mathbf{k}$ ([32], Chap. XIV, Sect. 2).

The theory in Sect. 6.3.2 can be applied to each cyclic torsion submodule $\mathbf{R}/\mathbf{R}\, b_i$ of T; using the primary decomposition of $\mathbf{R}/\mathbf{R}\, b_i$, one obtains the representation $\mathbf{R}/\mathbf{R}\, b_i = [X_i]_{\bar{\mathbf{k}}}$ where

$$\partial X_i = J_i X_i$$

and where the matrix J_i (with entries in $\bar{\mathbf{k}}$) is the Jordan canonical form of a cyclic matrix. Let

$$X = (X_i)_{1 \le i \le r}, \quad J = J_1 \bigoplus \ldots \bigoplus J_r.$$

One obtains $T = [X]_{\bar{\mathbf{k}}}$, $\partial X = J X$ and $A \approx J$ (considering A as a matrix over $\bar{\mathbf{k}}$). The matrix $J \in \bar{\mathbf{k}}^{r \times r}$ is called the *Jordan canonical form* of A.

To summarize, the Jordan canonical form of a matrix $A \in \mathbf{k}^{r \times r}$ is calculated as follows:

(i) Determine the Smith canonical form of $B(\partial) = \partial I_r - A$.

(ii) Calculate the elementary divisors of $B(\partial)$ (also called the elementary divisors of A) from the invariants $b_i(\partial)$ of A (found on the diagonal of the Smith canonical form).

(iii) Determine the Jordan blocks from the elementary divisors. The Jordan canonical form of A is the diagonal sum of all Jordan blocks.

Note that the *minimal polynomial* of A is $b_r(\partial)$ and that its *characteristic polynomial* is $\prod_{i=1}^{r} b_i(\partial)$ ([4], Chap. 7, Sect. 5).

Exercise 6.15. Let

$$A = \begin{bmatrix} 0 & 0 & 0 & 0 & 1 \\ 1 & 0 & 0 & 1 & 1 \\ -1 & 1 & 0 & -1 & 0 \\ 0 & 0 & 0 & 0 & -1 \\ 0 & 0 & 0 & 0 & 0 \end{bmatrix}.$$

(i) Show that the elementary divisors of A are ∂^3 and ∂^2. (ii) Deduce that $A \approx J_{3,0} \oplus J_{2,0}$ (with the notation in Sect. 6.3.2). (iii) Calculate the characteristic polynomial and the minimal polynomial of A. (iv) Calculate A^2, set $e_1 = [0 \;\; 0 \;\; 1 \;\; 0 \;\; 0]$ and show that $\{e_1 A^2, e_1 A, e_1\}$ is a basis of $E_1 \cong \mathbf{R}/\mathbf{R}\,\partial^3$. Setting $e_2 = [0 \;\; 0 \;\; 0 \;\; 1 \;\; 0]$, show that $\{e_2 A, e_2\}$ is a basis of $E_2 \cong \mathbf{R}/\mathbf{R}\,\partial^2$ and that $E_1 \cap E_2 = 0$. (v) Let Q be the matrix with rows $e_1 A^2, e_1 A, e_1, e_2 A, e_2$; check that $Q A Q^{-1} = J_{3,0} \oplus J_{2,0}$.

Smith zeros

The *Smith zeros* of the torsion $\mathbf{R}$-module T are defined as the roots over $\bar{\mathbf{k}}$ of the invariant factors of T ([9]).

Let $z \in \bar{\mathbf{k}}$ be a Smith zero of T and

$$(\partial - z)^{\nu_1}, ..., (\partial - z)^{\nu_j}$$

be the elementary divisors of T which are multiples of $\partial - z$, $1 \leq \nu_1 \leq ... \leq \nu_j$. The integers $\nu_1, ...\nu_j$ are the *structural indexes* of the Smith zero z of T, ν_j is its *order*, and $\sum_{i=1}^{j} \nu_i$ its degree ([26], Sect. 6.5).

If $\pi(\partial)$ is encountered l times ($l \geq 1$) in the list of all elementary divisors of T, l is called the *multiplicity* of the elementary divisor $\pi(\partial)$ ([4], Chap. 7, Sect. 4, n°8, def. 4).

Let $B(\partial)$ be a matrix over $\mathbf{R}$, M be the $\mathbf{R}$-module defined by $B(\partial)$ (i.e. $M \cong \text{coker} \bullet B(\partial)$) and $T = \mathcal{T}(M)$. The *Smith zeros* of $B(\partial)$ are defined as the Smith zeros of T, i.e. as the roots of the invariant factors of $B(\partial)$ over $\bar{\mathbf{k}}$ (with multiplicities defined as above).

Remark 6.6. If $\mathbf{R} = \mathbf{K}[\partial; \alpha, \delta]$ and $\mathbf{K}$ is *not a field of constants*, a nonzero f.g. torsion module is a cyclic module $T \cong \mathbf{R}/\mathbf{R}u$, where $u = u(\partial)$ is totally unbounded, by Corollary 6.2(ii) and Theorem 6.1. Assuming that $\mathbf{K}$ is algebraically closed, the Smith zeros of T can be defined as the *conjugacy classes* of the *right roots* of $u(\partial)$ (see Sect. 6.3.2). The Smith zeros of a matrix over $\mathbf{R}$ can be defined accordingly.

6.4 Linear Time-Varying Systems: A Module-Theoretic Setting

6.4.1 Basic Structural Properties

Some Definitions

From the mathematical point of view, a linear control system is an "object" characterized by a system of linear continuous- or discrete-time differential equations. As shown in Sect. 6.3.1 and Sect. 6.3.3, such a system of linear differential equations, e.g. (6.21), is a presentation (or definition) of a f.g. $\mathbf{R}$-module ($\mathbf{R} = \mathbf{K}[\partial; \alpha, \delta]$) by generators and relations. Other generators and relations can be chosen to describe the same linear control system. Therefore, we are led to the following definition [16]:

Definition 6.4. *A linear control system (or a linear system, for short) is a f.g.* $\mathbf{R}$*-module.*

The *input* of a linear system M is a finite sequence $u = (u_i)_{1 \le i \le m}$ of elements of M such that $M/[u]_{\mathbf{R}}$ is torsion [16]. Assuming that M is presented by the short exact sequence (6.23) (with $\mathbf{D} = \mathbf{R}$), $\operatorname{rk} M = k - r$, thus $\operatorname{rk} [u]_{\mathbf{R}} = k - r$ (by Proposition 6.9 and exercise 6.7(ii)). The input u is then said to be *independent* if $[u]_{\mathbf{R}}$ is free of rank m (i.e. if u is a basis of $[u]_{\mathbf{R}}$).

Without loss of generality, one can assume that u is independent. If not, indeed, the elements u_i $(1 \le i \le m)$ satisfy linear differential equations, i.e. are the variables of an "exogenous system" E, and M results from the interconnection of two subsystems: (1) a system M' with independent input $u' = (u'_i)_{1 \le i \le m}$; (2) the exogenous system E. The interconnection equation, resulting from a "fibered sum", is $u = u'$ (see [19]). Considering M as a whole, u is not its input; on the other hand, u' is an independent input of M'. –Therefore, in what follows, *the system input is assumed to be independent.*

The *output* of a linear system M is a finite sequence $y = (y_i)_{1 \le i \le p}$ of elements of M (from the engineering point of view, the elements of y are *measured*, using sensors). A linear system M with input u and output y is called an *input-output system* [16].

State-Space Representation

As the module $M/[u]_{\mathbf{R}}$ is torsion, it is defined by an equation having the form (6.24), e.g. $\partial \bar{\eta} = F \bar{\eta}$, where $\bar{\eta} = (\bar{\eta}_i)_{1 \le i \le n}$ is a basis of the $\mathbf{K}$-vector space $M/[u]_{\mathbf{R}}$ and where $F \in \mathbf{K}^{n \times n}$. There exist n elements $\eta_i \in M$ such that $\bar{\eta}_i = \eta_i + [u]_{\mathbf{R}}$ $(1 \le i \le n)$. Therefore, there exist matrices $G_j \in \mathbf{K}^{n \times m}$ $(0 \le j \le s$, s finite) such that

$$\partial \eta = F \eta + \sum_{j=0}^{s} G_j \, \partial^j \, u.$$

Assume that $s \ge 1$ and $G_s \neq 0$, and set

$$\eta^* = \eta - G_s^{\beta} \, \partial^{s-1} \, u$$

(where $\beta = \alpha^{-1}$).

Exercise 6.16. (i) Using the commutation rule (6.7), show that

$$\partial \eta^* = F \eta^* + \left(\sum_{j=0}^{s-2} G_j \, \partial^j + G'_{s-1} \, \partial^{s-1} \right) u$$

and calculate G'_{s-1} in function of G_{s-1}, G_s^{β} and G_s^{δ}. (ii) By induction, show that this elimination procedure yields a *Kalman representation* which does no longer contain derivatives in the control variables [16]:

$$\partial x = F\,x + G\,u \tag{6.25}$$

where $x_i + [u]_{\mathbf{R}} = \eta_i + [u]_{\mathbf{R}}$ $(1 \le i \le n)$.

Let $\bar{y}_i = y_i + [u]_{\mathbf{R}}$ $(1 \le i \le p)$. As $\bar{y}_i \in M/[u]_{\mathbf{R}}$, there exist matrices $H \in \mathbf{K}^{p \times n}$ and $L_j \in \mathbf{K}^{p \times n}$ $(0 \le j \le l)$ such that

$$y = H\,x + \sum_{j=0}^{l} L_j\,\partial^j\,u. \tag{6.26}$$

The input-output system is said to be *proper* (resp. *strictly proper*) if $L_j = 0$ for $j \ge 1$ (resp. $j \ge 0$).

For short, the state-space system (6.25), (6.26) is denoted by $\left(F,\ G,\ H,\ (L_j)_{0 \le j \le l}\right)$.

Exercise 6.17. Let $\zeta = V\,x$ where $V \in \mathbf{K}^{n \times n}$ is a change of basis matrix. Show that the new state ζ satisfies the state equation $\partial\zeta = \check{F}\,\zeta + \check{G}\,u$ where $\check{F}\,V = V^{\alpha}\,F + V^{\delta}$ (i.e. $\check{F}$ is (α, δ)-conjugate to F by V: see Sect. 6.3.2) and $\check{G} = V^{\alpha}\,G$.

Polynomial Matrix Description

Let $\xi = (\xi_i)_{1 \le i \le r}$ be such that $M = [\xi, u]_{\mathbf{R}}$ (where $[\xi, u]_{\mathbf{R}} = [\xi]_{\mathbf{R}} + [u]_{\mathbf{R}}$, i.e. $[\xi, u]_{\mathbf{R}}$ is the $\mathbf{R}$-module generated by the components of ξ and u). Equation (6.21), with $w = \begin{bmatrix} \xi \\ u \end{bmatrix}$, can be written

$$\begin{bmatrix} D(\partial) & \vdots & -N(\partial) \end{bmatrix} \begin{bmatrix} \xi \\ u \end{bmatrix} = 0 \tag{6.27}$$

where $D(\partial) \in \mathbf{R}^{r \times r}$ and $N(\partial) \in \mathbf{R}^{r \times m}$. As $y_i \in [\xi, u]_{\mathbf{R}}$ $(1 \le i \le p)$, there exist matrices $Q(\partial) \in \mathbf{R}^{p \times r}$ and $W(\partial) \in \mathbf{R}^{p \times m}$ such that

$$y = \begin{bmatrix} Q(\partial) & \vdots & W(\partial) \end{bmatrix} \begin{bmatrix} \xi \\ u \end{bmatrix}. \tag{6.28}$$

Equations (6.27), (6.28) are called a *polynomial matrix description* of the input-output system M (this notion is due to Rosenbrock [47]). The finite sequence $\xi = (\xi_i)_{1 \le i \le r}$ is called a *pseudo-state* of M.

Exercise 6.18. Show that $D(\partial)$ is invertible over the quotient field $\mathbf{Q}$ of $\mathbf{R}$. (Hint: $M/[u]_{\mathbf{R}}$ is torsion, thus...)

Controllability

Definition 6.5. *A linear system M is* controllable *if it is a free* **R**-*module [16].*[12]

By Theorem 6.5(i), the above definition means that no system variable satisfies an autonomous linear differential equation.

According to Theorem 6.5 & Theorem 6.1, a system M is controllable if, and only if $\mathcal{T}(M) = 0$. As $M/\mathcal{T}(M)$ is torsion free (see Sect. 6.3.1), we are led to the following definition [8]:

Definition 6.6. *The quotient $M/\mathcal{T}(M)$ is the* controllable quotient *of M.*

Proposition 6.15. *Let M be the linear system defined by equation (6.21), where $B = B(\partial) \in \mathbf{R}^{r\times k}$ is full row rank. The following properties are equivalent:*

(i) M is controllable;

(ii) there exist unimodular matrices $U \in \mathbf{R}^{r\times r}$ and $V \in \mathbf{R}^{k\times k}$ such that $U\,B\,V^{-1} = [I_r \quad 0]$;

(iii) the equality $\lambda\,B(\partial) = 0$ (where λ is a row of r elements[13]*) implies $\lambda = 0$;*

(iv) $B(\partial)$ is right-invertible over **R**.

Proof. The **R**-module M is free if, and only if the Smith canonical form of $B(\partial)$ is $[I_r \quad 0]$, thus (i) and (ii) are equivalent; in addition, (ii), (iii) and (iv) are equivalent. ■

Exercise 6.19. Let M be the system defined by equation (6.21), where $B = B(\partial) \in \mathbf{R}^{r\times k}$ is full row rank, and assume that M is controllable. (i) Let $U \in \mathbf{R}^{r\times r}$ and $V \in \mathbf{R}^{k\times k}$ be unimodular matrices such that $U\,B\,V^{-1} = [I_r \quad 0]$ and set $v = V(\partial)\,w$. Show that $\{v_{r+1}, ..., v_k\}$ is a basis of M (called a "flat output" [20], although it is not necessarily measured). Note that $U(\partial)$ is not needed. (ii) *Application*: use this *systematic method* to find a flat output of the system $\partial x + x = \partial u$. (Answer for (ii): $\varsigma = u - x$ is a flat output.)

Exercise 6.20. Let M be given by the polynomial matrix description (6.27), (6.28). The system M is controllable if, and only if the matrices $D(\partial)$ and $N(\partial)$ are left-coprime. (Hint: use Proposition 6.15 and exercise 6.13.)

---

[12] In the case of infinite-dimensional systems, there are different notions of controllability [22]. In our case, all kinds of controllability are equivalent since **R** is a principal ideal domain.

[13] These elements belong to a *right* **R**-module.

Theorem 6.6. *Let M be the state-space system* (6.25). *This system is controllable if*

$$rk\left[G \;\vdots\; F_1\,G^{\beta} \;\vdots\; \cdots \;\vdots\; F_{n-1}\,G^{\beta^{n-1}}\right] = n \tag{6.29}$$

where $F_i = (F - \delta\,I_n) \ldots \left(F^{\beta^{i-1}} - \delta\,I_n\right), 1 \leq i \leq n-1$. *(The matrix in the left-hand side of* (6.29) *is denoted by* $\Gamma\,(F, G)$ *and is called the* controllability matrix *of the pair* (F, G) *.)*

Proof. Equation (6.25) can be put in the form (6.21), setting

$$B\,(\partial) = \left[\partial I_n - F \;\vdots\; -G\right]$$

and $w = \begin{bmatrix} x \\ u \end{bmatrix}$. According to Proposition 6.15(iii), M is controllable if, and only if the two equalities (a) $\lambda\,(\partial I_n - F) = 0$, (b) $\lambda\,G = 0$, imply $\lambda = 0$. (1) By (6.8), $G\,\partial = \partial\,G^{\beta} - G^{\beta\delta}$, thus (b) is equivalent to $\lambda\,\left(\partial\,G^{\beta} - G^{\beta\delta}\right) = 0$. Using (a), this yields $\lambda\,\left(F\,G^{\beta} - G^{\beta\delta}\right) = 0$, i.e. $\lambda\,(F - \delta\,I_n)\,G^{\beta} = 0$, i.e. $\lambda\,F_1\,G^{\beta} = 0$. (2) Similarly, the equalities (a) and $\lambda\,F_1\,G^{\beta} = 0$ imply $\lambda\,F_1\,\left(F_1\,G^{\beta}\right)^{\beta} = 0$, i.e. $\lambda\,F_2\,G^{\beta^2} = 0$. (3) The "generalized Kalman test" (6.29) follows now by an obvious induction: equalities (a) and (b) imply $\lambda\,\Gamma = 0$; if $\mathrm{rk}\,\Gamma = n$, this implies $\lambda = 0$, and the system is controllable. ∎

In the case of constant coefficients, one obtains the famous criterion due to Kalman [27]:

Proposition 6.16. *If* **K** *is a field of constants, the state-space system* (6.25) *is controllable if, and only if*

$$rk\left[G \;\vdots\; F\,G \;\vdots\; \cdots \;\vdots\; F^{n-1}\,G\right] = n. \tag{6.30}$$

Proof. (1) This condition is sufficient by Theorem 6.6. (2) If $\mathrm{rk}\,\Gamma = \rho < n$, let $Q = [Q_1 \quad Q_2]$, where Q_1 is formed from ρ **K**-linearly independent columns of Γ and Q_2 is chosen such that Q is invertible over **K** (by "completion of the basis"). The **K**-vector space generated by the columns of Q_1 is F-invariant and contains the image of G, thus the matrices $Q^{-1}\,F\,Q$ and $Q^{-1}\,G$ have the structure

$$Q^{-1}\,F\,Q = \begin{bmatrix} F_c & * \\ 0 & F_{\bar{c}} \end{bmatrix}, \quad Q^{-1}\,G = \begin{bmatrix} G_c \\ 0 \end{bmatrix} \tag{6.31}$$

where $F_{\bar{c}}$ is an $(n-\rho) \times (n-\rho)$ matrix. Set $x = Q\,\chi$ where $\chi = \begin{bmatrix} \chi_c \\ \chi_{\bar{c}} \end{bmatrix}$. As $(\partial I_{n-r} - F_{\bar{c}})\,\chi_{\bar{c}} = 0$, $[\chi_{\bar{c}}]_{\mathbf{R}}$ is a nonzero torsion submodule of M, and the system M is not controllable. ∎

Remark 6.7. (i) Proposition 6.15(iii) and the result to be proved in exercise 6.20 are generalizations of the "Popov-Belevitch-Hautus test" for controllability (see, e.g., [26]). (ii) In the continuous-time case, the "generalized Kalman test" (6.29) was proved to be a *necessary and sufficient condition* for system (6.25) to be controllable, assuming that the entries of the matrices F and G are analytic functions ([52]; [25]; [53], Sect. 3.5). See also exercise 6.22. (iii) The *pair* (F, G), with coefficients in a field of constants $\mathbf{k}$, is said to be *controllable* if it satisfies the rank condition (6.30). The pair (F_c, G_c) in (6.31) is controllable (see, e.g., [45], Sect. 5.2, Corollary 5.2.25).

Definition 6.7. *A linear state-space system satisfying the rank condition* (6.29) *is* hypercontrollable.

Exercise 6.21. In the discrete-time case, write (6.25) in the form of a *difference equation*

$$\mathbf{q}x = A\,x + G\,u, \quad A = F + I_n, \tag{6.32}$$

and show that this system is hypercontrollable if, and only if

$$\mathrm{rk}\left[G \;\vdots\; \breve{A}_1\, G^{\beta} \;\vdots\; \cdots \;\vdots\; \breve{A}_{n-1}\, G^{\beta^{n-1}} \;\vdots \ldots\right] = n,$$

where $\breve{A}_i = A \ldots A^{\beta^{i-1}}, 1 \leq i \leq n-1$ (with $\beta = \alpha^{-1}$).

Definition 6.8. *The discrete-time state-space system* (6.32) *is said to be* reversible *if A is invertible over* $\mathbf{K}$ *([18]).*

Exercise 6.22. Consider the state-space system (6.25). (a) Assume that there exists a matrix $V \in \mathbf{GL}_n(\mathbf{K})$ such that

$$V^{\delta} = -V^{\alpha}\,F. \tag{6.33}$$

(i) Set $\eta = V\,x$, show that η satisfies a state-space equation of the form

$$\partial\eta = \breve{G}\,u \tag{6.34}$$

and determine $\breve{G}$ in function of V and G. (ii) Prove the following equality (similar to (6.9)):

$$b\,\partial^j = \sum_{i=0}^{j} (-1)^i \binom{j}{i}\, \partial^{j-i}\, b^{\beta^j\,\delta^i}. \tag{6.35}$$

(iii) Write condition (6.29) for system (6.34) and, using (6.35) and exercise 6.20, show that this condition is *necessary and sufficient* for system (6.34) to be controllable. (iv) Deduce that condition (6.29) is *necessary and sufficient* for system (6.25) to be controllable. (b) In the discrete-time case, show that there exists a matrix $V \in \mathbf{GL}_n(\mathbf{K})$ such that (6.33) holds if, and only if the state-space system (6.32) is *reversible.* (Hint for (iii): if system (6.34) is controllable, write that there exist matrices $X \in \mathbf{R}^{n\times n}$ and $Y = \sum_{0\leq j\leq q} \partial^j Y_j \in \mathbf{R}^{m\times n}$, $Y_j \in \mathbf{K}^{m\times n}$, such that $\partial X + \check{G} Y = I_n$; develop this expression and equalize the coefficients of same degree. See the proof of ([25], th. 6.4).)

Remark 6.8. In the continuous-time case, there does not generally exist a "fundamental matrix" $V \in \mathbf{GL}_n(\mathbf{K})$ satisfying (6.33). However, there exists a field extension $\check{\mathbf{K}}$ of $\mathbf{K}$, called a "Picard-Vessiot extension" of $\mathbf{K}$, such that (6.33) has solutions in $\mathbf{GL}_n\left(\check{\mathbf{K}}\right)$ ([54]).

Exercise 6.23. In part (iii) of Remark 6.7, determine the controllable quotient of the state-space system (6.25).

Remark 6.9. (i) According to Proposition 6.16, a state-space system with constant coefficients is controllable if, and only if it is hypercontrollable. (ii) Exercise 6.22 shows that systems for which there exists a matrix $V \in \mathbf{GL}_n(\mathbf{K})$ satisfying (6.33) (and, in particular, *reversible discrete-time systems*) are controllable if, and only if they are hypercontrollable.

Observability

Let M be an input-output system with input u and output y.

Definition 6.9. *The input-output system M is* observable *if $M = [u, y]_{\mathbf{R}}$ ([16]).*

Exercise 6.24. Let M be the input-output system given by the polynomial matrix description (6.27), (6.28). The system M is observable if, and only if the following equivalent conditions are satisfied: (a) $\begin{bmatrix} D(\partial) \\ Q(\partial) \end{bmatrix} \tilde{\xi} = 0$ implies $\tilde{\xi} = 0$ (meaning of $\tilde{\xi} = \left(\tilde{\xi}_i\right)_{1\leq i\leq r} : \tilde{\xi}_i = \xi_i + [u, y]_{\mathbf{R}}$); (b) the matrix $\begin{bmatrix} D(\partial) \\ Q(\partial) \end{bmatrix}$ is left-invertible over $\mathbf{R}$; (c) the matrices $D(\partial)$ and $Q(\partial)$ are right-coprime. (See [9, Proposition 2] and Chap. 5, Sect. 1 in the present text).

Let us consider a state space system $\left(F,\ G,\ H,\ (L_j)_{0\leq j\leq l}\right)$. With a slight modification of the proof of Theorem 6.6 (using condition (a) of exercise 6.24), one obtains the following result:

Theorem 6.7. *The state-space system under consideration is observable if*

$$rk\left[\tilde{G} \ \vdots\ \tilde{F}_1\,\tilde{G}^{\alpha} \ \vdots\ \cdots \ \vdots\ \tilde{F}_{n-1}\,\tilde{G}^{\alpha^{n-1}}\right] = n \tag{6.36}$$

where $\tilde{G} = {}^tH$ *and* $\tilde{F}_i = \left({}^tF + \delta\, I_n\right) \ldots \left({}^tF^{\alpha^{i-1}} + \delta\, I_n\right)$.[14] *(The transpose of the matrix in the left-hand side of* (6.36) *is denoted by* $\Omega\,(H, F)$ *and is called the* observability matrix *of the pair* (H, F)*.)*

Definition 6.10. *A linear state-space system satisfying the rank condition* (6.36) *is* hyperobservable.

Exercise 6.25. (i) Consider a state space system $(F,\ G,\ H,\ *)$ in the *continuous-time case* (where $*$ denotes any finite sequence of $p \times m$ matrices with entries in $\mathbf{K}$). This system is hypercontrollable (resp. hyperobservable) if, and only if a *dual system* is hyperobservable (resp. hypercontrollable). What is this dual system? (See [52] or [16]; for further details about this "duality", see [49] and ([46], Sect. 5.1).) (ii) In the *discrete-time case*, consider the state-space system (6.32) with $y = H\,x + \sum_{j=0}^{l} S_j\, \mathbf{q}^j\, u$. Show that this system is hyperobservable if and only if

$$\mathrm{rk}\left[\tilde{G} \ \vdots\ \hat{A}_1\,\tilde{G}^{\alpha} \ \vdots\ \cdots \ \vdots\ \hat{A}_{n-1}\,\tilde{G}^{\alpha^{n-1}}\right] = n,$$

$\hat{A}_i = {}^tA \ldots {}^tA^{\alpha^{i-1}}, 1 \leq i \leq n-1, \tilde{G} = {}^tH$. (iii) Assuming that the above discrete-time system is *reversible*, show that one can define a "dual system", as in the continuous-time case (this dual system is governed by a backward equation and is defined by other formulae).

Exercise 6.26. Reformulate Remark 6.9 to specify the connection between observability and hyperobservability.

6.4.2 Finite Poles and Zeros

The study of finite poles and zeros of linear systems is detailed in [9]. We develop here only a few points; some of them are complementary from those presented in the cited reference[15]. In what follows, M is an input-output system with input u and output y.

System Poles

Definition 6.11. *(i) The* module of system poles *is* $M/\,[u]_{\mathbf{R}}$. *(ii) The* system poles *are the Smith zeros of* $M/\,[u]_{\mathbf{R}}$.

[14] The symbol ${}^t\,(.)$ is used for the transpose.

[15] Infinite poles and zeros, which are extensively studied in [10], [37], [8], are not considered below, due to the lack of place.

Let $\mathbf{K} = \mathbf{k}$ be a field of constants.

Proposition 6.17. *(i) If M is given by the polynomial matrix description* $(6.27), (6.28)$, *the system poles of M are the Smith zeros of $D(\partial)$.*

(ii) If M is the state-space system $(6.25), (6.26)$, *the system poles are the eigenvalues of the "state matrix" F.*

Proof. (i) One has $M/[u]_{\mathbf{R}} = [\bar{\xi}]_{\mathbf{R}}$, where $\bar{\xi}_i = \xi_i + [u]_{\mathbf{R}}, 1 \leq i \leq r$, and $D(\partial)\,\bar{\xi} = 0$.

(ii) is a consequence of (i). ∎

Remark 6.10. The above can be generalized to the case where $\mathbf{K}$ is an *algebraically closed field which is not a field of constants* (see Remark 6.6). If M is the state-space system $(6.25), (6.26)$, the system poles are the conjugacy classes of the right eigenvalues of the state matrix F.

Invariant Zeros

Definition 6.12. *(i) The* module of invariant zeros *is* $\mathcal{T}(M/[y]_{\mathbf{R}})$. *(ii) The* invariant zeros *are the Smith zeros of* $\mathcal{T}(M/[y]_{\mathbf{R}})$.

The proof of the following result is similar to that of Proposition 6.17(i) (see [9]).

Proposition 6.18. *If M is given by the polynomial matrix description* (6.27), (6.28), *the invariant zeros of M are the Smith zeros of the "Rosenbrock matrix"*

$$R(\partial) = \begin{bmatrix} D(\partial) & -N(\partial) \\ Q(\partial) & W(\partial) \end{bmatrix}.$$

Let M be the state-space system $(6.25), (6.26)$, and assume that this system is *proper*. Let $u = u_c - K\,x$ be a state feedback and denote by M_c the closed-loop system[16]. Denoting by x_c and y_c, respectively, the state and the output of M_c, one obtains

$$\partial x_c = (F - G\,K)\,x_c + G\,u_c, \quad y_c = (H + L_0\,K)\,x_c + L_0\,u_c.$$

Exercise 6.27. Show that $\mathcal{T}(M/[y]_{\mathbf{R}}) \cong \mathcal{T}(M_c/[y_c]_{\mathbf{R}})$ and deduce that the invariant zeros of a proper state-space system are invariant by state feedback. (Hint: write the Rosenbrock matrix of the closed loop system and use elementary column operations.)

[16] The system M_c results from an interconnection, in the sense specified in [19].

Hidden Modes and Other Kinds of Poles and Zeros

The *module of uncontrollable modes* (also called the module of *input-decoupling* zeros) is $T = \mathcal{T}(M)$.

The *module of unobservable modes* (also called the module of *output-decoupling zeros*) is $M / [u, y]_{\mathbf{R}}$.

The module of *input-output-decoupling zeros* is $T / (T \cap [u, y]_{\mathbf{R}})$.

Assuming that $\mathbf{K} = \mathbf{k}$ is a field of constants, let $\{\text{i.d.z.}\}, \{\text{o.d.z.}\}$ and $\{\text{i.o.d.z.}\}$ be the sets of input-decoupling zeros, output-decoupling zeros and input-output-decoupling zeros, respectively (taking into account the multiplicities). The set of *hidden modes* is defined as:

$$\{\text{hidden modes}\} = \{\text{i.d.z.}\} + \{\text{o.d.z.}\} - \{\text{i.o.d.z.}\} \tag{6.37}$$

where the sign + is used for the "disjoint union" ([40], Sect. 1.8).

Other important kinds of poles and zeros are: the *transmission poles and zeros* (which are the poles and zeros of the *transfer matrix*, calculated from the Smith-MacMillan *form* of this matrix) and the *blocking zeros*. See [9], where the relations between the various poles and zeros are also studied. One of the main relations is:

$$\{\text{system poles}\} = \{\text{transmission poles}\} + \{\text{hidden modes}\} \tag{6.38}$$

Exercise 6.28. Let M be the input-output linear system defined by the polynomial matrix description $(6.27), (6.28)$ with

$$D(\partial) = \partial^2 (\partial - 1)^3 (\partial + 2)^2, \quad N(\partial) = \partial^2 (\partial - 1)(\partial + 1)$$
$$Q(\partial) = \partial^3, \quad W(\partial) = 0$$

(i) Calculate the various kinds of poles and zeros; show that $\{\text{i.o.d.z.}\} = \{0, 0\}$ using the equality (6.38). (ii) Calculate the elementary divisors of the module of invariant zeros and of the module of input-output-decoupling zeros. (Answers for (ii): the elementary divisors of $\mathcal{T}(M / [y]_{\mathbf{R}})$ are ∂^2, ∂^3, $\partial - 1$ and $\partial + 1$; the only elementary divisor of $T / (T \cap [u, y]_{\mathbf{R}})$ is ∂^2.)

6.5 Duality and Behaviors

6.5.1 The Functor Hom

Categories and Functors

The theory of categories and functors is well explained in [39]. We review below some essential points. Our presentation is also based on [50] and [48].

Categories

A *category* $\mathcal{C}$ consists of

(i) a class $\mathrm{Ob}(\mathcal{C})$, the objects of $\mathcal{C}$,

(ii) for each $X, Y \in \mathrm{Ob}(\mathcal{C})$, a set $\mathrm{Hom}_{\mathcal{C}}(X, Y)$, the morphisms (or "arrows") from X to Y,

(iii) for any $X, Y, Z \in \mathrm{Ob}(\mathcal{C})$, the composition $\mathrm{Hom}_{\mathcal{C}}(X, Y) \times \mathrm{Hom}_{\mathcal{C}}(Y, Z) \to \mathrm{Hom}_{\mathcal{C}}(X, Z)$.

The composition is associative and id_X, the identity of X, belongs to $\mathrm{Hom}_{\mathcal{C}}(X, X)$.

One often writes $X \in \mathcal{C}$ instead of $X \in \mathrm{Ob}(\mathcal{C})$ and $f : X \to Y$ instead of $f \in \mathrm{Hom}_{\mathcal{C}}(X, X)$.

A category $\mathcal{C}'$ is called a *subcategory* of $\mathcal{C}$ if $\mathrm{Ob}(\mathcal{C}') \subset \mathrm{Ob}(\mathcal{C})$ and for any $X, Y \in \mathrm{Ob}(\mathcal{C}')$, $\mathrm{Hom}_{\mathcal{C}'}(X, Y) \subset \mathrm{Hom}_{\mathcal{C}}(X, Y)$; the subcategory $\mathcal{C}'$ is *full* if the latter inclusion is an equality.

Let $\mathcal{C}$ be a category; its *opposite* $\mathcal{C}^{\mathrm{op}}$ is defined as follows: $\mathrm{Ob}(\mathcal{C}^{\mathrm{op}}) = \mathrm{Ob}(\mathcal{C})$; $\mathrm{Hom}_{\mathcal{C}^{\mathrm{op}}}(X, Y) = \mathrm{Hom}_{\mathcal{C}}(Y, X)$.

A *preadditive* category is a category in which each Hom set $\mathrm{Hom}_{\mathcal{C}}(X, Y)$ is an abelian group and for which composition is bilinear.

An *additive* category is a preadditive category which has a zero object 0 and a coproduct (or a product) of each pair of its objects.

An additive category $\mathcal{C}$ is *abelian* if: (i) every arrow in $\mathcal{C}$ has a kernel and a cokernel, (ii) every monic arrow is a kernel and every epic arrow is a cokernel ([39], Chap. VIII).

In what follows, we consider categories of modules, which are abelian. Let $\mathbf{A}$ be a ring; the category of left $\mathbf{A}$-modules is denoted by ${}_{\mathbf{A}}\mathbf{Mod}$. The objects are the left $\mathbf{A}$-modules and the arrows are the $\mathbf{A}$-linear morphisms. Let $X, Y \in {}_{\mathbf{A}}\mathbf{Mod}$; the set of all ($\mathbf{A}$-linear) morphisms from X to Y is denoted by $\mathrm{Hom}_{\mathbf{A}}(X, Y)$.

The category of *finitely generated* left $\mathbf{A}$-modules, denoted by ${}_{\mathbf{A}}\mathbf{Mod}^f$, is a full subcategory of ${}_{\mathbf{A}}\mathbf{Mod}$.

If $\mathbf{A} = \mathbb{Z}$, the ring of all integers, then ${}_{\mathbf{A}}\mathbf{Mod} = \mathbf{Ab}$, the category of abelian groups.

Functors

Definition Let $\mathcal{C}, \mathcal{D}$ be two categories.

A *covariant* (resp. *contravariant*) *functor* $F : \mathcal{C} \to \mathcal{D}$ consists of an "object-map" $F : \mathrm{Ob}(\mathcal{C}) \to \mathrm{Ob}(\mathcal{D})$ and for all $X, Y \in \mathcal{C}$, of an "arrow-map"[17] $F: \mathrm{Hom}_{\mathcal{C}}(X,Y) \to \mathrm{Hom}_{\mathcal{D}}(F(X), F(Y))$ (resp. $\mathrm{Hom}_{\mathcal{C}}(X,Y) \to \mathrm{Hom}_{\mathcal{D}}(F(Y), F(X))$) such that:

$$F(id_X) = id_{F(X)},$$

$$F(f \circ g) = F(f) \circ F(g) \quad (\text{resp. } F(f \circ g) = F(g) \circ F(f)).$$

For short, a contravariant functor "reverses the arrows". A contravariant functor $F: \mathcal{C} \to \mathcal{D}$ uniquely defines a covariant functor $\bar{F}: \mathcal{C}^{\mathrm{op}} \to \mathcal{D}$ and conversely. In what follows, we are mainly interested in contravariant functors.

Assuming that $\mathcal{C}$ and $\mathcal{D}$ are preadditive categories, a functor F is said to be *additive* if its arrow-map is $\mathbb{Z}$-linear.

Exactness Let $\mathcal{C}$ and $\mathcal{D}$ be abelian categories. A contravariant functor $F: \mathcal{C} \to \mathcal{D}$ is *left exact* if exactness of

$$X \xrightarrow{f} Y \xrightarrow{g} Z \longrightarrow 0 \tag{6.39}$$

implies exactness of

$$0 \longrightarrow F(Z) \xrightarrow{F(g)} F(Y) \xrightarrow{F(f)} F(X). \tag{6.40}$$

A contravariant functor F is *right exact* if exactness of

$$0 \longrightarrow X \xrightarrow{f} Y \xrightarrow{g} Z \tag{6.41}$$

implies exactness of

$$F(Z) \xrightarrow{F(g)} F(Y) \xrightarrow{F(f)} F(X) \longrightarrow 0. \tag{6.42}$$

A contravariant functor F is *exact* if it is both left and right exact.

Faithfulness A functor $F: \mathcal{C} \to \mathcal{D}$ is *faithful* if its arrow-map is monic. Assuming that $\mathcal{C}$ and $\mathcal{D}$ are preadditive categories and that the functor F is additive, this means that for every pair $X, Y \in \mathrm{Ob}(\mathcal{C})$ and every $f \in \mathrm{Hom}_{\mathcal{C}}(X,Y)$, the equality $F(f) = 0$ implies $f = 0$.

Fullness and surjectivity A covariant functor $F: \mathcal{C} \to \mathcal{D}$ is *full* if for every pair $X, Y \in \mathrm{Ob}(\mathcal{C})$ and for every arrow $g: F(X) \to F(Y)$, there exists an arrow $f: X \to Y$ such that $g = F(f)$. The functor F is *surjective* if for every $Z \in \mathrm{Ob}(\mathcal{D})$, there exists $X \in \mathrm{Ob}(\mathcal{C})$ such that $Z = F(X)$.

Isomorphism and strict duality A covariant functor $F: \mathcal{C} \to \mathcal{D}$ is an *isomorphism* if it is a bijection, both on objects and on arrows, that is, if it is full and faithful (i.e. "fully faithful") and bijective on objects. A contravariant functor $F: \mathcal{C} \to \mathcal{D}$ is a *strict duality* (or an *anti-isomorphism*) if the associated covariant functor $\bar{F}: \mathcal{C}^{\mathrm{op}} \to \mathcal{D}$ is an isomorphism[18].

[17] The "object map" and the "arrow map" are usually denoted by the same symbol.
[18] The notions of *equivalence* and of *duality* are weaker [39].

Image category Let $F : \mathcal{C} \to \mathcal{D}$ be a functor. Its *image* $F(\mathcal{C})$ consists of two classes: the class of objects $F(X), X \in \mathrm{Ob}(\mathcal{C})$, and of the class of arrows $F(f)$, f belonging to the class of arrows of $\mathcal{C}$. The image $F(\mathcal{C})$ is a subcategory of $\mathcal{D}$ if the functor F is *univalent* (i.e. monic) *with respect to objects* ([41], Chap. II, Sect. 10). We have the following result, based on ([41], Chap. II, Sect. 10, Prop. 10.4).

Proposition 6.19. *If $\mathcal{D}$ is a category of modules, a functor $F : \mathcal{C} \to \mathcal{D}$ is* naturally equivalent[19] *to a functor $F' : \mathcal{C} \to \mathcal{D}$ which is* univalent with respect to objects.

Proof . For $X \in \mathrm{Ob}(\mathcal{C})$, define $F'(X)$ to be the module whose elements are ordered pairs $(X, \tilde{x})$ where $\tilde{x}$ runs through all elements of the module $F(X)$. Addition and left multiplication by a scalar λ are defined in $F'(X)$ by $(X, \tilde{x}_1) + (X, \tilde{x}_2) = (X, \tilde{x}_1 + \tilde{x}_2)$ and $\lambda\,(X, \tilde{x}) = (X, \lambda\,\tilde{x})$, respectively. For a morphism $f : X \to Y$ in $\mathcal{C}$ we define $F'(f)(X, \tilde{x}) = (Y, F(f)(\tilde{x}))$. By construction, F' is univalent with respect to objects, and $\theta : F \to F'$, given by $\theta_X(\tilde{x}) = (X, \tilde{x})$, is a natural equivalence. (Note that both a zero object and a zero morphism in $F'(\mathcal{C})$ have the form $(X, 0) = 0_X$, $X \in \mathrm{Ob}(\mathcal{C})$.) ∎

In everything that follows, we consider only *module valued functors.*

Remark 6.11. Let $F : \mathcal{C} \to \mathcal{D}$ be a module valued functor. By Proposition 6.19, F can be (and, in the sequel, is) identified with a functor $F' : \mathcal{C} \to \mathcal{D}$ which is univalent with respect to objects. Therefore, $F(\mathcal{C})$ becomes a subcategory of $\mathcal{D}$; this subcategory is *generally not full.*

The following result is obvious:

Proposition 6.20. *A covariant (resp. contravariant) functor $F : \mathcal{C} \to \mathcal{D}$ is* faithful *if, and only if $F : \mathcal{C} \to F(\mathcal{C})$ is an* isomorphism *(resp. a* strict duality*).*

Exercise 6.29. Show that two objects of $F(\mathcal{C})$ which are isomorphic in $F(\mathcal{C})$ are isomorphic in $\mathcal{D}$, but that the converse does not generally hold.

Notation 1 *We write $F(X) \sim F(Y)$ to mean that $F(X)$ and $F(Y)$ are isomorphic in $F(\mathcal{C})$ (the symbol $\cong$ denotes an isomorphism in $\mathcal{C}$ or in $\mathcal{D}$).*

[19] For details regarding this notion, see ([41], Chap. II, Sect. 9) or ([39], Chap. I, Sect. 4). The main point is that two naturally equivalent functors can be identified.

Definition and General Properties of the Functor Hom

Let $W \in {}_{\mathbf{A}}\mathbf{Mod}$ and consider $F = \mathrm{Hom}_{\mathbf{A}}(., W)$ defined as follows:

(i) for any $M \in {}_{\mathbf{A}}\mathbf{Mod}$, $F(M) = \mathrm{Hom}_{\mathbf{A}}(M, W)$;

(ii) for any $X, Y \in {}_{\mathbf{A}}\mathbf{Mod}$ and $f : X \to Y$, $F(f) = Ff$ is the right-composition by f, i.e. for any $\gamma : Y \to W$, $(Ff)(\gamma) = \gamma \circ f$ (also written γf). This is represented by the commutative diagram below.

$$\begin{array}{ccc} X & \xrightarrow{f} & Y \\ (Ff)(\gamma) & \searrow & \downarrow \gamma \\ & & W \end{array}$$

Exercise 6.30. (i) Ff is a $\mathbb{Z}$-linear map $\mathrm{Hom}_{\mathbf{A}}(Y, W) \longrightarrow \mathrm{Hom}_{\mathbf{A}}(X, W)$ (i.e. F is an "additive functor"). (ii) $\mathrm{Hom}_{\mathbf{A}}(., W)$ is a contravariant functor. (iii) $\mathrm{Hom}_{\mathbf{A}}\left(\coprod_{i\in I} X_i, W\right) \cong \prod_{i\in I} \mathrm{Hom}_{\mathbf{A}}(X_i, W)$ (where the isomorphism is $\mathbb{Z}$-linear, i.e. is an isomorphism of abelian groups).

(The solution of the above exercise can be found in ([48], Chap. 2).)

Proposition 6.21. *The contravariant functor* $\mathrm{Hom}_{\mathbf{A}}(., W)$ *is left exact.*

Proof. Consider the exact sequence (6.39) and let us show that the sequence (6.40) is exact. (i) Fg is monic: if $\varepsilon : Z \to W$ is such that $(Fg)(\varepsilon) = 0$, one has $(\varepsilon g)(y) = 0$ for any $y \in Y$; as g is epic, this implies $\varepsilon = 0$. (ii) $\mathrm{Im}\, Fg \subset \ker Ff$: let $\gamma : Y \to W$ be such that $\gamma \in \mathrm{Im}\, Fg$. There exists $\varepsilon : Z \to W$ such that $\gamma = (Fg)(\varepsilon) = \varepsilon g$; therefore, $(Ff)(\gamma) = \varepsilon g f = 0$ since $gf = 0$. (iii) $\ker Ff \subset \mathrm{Im}\, Fg$: let $\gamma : Y \to W$ be such that $\gamma \in \ker Ff$, i.e. $(Ff)(\gamma) = \gamma f = 0$. Let $y \in \ker g$; as $\ker g = \mathrm{Im}\, f$, there exists $x \in X$ such that $y = f(x)$, thus $\gamma(y) = (\gamma f)(x) = 0$, hence $\ker g \subset \ker \gamma$. Let us identify Z with $Y/\ker g$ and g with the canonical epimorphism. Let $\bar{\gamma} : Y/\ker g \to W$ be the induced map (see exercise 6.4). One obtains $\gamma = (Fg)(\bar{\gamma}) \in \mathrm{Im}\, Fg$. ■

A very important consequence of Proposition 6.21 is the following one: consider the module $M \cong \mathrm{coker} \bullet B$, presented by the exact sequence (6.12). Using the functor $\mathrm{Hom}_{\mathbf{A}}(., W)$, one obtains the exact sequence (6.40) with $Z = M, Y = \mathbf{A}^k, X = \mathbf{A}^q$ and $f = \bullet B$. With the notation in Sect. 6.3.1, for any $\eta : \mathbf{A}^q \to W$ and any $\gamma : \mathbf{A}^k \to W$, set

$$\tilde{\eta} = (\eta(a_1), \ ..., \ \eta(a_q)) \in W^q, \quad \tilde{\gamma} = (\gamma(c_1), \ ..., \ \gamma(c_k)) \in W^k. \tag{6.43}$$

Exercise 6.31. The mapping $\mathrm{Hom}_{\mathbf{A}}\left(\mathbf{D}^k, W\right) \ni \gamma \longrightarrow \tilde{\gamma} \in W^k$ is an $\mathbf{A}$-monomorphism.

Invoking the result to be proved in exercise 6.31, $\mathrm{Hom}_{\mathbf{A}}(\mathbf{D}^q, W)$ and $\mathrm{Hom}_{\mathbf{A}}(\mathbf{D}^k, W)$ can be identified with W^q and W^k, respectively. Let $F(f)(\gamma) = \gamma f = \eta$; by (6.13),

$$\eta(a_i) = (F(f)(\gamma))(a_i) = \gamma\left(\sum_{j=1}^{k} b_{ij}\, c_j\right) = \sum_{j=1}^{k} b_{ij}\, \gamma(c_j), \quad 1 \le i \le q.$$

By the identification (6.43), $F(f)$ is identified with the mapping $W^k \ni \tilde{\gamma} \to \tilde{\eta} \in W^q$ such that

$$\tilde{\eta}_i = \sum_{j=1}^{k} b_{ij}\, \tilde{\gamma}_j, \quad 1 \le i \le q.$$

In other words, $F(f)$ is identified with the *left multiplication by* $B : W^k \to W^q$, written $B\bullet$. By exactness of (6.40), $F(M) = \mathrm{Hom}_{\mathbf{A}}(M, W)$ can be identified with the kernel of this mapping.

To summarize: the exact sequence (6.12) is a "cokernel representation", since $M \cong \mathrm{coker} \bullet B$. Using the functor $\mathrm{Hom}_{\mathbf{A}}(., W)$ and the above identifications, one obtains a "kernel representation", since $\mathrm{Hom}_{\mathbf{A}}(M, W) \cong \ker B\bullet$. The abelian group $\mathrm{Hom}_{\mathbf{A}}(M, W)$ is denoted by M^* and is called the "W-dual" of M ([30], Sect. 19D)[20]; this group is also called the kernel of $B\bullet$ in W^k, since (with the above identifications)

$$M^* = \left\{\tilde{w} \in W^k : B\,\tilde{w} = 0\right\}. \tag{6.44}$$

The structure of M^* can be further detailed. Let $\mathbf{E}$ be the endomorphism ring of W.

Proposition 6.22. *M^* is a left $\mathbf{E}$-module.*

Proof. We already know that M^* is an abelian group (see exercise 6.30(i)). Let $e \in \mathrm{Hom}_{\mathbf{A}}(W, W) = \mathbf{E}$ and $\tilde{w} \in \mathrm{Hom}_{\mathbf{A}}(M, W)$. Define $e\,\tilde{w} : M \to W$ as $(e\,\tilde{w})(m) = e(\tilde{w}(m))$ for any $m \in M$. It is easy to check that $e\,\tilde{w} \in \mathrm{Hom}_{\mathbf{A}}(M, W)$; thus M^*, equipped with the "left multiplication" by an element $e \in \mathbf{E}$, is a left $\mathbf{E}$-module. ■

Following [8], the image category $F\left({}_{\mathbf{A}}\mathbf{Mod}^f\right)$, where $F = \mathrm{Hom}_{\mathbf{A}}(., W)$, is denoted by $\left({}_{\mathbf{A}}\mathbf{Mod}^f\right)^*$ in the sequel. By Proposition 6.22, the objects of $\left({}_{\mathbf{A}}\mathbf{Mod}^f\right)^*$ are left $\mathbf{E}$-modules.

[20] The notation M^* is often used in the literature for the "algebraic dual" $\mathrm{Hom}_{\mathbf{A}}(M, \mathbf{A})$, which is not considered in this chapter.

Injective Modules

An $\mathbf{A}$-module W is *injective* if, for every $\mathbf{A}$-module Y and every submodule X of Y, every $\eta : X \to W$ can be extended to a map $\gamma : Y \to W$, as shown by the following commutative diagram:

$$\begin{array}{ccccc} & & W & & \\ & & \eta \uparrow & \nwarrow \gamma & \\ 0 & \longrightarrow & X & \xrightarrow{f} & Y \end{array}$$

Exercise 6.32. (i) Let $(W_j)_{j \in J}$ be a family of $\mathbf{A}$-modules; $\prod_{j \in J} W_j$ is injective if, and only if so is each W_j. (ii) Let $(W_j)_{1 \leq j \leq n}$ be a *finite* family of $\mathbf{A}$-modules; $\coprod_{j \in J} W_j$ is injective if, and only if so is each W_j.

A useful criterion to determine whether an $\mathbf{A}$-module W is injective is the "Baer criterion" ([48], th. 3.20):

Proposition 6.23. *(Baer criterion). An $\mathbf{A}$-module W is injective if, and only if every map $\eta : \mathbf{a} \to W$, where $\mathbf{a}$ is a left ideal of $\mathbf{A}$, can be extended to $\mathbf{A}$.*

The importance of injective modules is related to the following result:

Proposition 6.24. *An $\mathbf{A}$-module W is injective if, and only if the functor $F = Hom_{\mathbf{A}}(., W)$ is exact.*

Proof. (1) Assume that W is injective and let $f : X \to Y$ be a monomorphism (such as in the exact sequence (6.41)). Let $\eta : X \to W$; there exists $\gamma : Y \to W$ such that $\eta = \gamma f = (Ff)(\gamma)$, thus Ff is epic. (2) The converse is similar. ■

The following result can be proved ([48], th. 3.27, 3.30, 3.31):

Proposition 6.25. *Let V be an $\mathbf{A}$-module. (i) There exists an injective module containing V.*

(ii) Among all injective modules containing V, there exists a module $E(V)$, unique up to isomorphism fixing V pointwise, having the following property: if an injective module W is such that $V \subset W \subset E(V)$, then $W = E(V)$.

Definition 6.13. *The module $E(V)$ is called an* injective hull *of V.*[21]

[21] Also called *the* injective hull of W, by a mild abuse of language ([30], Sect. 3).

For explicit calculations, in particular in the case of modules over principal ideal domains, the notion of "divisible module" is very useful.

Definition 6.14. *Let* $\mathbf{D}$ *be an integral domain. A* $\mathbf{D}$*-module* W *is* divisible *if for each* $\tilde{v} \in W$ *and each* $0 \neq d \in \mathbf{D}$*, there exists* $\tilde{w} \in W$ *such that* $d\,\tilde{w} = \tilde{v}$.

Theorem 6.8. *(i) Every injective* $\mathbf{D}$*-module is divisible.*

(ii) Conversely, let $\mathbf{D}$ *be a* principal left ideal domain*; every divisible* $\mathbf{D}$*-module is injective.*

Proof. (i): Let W be an injective $\mathbf{D}$-module, $0 \neq d \in \mathbf{D}$ and $\tilde{v} \in W$. Let $\eta : \mathbf{D}\,d \to W$ be defined by $\eta\,(r\,d) = r\,\tilde{v}$ for any $r \in \mathbf{D}$, and let $\gamma : \mathbf{D} \to W$ extending η. It follows that $\tilde{v} = \eta\,(d) = \gamma\,(d) = d\,\tilde{w}$ with $\tilde{w} = \gamma\,(1)$, thus W is divisible. (ii) Let $\mathbf{a} = \mathbf{D}\,d$ be a nonzero left ideal of the principal left ideal domain $\mathbf{D}$ and W be a divisible $\mathbf{D}$-module. For any $\tilde{v} \in W$, there exists $\tilde{w} \in W$ such that $d\,\tilde{w} = \tilde{v}$; let $\eta : \mathbf{a} \to W$ be defined by $\eta\,(d\,r) = \tilde{v} = d\,\tilde{w}$. Then, $\gamma : \mathbf{D} \to W$, such that $\gamma\,(r) = \tilde{w}$, extends η ; as a result, W is injective by the Baer criterion (see Proposition 6.23). ■

Let $\mathbf{D}$ be a *commutative principal ideal domain* and $p \in \mathbf{D}$ be a prime. Set $\mathbf{D}\,(p^n) = \mathbf{D}/\mathbf{D}\,p^n$. By Proposition 6.10(iii), $\mathbf{D}\,(p^{n+1}) \cong \mathbf{D}\,(p^n)\,p \subset \mathbf{D}\,(p^n)$, thus $\mathbf{D}\,(p^n)$ can be identified with a submodule of $\mathbf{D}\,(p^{n+1})$ for any $n \geq 1$. Let

$$\mathbf{D}\,(p^\infty) = \bigcup_{n \geq 1} \mathbf{D}\,(p^n)$$

(called the *direct limit* of $(\mathbf{D}\,(p^n))_{n\geq 1}$, and written $\mathbf{D}\,(p^\infty) = \varinjlim \mathbf{D}\,(p^n)$ ([48], Chap. 2, exerc. 2.31)).

Proposition 6.26. $\mathbf{D}\,(p^\infty) = E\,(\mathbf{D}\,(p^n))$ *for any* $n \geq 1$.

Proof. (1) By definition of $\mathbf{D}\,(p^\infty)$, $\mathbf{D}\,(p^n) \subset \mathbf{D}\,(p^\infty)$ for any $n \geq 1$. (2) $\mathbf{D}\,(p^\infty)$ is divisible: let $0 \neq \tilde{v} \in \mathbf{D}\,(p^\infty)$, and let n be the least positive integer for which $\tilde{v} \in \mathbf{D}\,(p^n)$. For any $0 \neq d \in \mathbf{D}$, we have $d = p^k q$, where k is a natural integer and $q \in \mathbf{D}$ is not divisible by p. There exists $\tilde{x} \in \mathbf{D}\,(p+k)$ such that $\tilde{v} = p^k \tilde{x}$; in addition, $p^m \tilde{x} = 0$ if $m = p + k$. Since p^m and q are coprime, there exist elements $u, v \in \mathbf{D}$ such that $u\,p^m + v\,q = 1$. Therefore, $\tilde{x} = v\,q\,\tilde{x}$, thus $\tilde{v} = p^k v\,q\,\tilde{x} = d\,\tilde{w}$ with $\tilde{w} = v\,\tilde{x}$. (3) Therefore, $\mathbf{D}\,(p^\infty)$ is injective by Theorem 6.8(ii). (4) Denoting by (p^n) the ideal generated by p^n, $1 + (p^n)$ is identified with $p + (p^{n+1})$ for any $n \geq 1$. A divisible module containing $1 + (p^n)$ must contain an element $\tilde{w}$ such that $p\,\tilde{w} = p + (p^{n+1})$, thus it must contain $1 + (p^{n+1})$. Therefore, $E\,(\mathbf{D}\,(p^n)) \supset \mathbf{D}\,(p^{n+1})$, and by induction $E\,(\mathbf{D}\,(p^n)) \supset \mathbf{D}\,(p^\infty)$. It follows that $E\,(\mathbf{D}\,(p^n)) = \mathbf{D}\,(p^\infty)$. ■

Remark 6.12. The above construction of $E(\mathbf{D}(p))$ (where p is an atom) can be generalized to the case where $\mathbf{D} = \mathbf{K}\left[\partial; 1, \frac{d}{dt}\right]$ (i.e. where $\mathbf{D}$ is a ring of differential polynomials and ∂ is the continuous-time derivative), using "Picard-Vessiot rings" (see Remark 6.8 and [54]).

Exercise 6.33. (i) Let $\mathbf{D}$ be an integral domain; any product and any coproduct of divisible $\mathbf{D}$-modules is divisible. (ii) Let $\mathbf{D}$ be a principal left ideal domain; any coproduct of injective $\mathbf{D}$-modules is divisible. Compare with exercise 6.32(ii).

Cogenerators

Definition 6.15. *An* $\mathbf{A}$*-module* W *is a* cogenerator *(of the category* ${}_{\mathbf{A}}\boldsymbol{Mod}$*) if the functor* $F = Hom_{\mathbf{A}}(., W)$ *is faithful.*

As a consequence of this definition, any module containing a cogenerator is a cogenerator. The following result can be proved ([30], Sect. 19, Proposition 19.6 and Theorem 19.8):

Proposition 6.27. *The following conditions are equivalent:*

(i) W *is a cogenerator.*

(ii) For any $M \in {}_{\mathbf{A}}\boldsymbol{Mod}$ *and* $0 \neq m \in M$*, there exists* $\eta : M \to W$ *such that* $\eta(m) \neq 0$.

(iii) Any $M \in {}_{\mathbf{A}}\boldsymbol{Mod}$ *can be embedded into some direct product* $\prod_i W$.

(iv) For every simple module $S \in {}_{\mathbf{A}}\boldsymbol{Mod}$, W *contains a copy of* $E(S)$ *(the injective hull of* S*).*

Cogenerators are easier to characterize when they are *injective*: see ([6], Sect. 1, n°8), ([30], Sect. 19). The next result gives a *constructive proof* of existence of cogenerator modules (for a very concise but *not constructive proof* of existence of *injective cogenerators*, see [48], Lemma 3.37).

Theorem 6.9. *Let* $(S_i)_{i \in I}$ *be a representative system of simple* $\mathbf{A}$*-modules (see Definition 6.2). (i)* $W_0 = \bigoplus_{i \in I} E(S_i)$ *is a cogenerator, called a* canonical cogenerator *of* ${}_{\mathbf{A}}\boldsymbol{Mod}$.[22]

(ii) A module W *is a cogenerator if, and only if* W_0 *can be embedded in* W.

(iii) If $\mathbf{A}$ *is a principal ideal domain, the canonical cogenerator* W_0 *is injective.*

[22] By a mild abuse of language, W_0 is often called "the" canonical cogenerator, since it is uniquely defined up to isomorphism.

Proof. (i): W_0 satisfies condition (iv) of Proposition 6.27. (ii) Let W be a cogenerator. By Proposition 6.27(iv), we may assume that $E(S_i) \subset W$ for all i, thus $\sum_{i \in I} E(S_i) \subset W$. If $i \neq j$, $S_i \cap S_j = 0$ (as the reader can easily check), therefore $E(S_i) \cap E(S_j) = 0$ ([30], p. 114, exercise 8), thus the sum $\sum_{i \in I} E(S_i)$ is direct. (iii) is a consequence of exercise 6.33(ii). ∎

The following notation, introduced in [42], is useful for what follows: let $B \in \mathbf{A}^{q \times k}$, $N = \operatorname{Im} \bullet B = \mathbf{A}^q \bullet B \subset \mathbf{A}^k$, and $M \cong \operatorname{coker} \bullet B = \mathbf{A}^k / N$. Then $\ker B\bullet \subset W^k$ is denoted by $N^{\perp}$. Let $M^* \in \left({}_{\mathbf{A}}\mathbf{Mod}^f\right)^*$, $M^* \subset W^k$. The rows $b \in \mathbf{A}^k$ such that $b\, M^* = 0$ form a submodule of $\mathbf{A}^k$; this submodule is denoted by $(M^*)^{\perp}$ (it is finitely generated if the ring $\mathbf{A}$ is noetherian, and in particular if $\mathbf{A}$ is a principal ideal domain). Part (i) of the following theorem is classic ([30], Lemma 19.45(2)), whereas part (ii) and its proof given below are due to Oberst ([42], [44]).

Theorem 6.10. *Let W be a cogenerator. (i) For any $\mathbf{A}$-submodule N of $\mathbf{A}^k$, $N = N^{\perp\perp}$.*

(ii) Let $B \in \mathbf{A}^{q \times k}$ and $B' \in \mathbf{A}^{q' \times k}$ be such that $\ker B\bullet = \ker B'\bullet \subset W^k$. Then, $q = q'$, there exists a unimodular matrix $C \in \mathbf{A}^{q \times q}$ such that $B = C\, B'$, and coker $\bullet B =$ coker $\bullet B'$.

Proof. (i): By definition, $N \subset N^{\perp\perp}$. Let us prove the reverse inclusion. If $m \in \mathbf{A}^k$ does not belong to N, let $\bar{m} = m + N \in \mathbf{A}^k/N$. As W is a cogenerator and $\bar{m} \neq 0$, there exists $\bar{\eta} : \mathbf{A}^k/N \to W$ such that $\bar{\eta}(m) \neq 0$. The $\mathbf{A}$-morphism $\bar{\eta}$ is induced by an $\mathbf{A}$-morphism $\eta : \mathbf{A}^k \to W$ such that $\eta(N) = 0$ (see exercise 6.4), i.e. $\eta \in N^{\perp}$. As $\eta(m) \neq 0$, $m \notin N^{\perp\perp}$. (ii): Set $N = \operatorname{Im} \bullet B$ and $N' = \operatorname{Im} \bullet B' \subset W^k$. If $N'^{\perp} \subset N^{\perp}$, then $N^{\perp\perp} \subset N'^{\perp\perp}$. By (i), $N = N^{\perp\perp}$ and $N' = N'^{\perp\perp}$, thus $N \subset N'$, i.e. $\operatorname{Im} f \subset \operatorname{Im} f'$ where $f = \bullet B$ and $f' = \bullet B'$. As the $\mathbf{A}$-morphism f' is an epimorphism $\mathbf{A}^{q'} \to \operatorname{Im} f'$, there exists an $\mathbf{A}$-morphism $\gamma : \mathbf{A}^{q'} \to \mathbf{A}^q$ such that $f = f' \circ \gamma$ ([40], Chap. VI, th. 12)[23]. Denoting by C be matrix of γ in the canonical bases of $\mathbf{A}^{q'}$ and $\mathbf{A}^q$, one obtains $B = C\, B'$. If, moreover, $N^{\perp} \subset N'^{\perp}$, the matrix C is unimodular. ∎

6.5.2 Behaviors

The Category of Behaviors

Let $\mathbf{R} = \mathbf{K}[\partial; \alpha, \delta]$ and W be an $\mathbf{R}$-module. From Sect. 6.4.1 and Sect. 6.5.1, as well as Remark 6.11, we are led to the following definition.

[23] In other words: the free module $\mathbf{A}^{q'}$ is *projective*.

Definition 6.16. *(i) The* category of linear systems *is* ${}_{\mathbf{R}}\mathbf{Mod}^f$. *(ii) The* category of behaviors *in a power of* W *is* $\left({}_{\mathbf{R}}\mathbf{Mod}^f\right)^* = F\left({}_{\mathbf{R}}\mathbf{Mod}^f\right)$, *where* $F = \mathrm{Hom}_{\mathbf{R}}(.,W)$. *(iii) Let* M *be a linear system, i.e.* $M \cong \mathrm{coker} \bullet B(\partial)$, *assumed to be presented by the short exact sequence* (6.23) *(where* $\mathbf{D} = \mathbf{R}$ *and* $B = B(\partial)$*) and with equation* (6.21). *The* associated behavior *(in a power of* W*) is* $M^* = \mathrm{Hom}_{\mathbf{R}}(M,W)$, *identified with* $\ker B(\partial) \bullet = \left\{w \in W^k : B(\partial)\tilde{w} = 0\right\}$ *(see* (6.44)*).*

The category of behaviors can be very poor, depending on the "signal space" W under consideration. For example, if $W = 0$, all objects and morphisms of $\left({}_{\mathbf{R}}\mathbf{Mod}^f\right)^*$ are zero. However, by Proposition 6.20 and Definition 6.15, we have the following:

Theorem 6.11. *Let* W *be a cogenerator of* ${}_{\mathbf{R}}\mathbf{Mod}^f$*; then,* $\mathrm{Hom}_{\mathbf{R}}(.,W)$: ${}_{\mathbf{R}}\mathbf{Mod}^f \to \left({}_{\mathbf{R}}\mathbf{Mod}^f\right)^*$ *is a* strict duality.

The above theorem shows that the module-theoretic setting of linear systems, developed by Fliess [16], and the "behavioral theory", developed by Willems ([55], [45]), are *strictly dual* if the "signal space" W is a cogenerator. Under this condition (often implicit), several connections between Fliess' module theoretic setting and Willems' behavioral approach have been pointed out ([17], [9]). Theorem 6.10(ii) is called the "quasi-uniqueness of the representing matrix" [42]. By Theorem 6.9, we know that a cogenerator exists (more specifically, an *injective canonical cogenerator* exists).

Exercise 6.34. (i) A behavior M^* is a $\mathbf{k}$-vector space (where $\mathbf{k}$ is the subfield of constants of $\mathbf{K}$). (ii) Let $\mathbf{K} = \Re(t)$ and ∂ be the continuous-time derivative. Show that $(\mathbf{R}/\mathbf{R}\,\partial)^*$ is *not* a $\mathbf{K}$-vector space if W is a cogenerator.

Sub-Behaviors

Let W be an $\mathbf{R}$-module and $\mathbf{E}$ be the endomorphism ring of $\mathbf{W}$. Let $M \in_{\mathbf{R}}\mathbf{Mod}^f$ be a linear system and M/N be a quotient of M. This yields the exact sequence

$$N \longrightarrow M \longrightarrow M/N \longrightarrow 0.$$

By Proposition 6.21, one obtains the exact sequence

$$0 \longrightarrow (M/N)^* \longrightarrow M^* \longrightarrow N^*$$

and $(M/N)^*$ can be embedded in M^*. We are led to the following definition [8]:

Definition 6.17. *A sub-behavior of the behavior M^* is an* **E***-submodule $(M/N)^*$ (with the above embedding).*

The following proposition is a generalization of a result due to Willems ([55], prop. V.8), ([45], Chap. 5, th. 5.2.14 & exerc. 5.6). See also [8] where the case of "impulsive behaviors" is considered.

Proposition 6.28. *Let M be a linear system and $M/\mathcal{T}(M)$ be its controllable quotient. (i) There exist sub-behaviors $\mathcal{B}_c$ and $\mathcal{B}_u$ of $\mathcal{B} = M^*$ such that $\mathcal{B} = \mathcal{B}_c \oplus \mathcal{B}_u$, $\mathcal{B}_c \sim (M/\mathcal{T}(M))^*$ and $\mathcal{B}_u \sim (\mathcal{T}(M))^*$ (using notation 1); such a sub-behavior $\mathcal{B}_u$ (which is such that $\mathcal{B}_u \cong \mathcal{B}/\mathcal{B}_c$) is said to be* uncontrollable *(or* autonomous*).*

(ii) The sub-behavior $\mathcal{B}_c$ (called the controllable sub-behavior*) is unique if W is a cogenerator.*

Proof. (i): Let $B = B(\partial) \in \mathbf{R}^{q\times k}$ be a full row rank matrix and $M \cong \operatorname{coker} \bullet B$. Let U and V be unimodular matrices over $\mathbf{R}$ such that $U\,B\,V^{-1} = \begin{bmatrix} \Delta & 0 \end{bmatrix}$ is the Smith form of B. We know that

$$\mathcal{B} = \left\{w \in W^k : B\,w = 0\right\} = \left\{w \in W^k : \begin{bmatrix} \Delta & 0 \end{bmatrix} V\,w = 0\right\}.$$

Let

$$\mathcal{B}_c = \begin{bmatrix} \Delta & 0 \end{bmatrix} V \bullet,\ \mathcal{B}_u = \ker \begin{bmatrix} \Delta & 0 \\ 0 & I \end{bmatrix} V \bullet.$$

We have $\operatorname{coker} \bullet \begin{bmatrix} \Delta & 0 \end{bmatrix} V \cong M/\mathcal{T}(M)$, $\operatorname{coker} \bullet \begin{bmatrix} \Delta & 0 \\ 0 & I \end{bmatrix} V \cong \mathcal{T}(M)$ and $\mathcal{B} = \mathcal{B}_c \oplus \mathcal{B}_u$. Obviously, $\mathcal{B}_u \cong \mathcal{B}/\mathcal{B}_c$ and (i) is proved.

(ii) Let W be a cogenerator and $\mathcal{B} = \ker B \bullet = \ker B' \bullet$. By Proposition 6.10(ii), there exists a unimodular matrix C such that $B = C\,B'$. Let $U'\,B'\,V'^{-1} = \begin{bmatrix} \Delta & 0 \end{bmatrix}$ be the Smith form of B'. We have

$$\begin{bmatrix} C\,U^{-1}\Delta & 0 \end{bmatrix} = \begin{bmatrix} U'^{-1}\Delta & 0 \end{bmatrix} V'\,V^{-1}, \tag{6.45}$$

therefore there exists a unimodular matrix H such that $C\,U^{-1}\,\Delta = U'^{-1}\,\Delta\,H$, which implies that

$$C\,U^{-1}\,\Delta \begin{bmatrix} I & 0 \end{bmatrix} V = U'^{-1}\,\Delta H \begin{bmatrix} I & 0 \end{bmatrix} V$$

and by (6.45) the left-hand side member of the above equality is equal to $U'^{-1}\Delta \begin{bmatrix} I & 0 \end{bmatrix} V'$. Therefore, $H \begin{bmatrix} I & 0 \end{bmatrix} V = \begin{bmatrix} I & 0 \end{bmatrix} V'$. As a result, $\ker \begin{bmatrix} I & 0 \end{bmatrix} V \bullet = \ker \begin{bmatrix} I & 0 \end{bmatrix} V' \bullet$, and (ii) is proved. ■

Cogenerators Examples from Analysis

The construction of a canonical cogenerator provided by Theorem 6.9 is explicit but purely algebraic. We show below that in the case of systems with real or complex coefficients, there exists a canonical cogenerator which is classical in Analysis. The following result was proved by Oberst [43] in a more general context:

Theorem 6.12. *Let* $\mathbf{K} = \mathbb{C}$ *and* ∂ *be the continuous-time derivative. Let* $z \in \mathbb{C}$, $p = \partial - z$ *and* $C_{n,z}$ *be the* $\mathbb{C}$*-vector space with basis* $e^{z\,t}, t\,e^{z\,t}, ..., t^{n-1}e^{z\,t}$ *(with a slight abuse of notation). (i) There exists an* $\mathbf{R}$*-isomorphism* $C_{n,z} \cong \mathbf{R}\,(p^n)$ *(with the notation after Theorem 6.8).*

(ii) $W_0' = \sum_{n \geq 1,\ z \in \mathbb{C}} C_{n,z}$ *is a canonical cogenerator of* ${}_{\mathbf{R}}\mathbf{Mod}$*.*

Proof. (i): First note that $C_{n,z}$ is the $\mathbf{R}$-module generated by $t^{n-1}e^{z\,t}$. Let $\psi : \mathbf{R} \to C_{n,z}$ be the mapping defined by $\psi\,(r) = r\,t^{n-1}\,e^{z\,t}$ where $r = r\,(\partial) \in \mathbf{R}$. This mapping is an $\mathbf{R}$-epimorphism, and its kernel is $\mathrm{Ann}(C_{n,z}) = \mathbf{R}\,(p^n)$; this proves (i). (ii) is an obvious consequence of (i). ■

Exercise 6.35. Let $\mathbf{K}$ and ∂ be as in Theorem 6.12. (i) Let I be a nonempty open interval of $\Re$, and let $C^\infty\,(I)$ (resp. $\mathcal{D}'\,(I)$) be the space of indefinitely differentiable functions (resp. of distributions) on I, with complex values. Show that $C^\infty\,(I)$ and $\mathcal{D}'\,(I)$ are injective cogenerators of ${}_{\mathbf{R}}\mathbf{Mod}$. (ii) With $W = W_0'$ as defined in Theorem 6.12, show that a sub-behavior $\mathcal{B}_u$, as defined in Proposition 6.28, is a $\mathbb{C}$-vector space whose dimension is the sum of the degrees of the Smith zeros of $\mathcal{T}\,(M)$; show that this property still holds if W is any cogenerator. (Hint: for injectivity in question (i), use Theorem 6.8 and, regarding $\mathcal{D}'\,(I)$, see ([51], Chap. VI, Sect. 10, p. 213); for (iii), see, e.g., ([7], Chap. IV, Sect. 2, n°9) and use the fact that there exists a monomorphism $W_0' \to W$.)

Exercise 6.36. Reformulate Theorem 6.12 in the discrete-time case.

6.6 Concluding Remarks

This chapter presents a synthesis of some existing results, among the most important ones, in the theory of linear systems. Some points are original (or not well-known in the control community):

(i) Our presentation, where continuous- and discrete-time linear systems are merged into one general framework, is not classical (the same presentation is adopted in [9] for the study of "impulsive systems", in connection with the structure at infinity).

(ii) It is emphasized in Sect. 6.3.2 that in the case of time-varying coefficients, roots of differential polynomials and eigenvalues of matrices are not intrinsic notions: the relevant notions are the *conjugacy classes* of these roots and eigenvalues. This remark is of importance regarding stability: it is well-known that the stability of a linear system with real or complex coefficients is connected with the location of its poles in the complex plane. On the other hand, a continuous-time linear system with varying coefficients $\dot{x} = F(t)\, x$, where $F(t)$ has constant eigenvalues belonging to the open left half plane, is not necessarily stable ([53], Sect. 4.7, exerc. 4.6.17). This is not really paradoxical since these eigenvalues are not significant, contrary to their conjugacy classes. Even the latter do not completely characterize stability, as shown by the case of discrete-time N-periodic systems (see [14]).

(iii) Theorems 6.6 and 6.7, completed by Remark 6.9(ii), generalize classical results. Hypercontrollability and hyperobservability are new notions which, according to exercises 6.22 and 6.26, are respectively equivalent to controllability and observability only in the case of specific state-space systems. In the discrete-time case, we have called "controllability" the notion also called "reachability" by many authors. As suggested by exercise 6.25, the controllability $\longleftrightarrow$ observability of linear time-varying systems duality, which is classical in the continuous-time case, does not hold in the discrete-time one, except for "reversible systems".

(iv) There were several attempts to define the *category of behaviors* (e.g., [34]). Definition 6.16, based on Remark 6.11, seems to solve the main difficulty, which is pointed out in exercise 6.29: let $\mathcal{B}$ be a behavior (therefore an $\mathbf{E}$-module, using the notation in Sect. 6.5.2), and let $\mathcal{B}'$ be an $\mathbf{E}$-module such that $\mathcal{B}' \cong \mathcal{B}$; $\mathcal{B}'$ is generally not a behavior, since $\mathcal{B}'$ is a behavior only if $\mathcal{B}' \sim \mathcal{B}$ (using Notation 1).

(v) Proposition 6.28 (which is valid in the time-varying case) is a good illustration of the notion of cogenerator and of its importance.

For further reading, the following papers are recommended:

- about *poles and zeros (finite and infinite)*, see [9], [10] and [8];
- extensions to *infinite dimensional linear systems* can be found in [1], [21], [22] and related references.

References

1. Aoustin Y., Fliess M., Mounier H., Rouchon P., Rudolph J. (Sept. 1997) "Theory and Practice in the Motion Planning and Control of a Flexible Robot Arm Using Mikusiński Operators", *Proc. IFAC - SYROCO*, Nantes, France.

2. Blomberg H., Ylinen R. (1983) *Algebraic Theory for Multivariable Linear Systems.* Academic Press.
3. Bourbaki N. (1970) *Algèbre, Chapitres 1 à 3*, Hermann.
4. Bourbaki N. (1970) *Algèbre, Chapitres 4 à 7*, Hermann.
5. Bourbaki N. (1973) *Algèbre, Chapitre 8*, Hermann.
6. Bourbaki N. (1980) *Algèbre, Chapitre 10 – Algèbre homologique*, Masson.
7. Bourbaki N. (1976) *Fonctions d'une variable réelle*, Hermann.
8. Bourlès H. (2005, to appear) "Impulsive systems and behaviors in the theory of linear dynamical systems", *Forum Mathematicum.*
9. Bourlès, H., Fliess, M. (1997) "Finite poles and zeros of linear systems: an intrinsic approach", *Internat. J. Control*, vol. 68, pp. 897-922.
10. Bourlès H., Marinescu B. (1999) "Poles and Zeros at Infinity of Linear Time-Varying Systems", IEEE Trans. on Automat. Control, vol. 44, pp. 1981-1985.
11. Cohn P.M. (1977) *Skew field constructions*, London Math. Soc. Lecture Notes N° 27, Cambridge University Press.
12. Cohn P. M (1985) *Free Rings and their Relations*, Academic Press.
13. Dieudonné J. (1943) "Les déterminants sur un corps non commutatif", Bulletin de la S.M.F., tome 71, n°2, pp. 27-45.
14. El Mrabet Y., Bourlès H. (1998) "Periodic-Polynomial Interpretation for Structural Properties of Linear Periodic Discrete-Time Systems", *Systems & Control Letters*, vol. 33, pp. 241-251.
15. Fliess, M. (1989) "Generalized linear systems with lumped or distributed parameters and differential vector spaces", *Int. J. Control*, vol. 49, n°6, pp. 1989-1999.
16. Fliess M. (1990) "Some Structural Properties of Generalized Linear Systems", *Systems & Control Letters*, vol. 15, pp. 391-396.
17. Fliess M. (1992) "A remark on Willems' trajectory characterization of linear controllability", *Systems & Control Letters*, vol. 19, pp. 43-45.
18. Fliess M. (1993) "Reversible Linear and Nonlinear Discrete-Time Dynamics", IEEE Trans. on Automat. Control, vol. 37, N°8, pp. 1144-1153.
19. Fliess M., Bourlès H. (1996) "Discussing some examples of linear system interconnections", Systems & Control Letters, vol. 27, pp. 1-7.
20. Fliess M., Lévine J., Martin M., Rouchon P. (1995) "Flatness and defect of non-linear systems: introducting theory and examples", Int. J. Control, vol. 61, n°6, pp. 1327-1361.
21. Fliess M., Mounier M., Rouchon P., Rudolph J. (April 1998) "Controlling the transient of a chemical reactor: a distributed parameter approach", *Proc. CESA'98 IMACS-IEEE Multiconf.*, Hammamet (Tunisia).
22. Fliess M., Mounier H. (1999) "Tracking Control and π-Freeness of Infinite Dimensional Linear Systems", in: *Dynamical Systems, Control, Coding, Computer Vision*, Picci G., Gilliam D.S (edts), Birkhäuser, pp. 45-68.
23. Fröhler S., Oberst U. (1998) "Continuous time-varying linear systems", *Systems & Control Letters*, vol. 35, pp. 97-110.
24. Guéridon J., Dieudonné J. (1978) "L'algèbre depuis 1840", in *Abrégé d'histoire des mathématiques 1700-1900*, J. Dieudonné (edt), vol. 1, Chap. III, pp. 91-127, Hermann.
25. Ilchmann A., Nürnberger I., Schmale W. (1984) "Time-varying polynomial matrix systems", *Int. J. Control*, vol. 40, n°2, pp. 329-362.
26. Kailath T. (1980) *Linear Systems*, Prentice-Hall.

27. Kalman R.E. (1960) "On the general theory of control systems", Proc. 1st World Congress of IFAC, Moscou, pp. 481-493.
28. Kalman R.E. (1969) "Advanced theory of linear systems", in *Topics in Mathematical System Theory*, Kalman R.E., Falb P.L., Arbib M.A. (edts), McGraw-Hill, Chap. 10, pp. 245-339.
29. Lam T.Y. (2001) *A First Course in Noncommutative Rings* (2nd ed.), Springer.
30. Lam T.Y. (1999) *Lectures on Modules and Rings*, Springer.
31. Lam T.Y., Leroy A. (2000) "Principal One-Sided Ideals in Ore Polynomial Rings", *Contemporary Math.*, vol. 259, pp. 333-352.
32. Lang S. (2002) *Algebra*, Springer Verlag.
33. Leroy A. (1995) "Pseudo-linear transformations and evaluation in Ore extensions", *Bull. Belg. Math. Soc.*, vol. 2, pp. 321-347.
34. Lomadze V. (2004) "On duality of partial differential (and difference) equations", *J. of Algebra*, vol. 275, pp. 791-800.
35. Maisonobe, P., Sabbah, C. (edts) (1993) *$\mathcal{D}$-modules cohérents et holonomes*, Hermann.
36. Malgrange B. (1963) "Sur les systèmes différentiels à coefficients constants", Coll. CNRS, Paris, pp. 113-122.
37. Marinescu B., Bourlès H (2003) "The Exact Model Matching Problem for Linear Time-Varying Systems: an Algebraic Approach", IEEE Trans. on Automat. Control, vol. 48, pp. 166-169.
38. McConnel J.C., Robson J.C. (2001) *Noncommutative Noetherian Rings*, American Mathematical Society.
39. McLane S. (1971) *Categories for the Working Mathematician*, Springer Verlag.
40. McLane S., Bikhoff G. (1968) *Algebra*, McMillan.
41. Mitchell B. (1965) *Theory of Categories*, Academic Press.
42. Oberst U. (1990) "Multidimensional Constant Linear Systems", *Acta Applicandae Mathematicae*, vol. 20, pp 1-175.
43. Oberst U. (1995) "Variations on the Fundamental Principle for Linear Systems of Partial Differential and Difference Equations with Constant Coefficinets", *Applic. Alg. in Eng., Comm. and Comp.*, vol. 6, n°4/5, pp. 211-326.
44. Oberst U. (September 24, 2003) Personal communication.
45. Polderman J.W., Willems J.C. (1998) *Introduction to Mathematical Systems Theory: A Behavioral Approach.* Springer.
46. Pommaret J.-F. (2001) *Partial Differential Control Theory*, vol. 1 and 2, Kluwer Academic Pub.
47. Rosenbrock H.H. (1970) *State-space and Multivariable Theory*, Nelson.
48. Rotman J.J. (1979) *An Introduction to Homological Algebra*, Academic Press.
49. Rudolph J. (1996) "Duality in time-varying linear systems: a module theoretic approach", *Linear Algebra Applications*, vol. 245, pp. 83-105.
50. Schapira P. (Sept. 2000) *Elementary Homological Algebra*, Course at Paris VI University.
51. Schwartz L. (1966) *Théorie des distributions*, Hermann.
52. Silverman I.M., Meadows H.E. (1967) "Controllability and observability in time-variable linear systems", *SIAM J. Control*, vol. 5, pp. 64-73.
53. Sontag E.D. (1990) *Mathematical Control Theory*, Springer Verlag.
54. van der Put M., Singer M.F. (2003) *Differential Galois Theory*, Springer.
55. Willems J. C. (1991) Paradigms and Puzzles in the Theory of Dynamical Systems, *IEEE Trans. on Automat. Control*, vol. 36, pp. 259-294.

Lecture Notes in Control and Information Sciences

Edited by M. Thoma and M. Morari

Further volumes of this series can be found on our homepage: springeronline.com

Vol. 310: Janczak, A.
Identification of Nonlinear Systems Using Neural Networks and Polynomial Models
197 p. 2005 [3-540-23185-4]

Vol. 309: Kumar, V.; Leonard, N.; Morse, A.S. (Eds.)
Cooperative Control
301 p. 2005 [3-540-22861-6]

Vol. 308: Tarbouriech, S.; Abdallah, C.T.; Chiasson, J. (Eds.)
Advances in Communication Control Networks
358 p. 2005 [3-540-22819-5]

Vol. 307: Kwon, S.J.; Chung, W.K.
Perturbation Compensator based Robust Tracking Control and State Estimation of Mechanical Systems
158 p. 2004 [3-540-22077-1]

Vol. 306: Bien, Z.Z.; Stefanov, D. (Eds.)
Advances in Rehabilitation
472 p. 2004 [3-540-21986-2]

Vol. 305: Nebylov, A.
Ensuring Control Accuracy
256 p. 2004 [3-540-21876-9]

Vol. 304: Margaris, N.I.
Theory of the Non-linear Analog Phase Locked Loop
303 p. 2004 [3-540-21339-2]

Vol. 303: Mahmoud, M.S.
Resilient Control of Uncertain Dynamical Systems
278 p. 2004 [3-540-21351-1]

Vol. 302: Filatov, N.M.; Unbehauen, H.
Adaptive Dual Control: Theory and Applications
237 p. 2004 [3-540-21373-2]

Vol. 301: de Queiroz, M.; Malisoff, M.; Wolenski, P. (Eds.)
Optimal Control, Stabilization and Nonsmooth Analysis
373 p. 2004 [3-540-21330-9]

Vol. 300: Nakamura, M.; Goto, S.; Kyura, N.; Zhang, T.
Mechatronic Servo System Control
Problems in Industries and their Theoretical Solutions
212 p. 2004 [3-540-21096-2]

Vol. 299: Tarn, T.-J.; Chen, S.-B.; Zhou, C. (Eds.)
Robotic Welding, Intelligence and Automation
214 p. 2004 [3-540-20804-6]

Vol. 298: Choi, Y.; Chung, W.K.
PID Trajectory Tracking Control for Mechanical Systems
127 p. 2004 [3-540-20567-5]

Vol. 297: Damm, T.
Rational Matrix Equations in Stochastic Control
219 p. 2004 [3-540-20516-0]

Vol. 296: Matsuo, T.; Hasegawa, Y.
Realization Theory of Discrete-Time Dynamical Systems
235 p. 2003 [3-540-40675-1]

Vol. 295: Kang, W.; Xiao, M.; Borges, C. (Eds)
New Trends in Nonlinear Dynamics and Control, and their Applications
365 p. 2003 [3-540-10474-0]

Vol. 294: Benvenuti, L.; De Santis, A.; Farina, L. (Eds)
Positive Systems: Theory and Applications (POSTA 2003)
414 p. 2003 [3-540-40342-6]

Vol. 293: Chen, G. and Hill, D.J.
Bifurcation Control
320 p. 2003 [3-540-40341-8]

Vol. 292: Chen, G. and Yu, X.
Chaos Control
380 p. 2003 [3-540-40405-8]

Vol. 291: Xu, J.-X. and Tan, Y.
Linear and Nonlinear Iterative Learning Control
189 p. 2003 [3-540-40173-3]

Vol. 290: Borrelli, F.
Constrained Optimal Control of Linear and Hybrid Systems
237 p. 2003 [3-540-00257-X]

Vol. 289: Giarré, L. and Bamieh, B.
Multidisciplinary Research in Control
237 p. 2003 [3-540-00917-5]

Vol. 288: Taware, A. and Tao, G.
Control of Sandwich Nonlinear Systems
393 p. 2003 [3-540-44115-8]

Vol. 287: Mahmoud, M.M.; Jiang, J.; Zhang, Y.
Active Fault Tolerant Control Systems
239 p. 2003 [3-540-00318-5]

Vol. 286: Rantzer, A. and Byrnes C.I. (Eds)
Directions in Mathematical Systems Theory and Optimization
399 p. 2003 [3-540-00065-8]

Vol. 285: Wang, Q.-G.
Decoupling Control
373 p. 2003 [3-540-44128-X]

Vol. 284: Johansson, M.
Piecewise Linear Control Systems
216 p. 2003 [3-540-44124-7]

Vol. 283: Fielding, Ch. et al. (Eds)
Advanced Techniques for Clearance of
Flight Control Laws
480 p. 2003 [3-540-44054-2]

Vol. 282: Schröder, J.
Modelling, State Observation and
Diagnosis of Quantised Systems
368 p. 2003 [3-540-44075-5]

Vol. 281: Zinober A.; Owens D. (Eds)
Nonlinear and Adaptive Control
416 p. 2002 [3-540-43240-X]

Vol. 280: Pasik-Duncan, B. (Ed)
Stochastic Theory and Control
564 p. 2002 [3-540-43777-0]

Vol. 279: Engell, S.; Frehse, G.; Schnieder, E. (Eds)
Modelling, Analysis, and Design of Hybrid Systems
516 p. 2002 [3-540-43812-2]

Vol. 278: Chunling D. and Lihua X. (Eds)
H_∞ Control and Filtering of
Two-dimensional Systems
161 p. 2002 [3-540-43329-5]

Vol. 277: Sasane, A.
Hankel Norm Approximation
for Infinite-Dimensional Systems
150 p. 2002 [3-540-43327-9]

Vol. 276: Bubnicki, Z.
Uncertain Logics, Variables and Systems
142 p. 2002 [3-540-43235-3]

Vol. 275: Ishii, H.; Francis, B.A.
Limited Data Rate in Control Systems with Networks
171 p. 2002 [3-540-43237-X]

Vol. 274: Yu, X.; Xu, J.-X. (Eds)
Variable Structure Systems:
Towards the 21st Century
420 p. 2002 [3-540-42965-4]

Vol. 273: Colonius, F.; Grüne, L. (Eds)
Dynamics, Bifurcations, and Control
312 p. 2002 [3-540-42560-9]

Vol. 272: Yang, T.
Impulsive Control Theory
363 p. 2001 [3-540-42296-X]

Vol. 271: Rus, D.; Singh, S.
Experimental Robotics VII
585 p. 2001 [3-540-42104-1]

Vol. 270: Nicosia, S. et al.
RAMSETE
294 p. 2001 [3-540-42090-8]

Vol. 269: Niculescu, S.-I.
Delay Effects on Stability
400 p. 2001 [1-85233-291-316]

Vol. 268: Moheimani, S.O.R. (Ed)
Perspectives in Robust Control
390 p. 2001 [1-85233-452-5]

Vol. 267: Bacciotti, A.; Rosier, L.
Liapunov Functions and Stability in Control Theory
224 p. 2001 [1-85233-419-3]

Vol. 266: Stramigioli, S.
Modeling and IPC Control of Interactive Mechanical
Systems – A Coordinate-free Approach
296 p. 2001 [1-85233-395-2]

Vol. 265: Ichikawa, A.; Katayama, H.
Linear Time Varying Systems and Sampled-data Systems
376 p. 2001 [1-85233-439-8]

Vol. 264: Baños, A.; Lamnabhi-Lagarrigue, F.;
Montoya, F.J
Advances in the Control of Nonlinear Systems
344 p. 2001 [1-85233-378-2]

Vol. 263: Galkowski, K.
State-space Realization of Linear 2-D Systems with
Extensions to the General nD (n$>$2) Case
248 p. 2001 [1-85233-410-X]

Vol. 262: Dixon, W.; Dawson, D.M.; Zergeroglu, E.;
Behal, A.
Nonlinear Control of Wheeled Mobile Robots
216 p. 2001 [1-85233-414-2]

Vol. 261: Talebi, H.A.; Patel, R.V.; Khorasani, K.
Control of Flexible-link Manipulators
Using Neural Networks
168 p. 2001 [1-85233-409-6]

Vol. 260: Kugi, A.
Non-linear Control Based on Physical Models
192 p. 2001 [1-85233-329-4]

Vol. 259: Isidori, A.; Lamnabhi-Lagarrigue, F.;
Respondek, W. (Eds)
Nonlinear Control in the Year 2000 Volume 2
640 p. 2001 [1-85233-364-2]

Vol. 258: Isidori, A.; Lamnabhi-Lagarrigue, F.;
Respondek, W. (Eds)
Nonlinear Control in the Year 2000 Volume 1
616 p. 2001 [1-85233-363-4]

Vol. 257: Moallem, M.; Patel, R.V.; Khorasani, K.
Flexible-link Robot Manipulators
176 p. 2001 [1-85233-333-2]